AF595347

Sustainable and Environmental Catalysis

Sustainable and Environmental Catalysis

Editors

Raffaele Cucciniello
Daniele Cespi
Tommaso Tabanelli

MDPI • Basel • Beijing • Wuhan • Barcelona • Belgrade • Manchester • Tokyo • Cluj • Tianjin

Editors
Raffaele Cucciniello
University of Salerno
Italy

Daniele Cespi
NIER Ingegneria SpA
Italy

Tommaso Tabanelli
Alma Mater Studiorum Università di Bologna
Italy

Editorial Office
MDPI
St. Alban-Anlage 66
4052 Basel, Switzerland

This is a reprint of articles from the Special Issue published online in the open access journal *Catalysts* (ISSN 2073-4344) (available at: https://www.mdpi.com/journal/catalysts/special_issues/SEC_catalysts).

For citation purposes, cite each article independently as indicated on the article page online and as indicated below:

LastName, A.A.; LastName, B.B.; LastName, C.C. Article Title. *Journal Name* **Year**, *Volume Number*, Page Range.

ISBN 978-3-0365-1980-7 (Hbk)
ISBN 978-3-0365-1981-4 (PDF)

Contents

 catalysts

MDPI

Editorial

Sustainable and Environmental Catalysis

Tommaso Tabanelli [1,*], Daniele Cespi [2,*] and Raffaele Cucciniello [3,*]

1 Dipartimento di Chimica Industriale "Toso Montanari", Università di Bologna, Viale del Risorgimento 4, 40136 Bologna, Italy
2 NIER Ingegneria SpA, Via C. Bonazzi 2-40013-Castel Maggiore, 40013 Bologna, Italy
3 Department of Chemistry and Biology, "Adolfo Zambelli" Università di Salerno, Via Giovanni 6 Paolo II, 132, 84084 Fisciano, Italy
* Correspondence: tommaso.tabanelli@unibo.it (T.T.); d.cespi@nier.it (D.C.); rcucciniello@unisa.it (R.C.)

Citation: Tabanelli, T.; Cespi, D.; Cucciniello, R. Sustainable and Environmental Catalysis. *Catalysts* **2021**, *11*, 225. https://doi.org/10.3390/catal11020225

Academic Editor: Jean-François Lamonier

Received: 1 February 2021
Accepted: 3 February 2021
Published: 9 February 2021

1. Introduction

Over the last few decades, an increasing amount of interest from academia and industry has been devoted to the application of the Twelve Principles of the Green Chemistry in order to pursue the Sustainable Development Goals (SDGs) recommended by the United Nations [1]. They are based on the fundamental idea of guiding research and innovation toward more environmental-friendly practices as well as socioeconomic sustainable solutions. Among them, catalysis plays a pivotal role able to address some important needs: (i) avoiding the use of stoichiometric amount of chemicals; (ii) increasing selectivity (i.e., decreasing the generation of waste), and (iii) decreasing energy demand. In this context, with the aim to eliminate environmental pollutants (NO_x, CO, SO_x, VOCs) and to design new catalytic routes for clean energy production, environmental and sustainable catalysis represents a noble and one of the most investigated research directions. Moreover, as recently suggested, sustainable and environmental catalysis allows for converting waste in value-added products, and the degradation of pollutants through catalytic combustion/photocatalysis are considered main topics in this context [2,3]. As a matter of fact, it is worth highlighting some recent research advances. The oxidation of chlorinated compounds by catalytic combustion on non-noble metals based oxides [4,5], NO_x reduction to nitrogen or by selective catalytic reduction (SCR) with ammonia in the presence of heterogeneous catalysts [6]; biofuel production from vegetable oils [7]; the development of new active catalysts for Fenton and photo-Fenton processes [8]; the development of supported nanoparticle catalysts for the treatment of the emission of harmful substances from automobiles, and the use of waste as source of metals for catalysts preparation [9]; the design of new active catalytic systems for the preparation of olefins from polyols [10]; the catalytic hydrotreatment in the valorization of bio-oils obtained from lignocellulosic biomass pyrolysis [11]; and the design of active catalysts for the preparation of organic carbonates from CO_2 or bio-based compounds are some examples in this framework [12,13].

This Special Issue aims to cover new and promising examples of sustainable and environmental catalysis with particular emphasis on the adoption of green chemistry principles at both a laboratory and industrial scale.

2. The Contents of the Special Issue

In this Special Issue, several fields of research on the sustainable catalytic valorization of biomasses are discussed. Therefore, we would like to sincerely thank all the authors who contributed with their excellent contributions to this Special Issue, which includes fourteen articles.

Cucciniello et al. investigated the use of iron-doped mayenite catalysts for trichloroethylene (TCE) oxidation in the gas phase. All the synthesized catalysts showed good performances for TCE oxidation, totally converted into CO_2, CO, and HCl. Mayenite loaded with 2% iron was found to be the best catalyst in terms of T_{50} (300 °C). This result was correlated with an optimum combination of the oxidative properties of the mayenite active support with the redox properties of iron, as Temperature Programmed Reduction (TPR), X-ray photoelectron spectroscopy (XPS) and Raman results have shown [14].

Chevez Rivas et al. reported the preparation and characterization of natural purified mordenite treated with a hydrothermal ion exchange process in an acid medium with Fe^{2+} or Fe^{3+} salts. The set of samples was characterized regarding their textural properties, morphology, and crystallinity, and tested in the NO reduction with CO/C_3H_6. The activity in hydrocarbons-selective catalytic reduction (HC–SCR) (used as a reaction test) was found to depend on the ion exchange process [15].

Husnain et al. discussed the selective catalytic reduction of NO with NH_3 at low temperatures with natural iron ore catalysts. The results showed that the sample based on α-Fe_2O_3 and γ-Fe_2O_3 calcined at 250 °C, achieved excellent selective catalytic reduction (SCR) activity (above 80% at 170–350 °C) and N_2 selectivity (above 90%, up to 250 °C) at low temperatures [16].

Ngoie et al. investigated the potential of a heterogeneous catalyst produced from mineral processing waste for biodiesel production. Tailings from the concentration of cupriferous minerals served as the starting material for synthesis of the catalyst. The nano-magnetic catalysts were prepared using coprecipitation and sol–gel methods, combined with zero-valent iron nanoparticles to form a hydride catalyst. Catalyst properties were assessed using SEM, TEM, BET, and EDX. The maximum yield obtained with this catalyst was 88% and an average of 27% decrease in biodiesel yield observed after four reaction cycles [17].

Muhammad Habib et al. discussed the preparation of two alternative catalysts (Co_3O_4/TiO_2 and 8 wt.% ZrO_2–Co_3O_4/TiO_2) for CO, HC, and NO_x conversions. These systems were produced along with a wire mesh-based substrate and tested by mounting them at the exhaust tailpipe of a motorcycle engine. Zirconia-promoted catalyst showed more promising towards NO_x conversion. The cobalt supported by the titania Co_3O_4/TiO_2 catalyst showed a performance towards conversion of carbon monoxide, nitrogen oxides, and unburnt hydrocarbons to values of 78.1%, 61.9%, and 82.6% efficiency, respectively, at 1500 RPM; whereas, the conversion efficiency of zirconia promoted ZrO_2–Co_3O_4/TiO_2 catalyst was 81.3%, 78.6%, and 55.1% towards HCs, CO, and NO_x, respectively, at 1500 RPM value. Due to small crystalline size, thermal stability, and fouling inhibition, the ZrO_2-promoted Co_3O_4/TiO_2 catalyst showed better conversion efficiency towards CO and NO_x. The slightly lower efficiency of zirconia-promoted catalyst towards HCs is due to the nonavailability of the vacancies of oxygen for oxidation on Co_3O_4. Both catalysts showed selectivity towards CO, NO_x, and HC, and had comparable performance with respect to the activity of a conventional catalyst [18].

Zhou et al. proposed a copper-based Fenton-like catalyst ($Cu/Al_2O_3/g$–C_3N_4) to achieve high degradation efficiencies for Rhodamine B (Rh B) in a wide range of pH (4.9–11.0). The Cu/Al_2O_3 composite was first prepared via a hydrothermal method followed by a calcination process. The composite was then stabilized on graphitic carbon nitride (g–C_3N_4). The characterizations show that Cu species were immobilized on the Al_2O_3 framework in the form of Cu^{2+} and Cu^+, and the introduction of CN increased its specific surface area and adsorption capacity for Rh B. The $Cu/Al_2O_3/CN$ composite showed an excellent catalytic performance in a wide range of pH (4.9–11.0). The specific reaction rate constant of Rh B degradation was calculated as $(5.9 \pm 0.07) \times 10^{-9}$ m·s^{-1}, and the activation energy was calculated to be 71.0 kJ/mol. The recycling experiment demonstrated its durability for Rh B removal and proved that the degradation reaction was surface dominated, with a negligible leaching of copper species in solution [19].

A novel method for preparing highly efficient glass fiber-supported Pt nanoparticle catalysts for the treatment of the emission of harmful substances from automobiles was proposed by Sasaki et al. The Pt catalyst was prepared by depositing a reactive chemically adsorbed monomolecular film with a thiol group at the molecular terminal on the surface of a hydrophilized glass fiber. The monomolecular film enabled the uniform and dense grafting of Pt nanoparticles on the surface of the glass fiber as a single layer. Both catalysts exhibited catalytic performance at temperatures ≥200 °C. At 300 °C, the octane combustion rates (index for evaluating performances of the Pt and commercially available catalysts) of the catalysts were ≥80%. At 50–400 °C, the octane combustion rates of both catalysts were approximately the same. Pt catalysts were found to exhibit catalytic performance comparable to that of Pd catalysts under lean conditions, despite the lower price [20].

Research conducted by Villamarin-Barriga et al. showed that the residual sludge of galvanic industries can be used as catalytic materials (after thermal treatment) for the synthesis of materials. Researchers evaluated the sludge from three galvanic industries. Catalyst was obtained from a thermal process based on sludge dried between 100 and 120 °C, and calcination of sludges between 400 to 700 °C. Catalytic activity was analyzed by thermogravimetric analysis of a thermocatalytic decomposition of crude oil. The best conditions for catalyst synthesis were calcination between 400 and 500 °C, the temperature of reduction between 750 and 850 °C for 15 min. The catalytic material had mainly Fe as active phase and a specific surface between 17.68–96.15 $m^2 \cdot g^{-1}$; the catalysts promoted around 6% more weight-loss of crude oil in the thermal decomposition compared with assays without the catalyst. To determinate the environmental sustainability of this process, the E-factor (mass of waste/mass of product) was calculated by obtaining values between 0.1 and 0.3; the ideal E-factor is zero [21].

In the manuscript of Klein Gebbink et al., two new tetradentate N-donor ligand-supported ReO_2^+ complexes (cis-[(BPMEN)ReO_2]PF_6 and cis-[(BmdmPMEN)ReO_2]PF_6) were synthesized and fully characterized. They were found to be active in deoxydehydration (DODH) reactions of diols, suggesting that Re(V) dioxo complexes can be involved in DODH catalysis. Treatment of (N_2Py_2) ReO_2^+ with an excess amount of water generates an active species for DODH catalysis; use of the Re-product of this reaction shows a much shorter induction period compared to the pristine complex. No ligand is coordinated to the "water-treated" complex, indicating that the real catalyst is formed after ligand dissociation. IR analysis suggested this catalyst to be a rhenium-oxide/hydroxide oligomer. The monodentate pyridine ligand is much easier to dissociate from the metal center than a tetradentate N_2Py_2 ligand, which makes the $Py_4ReO_2^+$-initiated DODH reaction more efficient. For the $Py_4ReO_2^+$-initiated DODH of diols and biomass-based polyols, both PPh_3 and 3-pentanol could be used as a reductant. Excellent olefin yields are achieved [22].

The manuscript of Ray et al. reports on the photocatalytic and antibacterial activity of ZnO nanoparticles. This activity is strongly related to the amount of surficial reactive oxygen species (ROS) and metal ions (Zn^{2+}) which, in turn, is strictly dependent on the morphology of the nanomaterials. In this context, the authors managed to optimize the preparation of different ZnO nanospheres, nanopetals, and nanorods by varying the precipitation solvent ranging from PEG400, water, and toluene. These nanocatalysts (dimension ranging from 10 to few hundreds nanometers) were effective in killing 99% of bacteria at low concentrations in water while the photodegradation of caffeine using solar light was completed after 120 min using the optimized ZnO nanospheres. These results are promising; therefore, the mechanism of both the processes will be investigated in depth in future investigations [23].

The research work of Kaewkannetra and coworkers show the possibility to successfully valorize the triglyceride-rich palm-oil mill effluent (POME) through the combination of biological and thermochemical processes. In this complementary approach, the aqueous waste of the palm oil manufacture first undergoes an enzymatic hydrolysis with yield up to 90% in free fatty acids (FFA). Subsequently, FFA are efficiently converted into long chain hydrocarbon mixtures, suitable for jet fuel application, via the heterogeneously catalyzed hydrogenation reaction (Pd/Al_2O_3, 400 °C, 10 bar of H_2, 1 h). This innovative and complementary approach for the valorization of POME was never reported in literature and represents a new way to both alleviate the wastewater environmental issue and to obtain a high value-added product (jet fuels) for local biorefineries (e.g., Southeast Asia) [24].

Dutta et al. reported on the catalytic activity of Pd/SiO_2 for the selective oxidation of CO at low temperature. An extensive characterization of the catalytic materials (by FTIR, TGA–DSC, TEM, and XPS) allowed for investigating the influence of the thermal pretreatment on catalyst activity and stability. Interestingly, a structure–activity relationship was proposed, correlating the catalyst activation and CO oxidation hysteresis behavior with the information on both the palladium oxidation states and nanoparticle dispersion. In particular, the best low-temperature CO oxidation performance was achieved by calcining the catalyst at 450 °C [25].

In a comprehensive review, Paone et al. extensively discussed the potentials of catalytic hydrotreatment in the valorization of bio-oils obtained from either lignocellulosic biomass pyrolysis or by the physical pressing of oleaginous vegetable seeds. This treatment uses high-pressure hydrogen in order to both stabilize the bio-oil and decrease the amount of oxygenated compounds, in this way obtaining mixtures suitable for direct use as drop-in biofuels. In particular, the current state-of-the-art on the reaction pathways, catalyst formulations, and reactor technology (i.e., traditional vs. membrane) were analyzed in-depth [26].

Zhang et al. proposed an innovative series of binuclear aluminum complexes as highly active and selective homogeneous catalysts for the cycloaddition reaction of epoxides and carbon dioxide. The reaction is of interest because the market for organic carbonates is steadily increasing due to their utilization as solvents, electrolytes, and monomers for high-performance polymers (i.e., polycarbonates, polyurethane, and polyureas), as well as a way to valorize CO_2 as a feedstock. Interestingly, the bifunctional tridentate tetrammino phenoxyimino-phenoxy aluminum complex is also active in very mild conditions, i.e., 100 °C and low CO_2 pressure ($\leq$2 atm), which allows for obtaining good to excellent conversions of the parental epoxides. Kinetic investigations showed a first-order dependence on the concentration of the epoxide for the target reaction [27].

Funding: This research received no external funding.

Conflicts of Interest: The authors declare no conflict of interest.

References

1. Cespi, D.; Esposito, I.; Cucciniello, R.; Anastas, P.T. Beyond the beaker: Benign by design society. *Curr. Res. Green Sustain. Chem.* **2020**, *3*, 100028. [CrossRef]
2. Fasolini, A.; Cespi, D.; Tabanelli, T.; Cucciniello, R.; Cavani, F. Hydrogen from Renewables: A Case Study of Glycerol Reforming. *Catalysts* **2019**, *9*, 722. [CrossRef]
3. Bellè, A.; Tabanelli, T.; Fiorani, G.; Perosa, A.; Cavani, F.; Selva, M. A Multiphase Protocol for Selective Hydrogenation and Reductive Amination of Levulinic Acid with Integrated Catalyst Recovery. *ChemSusChem* **2019**, *12*, 3343–3354. [CrossRef]
4. Cucciniello, R.; Intiso, A.; Castiglione, S.; Genga, A.; Proto, A.; Rossi, F. Total oxidation of trichloroethylene over mayenite ($Ca_{12}Al_{14}O_{33}$) catalyst. *Appl. Catal. B Environ.* **2017**, *204*, 167–172. [CrossRef]
5. Intiso, A.; Martinez-Triguero, J.; Cucciniello, R.; Proto, A.; Palomares, A.E.; Rossi, F. A Novel Synthetic Route to Prepare High Surface. *Catalysts* **2019**, *9*, 27. [CrossRef]
6. Chen, Z.; Liu, Q.; Guo, L.; Zhang, S.; Pang, L.; Guo, Y.; Li, T. The promoting mechanism of in situ Zr doping on the hydrothermal stability of Fe-SSZ-13 catalyst for NH3-SCR reaction. *Appl. Catal. B Environ.* **2021**, *286*, 119816. [CrossRef]
7. Ricciardi, M.; Falivene, L.; Tabanelli, T.; Proto, A.; Cucciniello, R.; Cavani, F. Bio-Glycidol Conversion to Solketal over Acid Heterogeneous Catalysts: Synthesis and Theoretical Approach. *Catalysts* **2018**, *8*, 391. [CrossRef]
8. Prete, P.; Fiorentino, A.; Rizzo, L.; Proto, A.; Cucciniello, R. Review of aminopolycarboxylic acids-based metal complexes application to water and wastewater treatment by (photo-)Fenton process at neutral pH. *Curr. Opin. Green Sustain. Chem.* **2021**. [CrossRef]
9. Cova, C.M.; Zuliani, A.; Manno, R.; Sebastian, V.; Luque, R. Scrap waste automotive converters as efficient catalysts for the continuous-flow hydrogenations of biomass derived chemicals. *Green Chem.* **2020**, *22*, 1414–1423. [CrossRef]
10. Scioli, G.; Tonucci, L.; Di Profio, P.; Proto, A.; Cucciniello, R.; D'Alessandro, N. New green route to obtain (bio)-propene through 1,2-propanediol deoxydehydration. *Sustain. Chem. Pharm.* **2020**, *17*, 100273. [CrossRef]
11. Kang, S.; Li, X.; Fan, J.; Chang, J. Hydrothermal conversion of lignin: A review. *Ren. Sustain. Energy Rev.* **2013**, *27*, 546–558. [CrossRef]
12. Aomchad, V.; Cristofol, A.; Della Monica, F.; Limburg, B.; D'Elia, V.; Kleij, A. Recent progress in the catalytic transformation of carbon dioxide into biosourced organic carbonates. *Green Chem.* **2021**. [CrossRef]

13. Tabanelli, T.; Giliberti, C.; Mazzoni, R.; Cucciniello, R.; Cavani, F. An innovative synthesis pathway to benzodioxanes: The peculiar reactivity of glycerol carbonate and catechol. *Green Chem.* **2019**, *21*, 329. [CrossRef]
14. Cucciniello, R.; Intiso, A.; Siciliano, T.; Palomares, A.P.; Martinez-Triguero, J.; Cerrillo, J.L.; Proto, A.; Rossi, F. Oxidative Degradation of Trichloroethylene over Fe_2O_3-doped Mayenite: Chlorine Poisoning Mitigation and Improved Catalytic Performance. *Catalysts* **2019**, *9*, 747. [CrossRef]
15. Rivas, F.C.; Rodriguez-Iznaga, I.; Berlier, G.; Ferro, D.T.; Concepcion-Rosabal, B.; Petranovskii, V. Fe Speciation in Iron Modified Natural Zeolites as Sustainable Environmental Catalysts. *Catalysts* **2019**, *9*, 866. [CrossRef]
16. Husnain, N.; Wang, E.; Fareed, S. Low-Temperature Selective Catalytic Reduction of NO with NH_3 over Natural Iron Ore Catalyst. *Catalysts* **2019**, *9*, 956. [CrossRef]
17. Ngoie, W.I.; Weiz, P.J.; Ikhu-Omoregbe, D.; Oyekola, O.O. Heterogeneous Nanomagnetic Catalyst from Cupriferous Mineral Processing Gangue for the Production of Biodiesel. *Catalysts* **2019**, *9*, 1047. [CrossRef]
18. Rehman, M.H.U.; Noor, T.; Iqbal, N. Effect of Zirconia on Hydrothermally Synthesized Co_3O_4/TiO_2 Catalyst for NO_x Reduction from Engine Emissions. *Catalysts* **2020**, *10*, 209. [CrossRef]
19. Zhou, C.; Liu, Z.; Fang, L.; Guo, Y.; Feng, Y.; Yang, M. Kinetic and Mechanistic Study of Rhodamine B Degradation by H_2O_2 and $Cu/Al_2O_3/g\text{-}C_3N_4$ Composite. *Catalysts* **2020**, *10*, 317. [CrossRef]
20. Sasaki, T.; Horino, Y.; Ohtake, T.; Ogawa, K.; Suzaki, Y. A Highly Efficient Monolayer Pt Nanoparticle Catalyst Prepared on a Glass Fiber Surface. *Catalysts* **2020**, *10*, 472. [CrossRef]
21. Villamarin-Barriga, E.; Canacuan, J.; Londono-Larrea, P.; Solis, H.; De La Rosa, A.; Saldarriaga, J.F.; Montero, C. Catalytic Cracking of Heavy Crude Oil over Iron-Based Catalyst Obtained from Galvanic Industry Wastes. *Catalysts* **2020**, *10*, 736. [CrossRef]
22. Li, J.; Lutz, M.; Gebbink, R.J.M.K. N-Donor Ligand Supported "ReO^{2+}": A Pre-Catalyst for the Deoxydehydration of Diols and Polyols. *Catalysts* **2020**, *10*, 754. [CrossRef]
23. Thakur, S.; Neogi, S.; Ray, A.K. Morphology-Controlled Synthesis of ZnO Nanostructures for Caffeine Degradation and *Escherichia coli* Inactivation in Water. *Catalysts* **2021**, *11*, 63. [CrossRef]
24. Muanruksa, P.; Winterburn, J.; Kaewkannetra, P. Biojet Fuel Production from Waste of Palm Oil Mill Effluent through Enzymatic Hydrolysis and Decarboxylation. *Catalysts* **2021**, *11*, 78. [CrossRef]
25. Al Soubaihi, R.M.; Saoud, K.M.; Myint, M.T.Z.; Gothelid, M.A.; Dutta, J. CO Oxidation Efficiency and Hysteresis Behavior over Mesoporous Pd/SiO_2 Catalyst. *Catalysts* **2021**, *11*, 131. [CrossRef]
26. Bagnato, G.; Sanna, A.; Paone, E.; Catizzone, E. Recent Catalytic Advances in Hydrotreatment Processes of Pyrolysis Bio-Oil. *Catalysts* **2021**, *11*, 157. [CrossRef]
27. Zhang, Z.; Wang, T.; Xiang, P.; Du, Q.; Han, S. Syntheses, Characterization, and Application of Tridentate Phenoxyimino-Phenoxy Aluminum Complexes for the Coupling of Terminal Epoxide with CO_2: From Binary System to Single Component Catalyst. *Catalysts* **2021**, *11*, 145. [CrossRef]

Communication

Base-Free Synthesis of Furfurylamines from Biomass Furans Using Ru Pincer Complexes

Danielle Lobo Justo Pinheiro and Martin Nielsen *

Department of Chemistry, Technical University of Denmark, DK-2800 Kgs. Lyngby, Denmark; dane@kemi.dtu.dk
* Correspondence: marnie@kemi.dtu.dk; Tel.: +45-24651045

Abstract: We report the first example of employing homogeneous organometal-catalyzed transfer hydrogenation for the selective reductive amination of furfurals to furfurylamines. An efficient, chemoselective, and base-free method is described using Ru-MACHO-BH as catalyst and *i*PrOH as H donor. The method tolerates a range of substituents affording moderate to excellent yields.

Keywords: transfer hydrogenation; furfurals; furfurylamine; reductive amination; Ru-MACHO

Citation: Pinheiro, D.L.J.; Nielsen, M. Base-Free Synthesis of Furfurylamines from Biomass Furans Using Ru Pincer Complexes. *Catalysts* **2021**, *11*, 558. https://doi.org/10.3390/catal11050558

Academic Editors: Raffaele Cucciniello, Daniele Cespi and Tommaso Tabanelli

Received: 30 March 2021
Accepted: 26 April 2021
Published: 28 April 2021

1. Introduction

The development of sustainable techniques to transform biomass into useful compounds is one of the biggest challenges of modern chemistry [1]. The introduction of nitrogen in biomass-derived compounds adds value and expands their industry applicability [2]. Furfurals are aldehydes derived from biomass and are identified as one of the key chemicals produced by the lignocellulosic biorefineries. Around 280 kTon are produced globally per year [3]. Furfurylamines (amines derived from furfurals) present diverse applications in the industry, including the preparation of pharmaceutical compounds such as Furesomide, Furtrethonium, an anti-hepatitis-B, and Barmastine (Figure 1), as well as polymers, antiseptic agents, agrochemicals, pesticides, and synthetic resins [1,2,4].

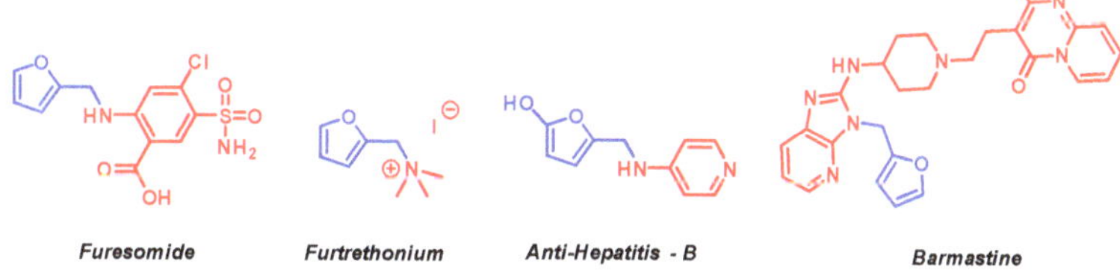

Figure 1. Pharmaceutical compounds containing furfurylamines.

The synthesis of furfurylamines from furfurals by reductive amination has been investigated using diverse reducing agents and catalysts. Studies involving hydrogen gas, silanes, borohydrides, and formic acid as reductants have been reported in the literature. Hydrogen gas as reductant is an interesting green tool; however, the method needs to operate under pressure of a highly flammable gas, increasing the operating cost. Nevertheless, there are many examples in the literature using H_2 as reductant for reductive amination with noble and non-noble metal catalysts such as Ru, Au, Ir, Pt, Ni, Co and Fe [5–11]. Although silane is obtained from waste residues of the silicon industry, their use is still in stoichiometric amounts, generating excessive amounts of waste [12–14]. The use of formic acid as H donor for the reductive amination of furfural was demonstrated as well. Cao and co-workers synthesized N-(furan-2-ylmethyl)aniline in 93% yield from nitrobenzene and furfural using Au/TiO_2-R as catalyst at 80 °C for 4 h [15]. Smith Jr and co-workers also employed formic acid as H donor, but used formamide as N source [16]. To the best of our knowledge, the only work involving an alcohol as H donor (*i*PrOH) for the synthesis of furfurylamines from furfural was reported by Yus [17]. In this work, the reaction between furfural and heptylamine using 20 mol% of NiNPs at 76 °C for 48 h afforded 30% yield of the furfurylamine.

One of the most powerful and robust methods for effective C–N bond formation of amines is the reductive amination of carbonyl compounds. [4,18–30]. This transformation features compelling advantages, such as simple operating setups, mild reaction conditions, direct use of available substrates, and inexpensive reagents [31]. The reductive amination using transfer hydrogenation for the synthesis of furfurylamines from furfurals is limited, even though this transformation as a synthetic tool is non-toxic, environmentally friendly, does not require flammable gasses, and employs a stable, easy to handle, and inexpensive source of hydrogen [4,32–37]. However, transfer hydrogenation catalysts typically require strong bases to be active, which can be detrimental for substrates that are base-sensitive [38]. Therefore, studies applying base-free conditions must be developed to avoid this drawback.

The use of homogenous metal catalysis has demonstrated great reactivity for transfer hydrogenation of carbonyl compounds and has been proven to hold many advantages [38–41]. In 2018, De Vries reported a base-free transfer hydrogenation of α,β-unsaturated ketones and aldehydes using the PNP pincer complex carbonylhydrido (tetrahydroborato)[bis(2-diphenylphosphinoethyl)amino]ruthenium(II) (Ru-MACHO-BH) as catalyst, in the presence of EtOH or *i*PrOH as H source and showed high activity and selectivity [42]. The amino-based Ru-PNP complexes are also very efficient catalysts for hydrogenation [43–49] and dehydrogenation [50–57] reactions. The high activity of these Ru PNP complexes in hydrogenations is often attributed to the presence of the Ru–H unit and N–H group [58].

Inspired by these works, we investigated the use of Ru-MACHO [59] (carbonylhydrido (tetrahydroborato)[bis(2-diphenylphosphinoethyl)amino]ruthenium(II)) and Ru-MACHO-BH complexes as potential catalysts for the transfer hydrogenation of the reductive amination in this work.

2. Results and Discussion

Our studies commenced with testing Ru-MACHO (1 mol%) as the catalyst for the transfer hydrogenation of the aldimine **1a** (Figure S1) in the presence of *i*PrOH (0.2 M of **1a**) as hydrogen source and KO*t*Bu (20 mol%) as additive at 90 °Cfor 3 h (Scheme 1). To our delight, the reaction afforded >99% conversion to furfurylamine **2a**. We then set out to evaluate the transfer hydrogenation of **1a** using varying catalyst loading, additives, temperatures, and reaction times with the aim of developing a mild protocol for this reaction.

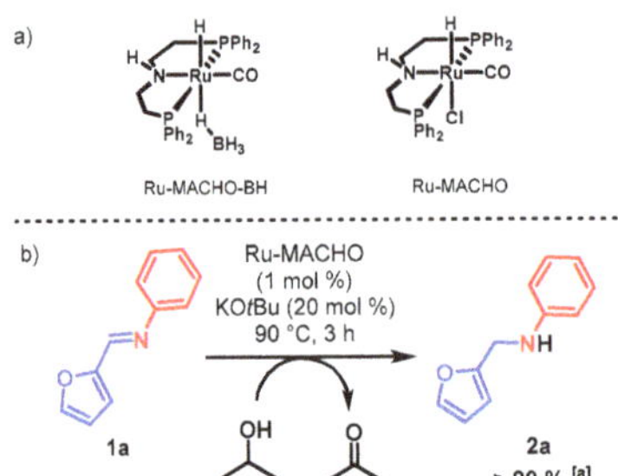

Scheme 1. (**a**) Ru-PNP catalysts used in this work. (**b**) Transfer hydrogenation of aldimine using Ru-MACHO. [a] Measured by ^{1}H NMR spectroscopy analysis of the crude reaction mixture.

Reducing the reaction time to 15 min, the catalyst loading of Ru-MACHO to 0.5 mol%, and the KO*t*Bu loading to 10 mol% still led to full conversion (Table 1, Entry 3). In fact, after 5 min, 51% was already converted (Entry 4). Changing the additive to NaOH had a detrimental effect, and only 18% conversion was observed. Likewise, lowering the catalyst loading to 0.1 mol% afforded less than 5% conversion. Changing the catalyst to Ru-MACHO-BH showed very low activity within 15 min, both with and without additive (Entries 6 and 7, respectively).

Table 1. Transfer hydrogenation of aldimines: Initial studies.

Entry [a]	Catalyst (mol%)	Additive [b]	Time	Conv. [c] (%)
1	Ru-MACHO (0.5)	KO*t*Bu	1 h	>99
2	Ru-MACHO (0.5)	KO*t*Bu	30 min	>99
3	Ru-MACHO (0.5)	KO*t*Bu	15 min	>99
4	Ru-MACHO (0.5)	KO*t*Bu	5 min	51
5	Ru-MACHO (0.5)	NaOH	15 min	18
6	Ru-MACHO (0.1)	KO*t*Bu	15 min	<5
7	Ru-MACHO-BH (0.5)	-	15 min	<5

[a] Reactions were carried out using 1.3 mmol of furfural and aniline in 7 mL *i*PrOH at 90 °C. [b] 10 mol% additive used. [c] Measured by ^{1}H NMR spectroscopy analysis of the crude reaction mixture.

Motivated by these initial positive results, the reductive amination of furfural with aniline was further investigated. Thus, in the presence of 10 mol% KO*t*Bu, 0.5 mol% of Ru-MACHO afforded >99% conversion after 18 h at 90 °C. However, the furfuryl alcohol (FA, **3**) appeared as a significant side product in a proportion of 7:3 (**2a**/**3**) (Scheme 2). Fortunately, introducing $MgSO_4$ as drying agent led to >99% conversion selectively to the desired product in 3 h (Table 2, Entry 2). Reducing the reaction time to 1 h decreased the selectivity to 93:7. Using Ru-MACHO-BH (0.5 mol%) and $MgSO_4$ but without the basic additive still resulted in 93% conversion after 1 h and with **2a** as the sole product by ^{1}H NMR analysis (Entry 3). Increasing the amount of aniline from 1.0 to 1.2 equivalent afforded >99% **2a** under otherwise identical conditions (Entry 5). Unfortunately, it was not possible to further reduce the reaction time without compromising the conversion and selectivity (Entries 6–8). Decreasing the amount of Ru-MACHO-BH to 0.25 mol% also led to a low conversion of 11% (Entry 9). Lowering the temperature to 70 °C resulted in practically no conversion (<5%, Entry 10). However, by increasing the temperature to 120 °C, it was possible to achieve exclusively **2a** with >99% conversion within 30 min (Entry 11).

A number of drying agents were then tested. Using Na_2SO_4 at 90 °C afforded >99% conversion in 1 h. However, the selectivity decreased to 97:3 (**2a**/**3**) (Entry 12). Decreasing the time further to 15 min maintained the full conversion but led to even lower selectivity, down to 57:42 (**2a**/**3**) (Entries 13–15). These observations suggest that the formation of **3** is highly reversible, and that **1a** is regenerated from **3** throughout the course of the reaction. Moreover, decreasing the reaction temperature to 70 °C led to merely 17% conversion (Entry 16). Molecular sieves (4 Å) were also evaluated and showed full conversion after 1 h, albeit with slightly lower selectivity (94:6 **2a**/**3**) (Entry 17). Decreasing the time further to 15 min maintained the full conversion but also led to lower selectivity, (71:29 **2a**/**3**) (Entry 18). The temperature was evaluated, and carrying out the reaction at 70 °C led to 71% conversion and 96:4 (**2a**/**3**) of selectivity (Entry 19).

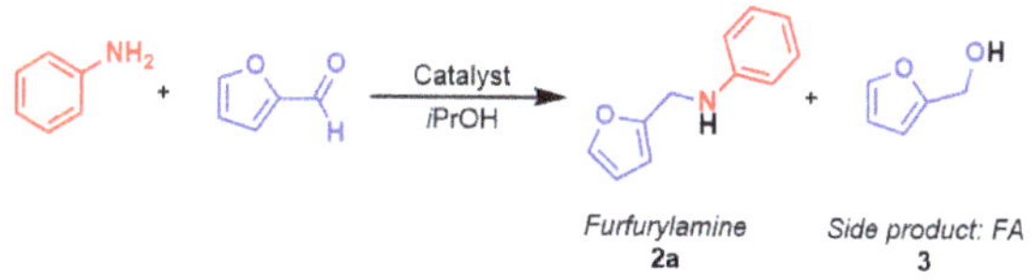

Scheme 2. Reductive amination between furfural and aniline.

Table 2. One-pot synthesis of furfurylamines: Optimization.

Entry [a]	Catalyst (mol%)	Additive [b]	Temperature (°C)	Time	Conversion [c] (%)	2a [c] (%)	3 [c] (%)
1	Ru-MACHO (0.5)	KO*t*Bu	90	18 h	>99	70	30
2	Ru-MACHO (0.5)	KO*t*Bu + $MgSO_4$	90	3 h	>99	>99	-
3	Ru-MACHO (0.5)	KO*t*Bu + $MgSO_4$	90	1 h	>99	93	7
4	Ru-MACHO-BH (0.5)	$MgSO_4$	90	1 h	93	>99	-
5 [d]	Ru-MACHO-BH (0.5)	$MgSO_4$	90	1 h	>99	>99	-
6 [d]	Ru-MACHO-BH (0.5)	$MgSO_4$	90	45 min	75	86	14
7 [d]	Ru-MACHO-BH (0.5)	$MgSO_4$	90	30 min	30	73	27
8 [d]	Ru-MACHO-BH (0.5)	-	90	30 min	15	52	48
9 [d]	Ru-MACHO-BH (0.25)	$MgSO_4$	90	1 h	11	-	>99
10 [d]	Ru-MACHO-BH (0.5)	$MgSO_4$	70	1 h	<5	-	-
11 [d]	Ru-MACHO-BH (0.5)	$MgSO_4$	120	30 min	>99	>99	-
12 [d]	Ru-MACHO-BH (0.5)	Na_2SO_4	90	1 h	>99	97	3
13 [d]	Ru-MACHO-BH (0.5)	Na_2SO_4	90	45 min	>99	90	10
14 [d]	Ru-MACHO-BH (0.5)	Na_2SO_4	90	30 min	>99	76	24
15 [d]	Ru-MACHO-BH (0.5)	Na_2SO_4	90	15 min	>99	57	42
16 [d]	Ru-MACHO-BH (0.5)	Na_2SO_4	70	1 h	17	72	28
17 [d]	Ru-MACHO-BH (0.5)	MS 4 Å	90	1 h	>99	94	6
18 [d]	Ru-MACHO-BH (0.5)	MS 4 Å	90	15 min	>99	71	29
19 [d]	Ru-MACHO-BH (0.5)	MS 4 Å	70	1 h	71	96	4

[a] Reactions were carried out using 1.3 mmol of furfural, aniline, and 1.3 mmol of drying agent in 7 mL *i*PrOH. [b] 10 mol% of KOtBu used. [c] Measured by ^{1}H NMR spectroscopy analysis of the crude reaction mixture. [d] Reactions were carried out using 1.2 equivalent of aniline. MS = Molecular sieves.

As seen in Figure 2, the levels of **1–3** differed significantly throughout the course of the reaction, depending on whether Na_2SO_4 or $MgSO_4$ was employed. Within 15 min, almost all **1a** had disappeared and 60% of **2a** had already been generated when using Na_2SO_4. Surprisingly, 35% of **3** was observed at this point. Hereafter, the reaction slowed significantly, and after 30 min, merely 70% of **2a** had been produced and **3** had only dropped to 22%. By contrast, with $MgSO_4$ the level of **3** did not exceed 15% throughout the entire course of the reaction, and after 30 min, it was 12%. At this time, there was still an ample amount of **1a** (45%) to undergo hydrogenation, and 43% of **2a** had been produced. This difference in amount of **1a** present during the course of the reaction might explain the superiority of $MgSO_4$ as drying agent after 60 min.

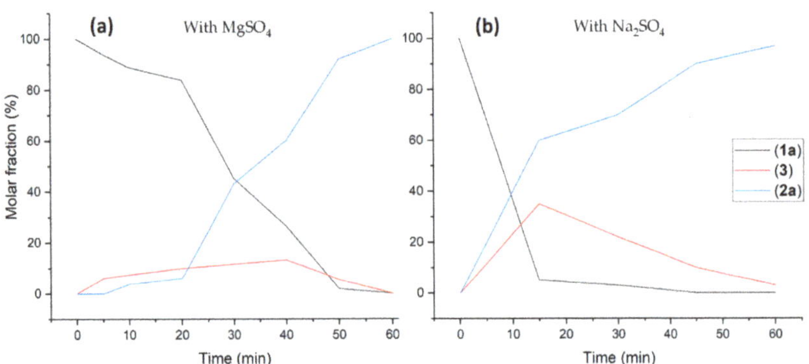

Figure 2. Monitoring the reaction of furfural with aniline using either $MgSO_4$ as drying agent (**a**) or Na_2SO_4 as drying agent (**b**). Reactions were carried out using 1.3 mmol of furfural, 1.2 equivalent aniline, and 1.3 mmol of drying agent in 7 mL *i*PrOH.

Therefore, although Na_2SO_4 and molecular sieves demonstrate higher conversion rates than $MgSO_4$, the latter drying agent was chosen due to the higher yield provided after 1 h of reaction time. Therefore, the conditions described in the Entry 5 in Table 2 were defined as standard conditions for the scope.

To assess the general applicability of the Ru-MACHO-BH as a catalyst for the one-pot synthesis of furfurylamines from furfurals and amines, various anilines were evaluated using the standard conditions (Scheme 3). Generally, moderate to excellent yields were obtained. The parent aniline afforded an excellent 93% of isolated product. Comparing the anilines containing either electron-donating or -withdrawing substituents, the latter group showed superior yield. As such, 4-F-$C_6H_4NH_2$, 4-CF_3-$C_6H_4NH_2$, and 4-aminopyridine generated the best yields of the substituted anilines with 74-95% of isolated products **2h–j**. The product **2j** is analogous to the anti-hepatitis-B compound shown in Figure 1, which demonstrates the direct applicability of the method for the synthesis of pharmacological activity compounds. On the other hand, a donating group (4-CH_3-$C_6H_4NH_2$) afforded lower yield of 61% of **2d**. This observation can perhaps be explained by the increased electronic deficiency of the imines when employing 4-CF_3-$C_6H_4NH_2$ as reagent [1]. Various halogens were tested as well and showed moderate to good yields (**2b**, **2e**, **2h**). Compounds with substituent in different positions, such as 3-Cl-$C_6H_4NH_2$ and 2-F-$C_6H_4NH_2$, showed good tolerance, yielding 60% and 56% of **2c** and **2f**, respectively. The method was also tested with the secondary amine *N*-methylaniline, which afforded the tertiary amine **2g** in high yield (89%). Unfortunately, no products were observed when employing various primary and secondary alkyl amines (*t*BuNH_2, *n*HepNH_2, Me_2NH, morpholine).

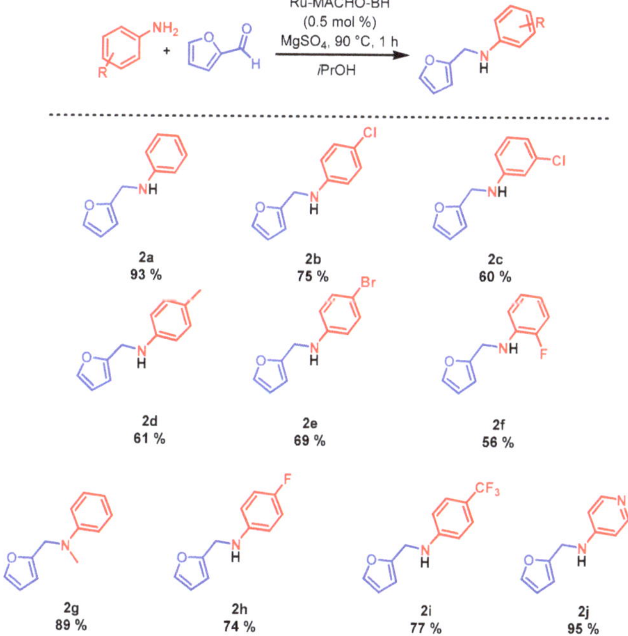

Scheme 3. One-pot synthesis of furfurylamines catalyzed by Ru-MACHO-BH. Reactions were carried out using 1.3 mmol of furfural, 1.56 mmol of aniline, and 1.3 mmol of $MgSO_4$ in 7 mL *i*PrOH. All yields are isolated.

5-(hydroxymethyl)furfural (HMF) and 5-methylfurfural are other important biomass-derived furans with industrial applications [60,61]. The furfurylamines derived from HMF are used in the synthesis of biopolymers (polyamides) and pharmaceuticals [4]. The *N*-(5-methylfurfuryl)aniline is a very important compound used in the synthesis of epoxyisonindoles and bioactive compounds such as anti-bacterial, anti-tuberculosis, anti-tumor, and anti-inflammatory entities [62–70]. Therefore, the method is an interesting alternative for the production of these valuable compounds. Hence, we also evaluated this compound as a potential substrate (Scheme 4). The reactions afforded a high yield of **4** (87%) and a moderate yield of **5** (54%).

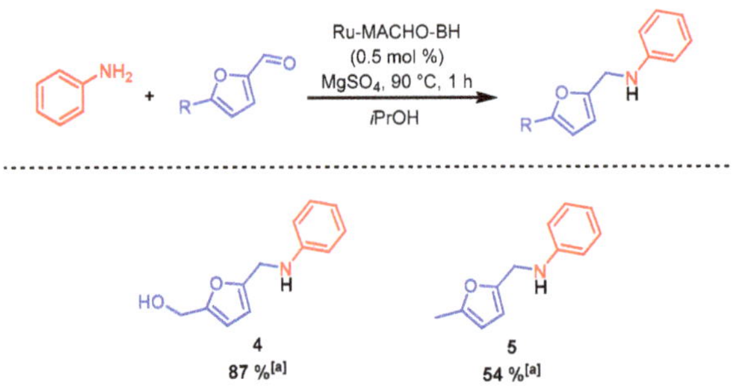

Scheme 4. One-pot synthesis of furfurylamines catalyzed by Ru-MACHO-BH. Reactions were carried out using 1.3 mmol of furfural, 1.56 mmol of aniline and 1.3 mmol of $MgSO_4$ in 7 mL of *i*PrOH. [a] Isolated yield.

3. Materials and Methods

3.1. Materials

Most chemicals were purchased from commercial suppliers and used without further purification unless otherwise stated. Hydroxymethylfurfural (HMF, 99%) (Sigma-Aldrich, St. Louis, MO, USA), furfural (99%) (Sigma-Aldrich, St. Louis, MO, USA), 5-methylfurfural (99%, Sigma-Aldrich, St. Louis, MO, USA), KO*t*Bu (99%, Sigma-Aldrich, St. Louis, MO, USA), *i*PrOH (anhydrous, 99.5%, Sigma-Aldrich, St. Louis, MO, USA), Ru-MACHO (Sigma-Aldrich, St. Louis, MO, USA), and Ru-MACHO-BH (Strem Chemicals, Newburyport, MA, USA) are commercially available and were used without further purification. All reactions dealing with air or moisture-sensitive compounds were performed using standard Schlenk techniques or in an argon-filled glovebox. The ^{1}H and ^{13}C NMR spectra were recorded on a Bruker Avance III 400 MHz spectrometer (Bruker, Billerica, MA, USA) and were referenced to the solvent peak. The software MestReNova version 11.0.0-17609 (Mestrelab, Escondido, CA, USA, 2016) was used for NMR analysis. The software OriginPro 2019 9.6.0.172 (Academic) (OriginLab, Northampton, MA, USA, 2019) was used for graphic plot. All the products are literature known compounds, and the experimental data (^{1}H and $^{13}C\{^{1}H\}$ NMR spectra) fit those reported.

3.2. Methods

3.2.1. Preparation of Aldimine 1a

A mixture of furfural (54 mmol), aniline (54 mmol) and methanol (0.5 M) in the presence of MS (4 Å) was stirred at room temperature for 3 h. After completion of the reaction, the crude mixture was filtered off and evaporated under reduced pressure. The product **1a** was obtained as a brown oil, 7.83 g, 85%.

3.2.2. General Procedure for Transfer Hydrogenation of Aldimine 1a Catalyzed by Ru-PNP Complexes

A Schlenk pressure vessel containing catalyst, additive and magnetic bar was sealed and flushed with argon (three times). The solvent and H-donor (*i*-PrOH) was introduced by a needle and stirred at 90 °C. After 10 min, the aldimine **1a** was added to the solution. After a certain reaction time (5–18 h), the reaction was stopped, and the crude was analyzed. The conversion was determined by spectroscopy ^{1}H NMR.

3.2.3. General Procedure for One-Pot Reductive Amination of Furfural

In a Schlenk pressure vessel containing Ru-MACHO-BH (0.5 mol %) and $MgSO_4$ (1.3 mmol), a magnetic stirring bar was added and the vessel was sealed and flushed with argon (three times). During argon flow, 4.5 mL of *i*PrOH was introduced by a needle and the solution was heated at 90 °C and stirred for 10 min. In a flame-dried screw-cap vial, aniline (1.56 mmol) and furfural (1.3 mmol) were mixed with 2.5 mL of *i*PrOH (to provide a solution with furfural concentration of 0.18 M) under argon flow. The atmosphere was replaced with argon and the solution was introduced to the Schlenk pressure vessel. The reaction mixture was kept at 90 °C for 1 h. The crude reaction mixture was evaporated under reduced pressure, and the product was obtained after purification through chromatography column (Ethyl acetate/pentane, 90:10). For the optimization process, the method of employing relative conversions as measured by NMR was confirmed with respect to absolute values by a single duplicate test reaction using mesitylene as internal standard.

4. Conclusions

In conclusion, we report the first example of an efficient base free one-pot transfer hydrogenative reductive amination of furfural for the synthesis of furfurylamines under mild conditions, employing low amounts of the commercially available catalyst Ru-MACHO-BH and *i*PrOH as H donor. The general applicability of the method is demonstrated by the use of furfural and various anilines with different substituents, which afforded yields that varied from moderate to excellent (56–93%). Furthermore, this chemoselective methodology established a high yield (83%) in the synthesis of the furfurylamine derived from HMF and a moderate yield (54%) from *N*-(5-methylfurfuryl)aniline.

Supplementary Materials: The following are available online at https://www.mdpi.com/article/10.3390/catal11050558/s1, Table S1: Monitoring the reaction of furfural and aniline using $MgSO_4$ as drying agent. Table S2: Monitoring the reaction of furfural and aniline using Na_2SO_4 as drying agent. Figure S1: ^{1}H NMR spectrum of **1a** (400 MHz, $CDCl_3$), Figure S2: ^{13}C NMR spectrum of **1a** (100 MHz, $CDCl_3$), Figure S3: ^{1}H NMR spectrum of **2a** (400 MHz, $CDCl_3$), Figure S4: ^{13}C NMR spectrum of **2a** (100 MHz, $CDCl_3$), Figure S5: ^{1}H NMR spectrum of **2b** (400 MHz, $CDCl_3$), Figure S6: ^{13}C NMR spectrum of **2b** (100 MHz, $CDCl_3$), Figure S7: ^{1}H NMR spectrum of **2c** (400 MHz, $CDCl_3$), Figure S8: ^{13}C NMR spectrum of **2c** (100 MHz, $CDCl_3$), Figure S9: ^{1}H NMR spectrum of **2d** (400 MHz, $CDCl_3$), Figure S10: ^{13}C NMR spectrum of **2d** (100 MHz, $CDCl_3$), Figure S11: ^{1}H NMR spectrum of **2e** (400 MHz, $CDCl_3$), Figure S12: ^{13}C NMR spectrum of **2e** (100 MHz, $CDCl_3$), Figure S13: ^{1}H NMR spectrum of **2f** (400 MHz, $CDCl_3$), Figure S14: ^{13}C NMR spectrum of **2f** (100 MHz, $CDCl_3$), Figure S15: ^{1}H NMR spectrum of **2g** (400 MHz, $CDCl_3$), Figure S16: ^{13}C NMR spectrum of **2g** (100 MHz, $CDCl_3$), Figure S17: ^{1}H NMR spectrum of **2h** (400 MHz, CD_3OD), Figure S18: ^{13}C NMR spectrum of **2h** (100 MHz, $CDCl_3$), Figure S19: ^{1}H NMR spectrum of **2i** (400 MHz, $CDCl_3$), Figure S20: ^{13}C NMR spectrum of **2i** (100 MHz, $CDCl_3$), Figure S21: ^{1}H NMR spectrum of **2j** (400 MHz, $CDCl_3$), Figure S22: ^{13}C NMR spectrum of **2j** (100 MHz, $CDCl_3$), Figure S23: ^{1}H NMR spectrum of **4** (400 MHz, $CDCl_3$), Figure S24: ^{13}C NMR spectrum of **4** (100 MHz, $CDCl_3$), Figure S25: ^{1}H NMR spectrum of **5** (400 MHz, $CDCl_3$), Figure S26: ^{13}C NMR spectrum of **5** (100 MHz, $CDCl_3$).

Author Contributions: D.L.J.P. did the experimental part. M.N. did funding acquisition and project administration. Everything else, from conceptualization to manuscript writing, D.L.J.P. and M.N. did equally. All authors have read and agreed to the published version of the manuscript.

Funding: This work was supported by a research grant (19049) from VILLUM FONDEN.

Conflicts of Interest: The authors declare no conflict of interest.

References

1. Caetano, J.A.T.; Fernandes, A.C. One-pot synthesis of amines from biomass resources catalyzed by $HReO_4$. *Green Chem.* **2018**, *20*, 2494–2498. [CrossRef]
2. Dunbabin, A.; Subrizi, F.; Ward, J.M.; Sheppard, T.D.; Hailes, H.C. Furfurylamines from biomass: Transaminase catalysed upgrading of furfurals. *Green Chem.* **2017**, *19*, 397–404. [CrossRef]

3. Mariscal, R.; Maireles-Torres, P.; Ojeda, M.; Sádabaa, I.; Granados, M.L. Furfural: A renewable and versatile platform molecule for the synthesis of chemicals and fuels. *Energy Environ. Sci.* **2016**, *9*, 1144–1189. [CrossRef]
4. He, J.; Chen, L.; Liu, S.; Song, K.; Yang, S.; Riisager, A. Sustainable access to renewable N-containing chemicals from reductive amination of biomass-derived platform compounds. *Green Chem.* **2020**, *22*, 6714–6747. [CrossRef]
5. Chieffi, G.; Braun, M.; Esposito, D. Continuous reductive amination of biomass-derived molecules over carbonized filter paper-supported FeNi alloy. *ChemSusChem* **2015**, *8*, 3590–3594. [CrossRef]
6. Deng, D.; Kita, Y.; Kamata, K.; Hara, M. Low-Temperature Reductive Amination of Carbonyl Compounds over Ru Deposited on $Nb_2O_5 \cdot nH_2O$. *ACS Sustain. Chem. Eng.* **2019**, *7*, 4692–4698. [CrossRef]
7. Laroche, B.; Ishitani, H.; Kobayashi, S. Direct Reductive Amination of Carbonyl Compounds with H_2 Using Heterogeneous Catalysts in Continuous Flow as an Alternative to N-Alkylation with Alkyl Halides. *Adv. Synth. Catal.* **2018**, *360*, 4699–4704. [CrossRef]
8. Gould, N.S.; Landfield, H.; Dinkelacker, B.; Brady, C.; Yang, X.; Xu, B. Selectivity Control in Catalytic Reductive Amination of Furfural to Furfurylamine on Supported Catalysts. *ChemCatChem* **2020**, *12*, 2106–2115. [CrossRef]
9. Murugesan, K.; Senthamarai, T.; Chandrashekhar, V.G.; Natte, K.; Kamer, P.C.J.; Beller, M.; Jagadeesh, R.V. Catalytic reductive aminations using molecular hydrogen for synthesis of different kinds of amines. *Chem. Soc. Rev.* **2020**, *49*, 6273–6328. [CrossRef] [PubMed]
10. Zhou, K.; Chen, B.; Zhou, X.; Kang, S.; Xu, Y.; Wei, J. Selective Synthesis of Furfurylamine by Reductive Amination of Furfural over Raney Cobalt. *ChemCatChem* **2019**, *11*, 5562–5569. [CrossRef]
11. Dong, C.; Wang, H.; Du, H.; Peng, J.; Cai, Y.; Guo, S.; Zhang, J.; Samart, C.; Ding, M. Ru/HZSM-5 as an efficient and recyclable catalyst for reductive amination of furfural to furfurylamine. *Mol. Catal.* **2020**, *482*, 110755. [CrossRef]
12. Carrillo, A.I.; Llanes, P.; Pericàs, M.A. A versatile, immobilized gold catalyst for the reductive amination of aldehydes in batch and flow. *React. Chem. Eng.* **2018**, *3*, 714–721. [CrossRef]
13. Maya, R.J.; Poulose, S.; John, J.; Varma, R.L. Direct Reductive Amination of Aldehydes via Environmentally Benign Bentonite-Gold Nanohybrid Catalysis. *Adv.Synth. Catal.* **2017**, *359*, 1177–1184. [CrossRef]
14. Mirza-Aghayan, M.; Kalantari, M.; Boukherroub, R. Palladium oxide nanoparticles supported on graphene oxide: A convenient heterogeneous catalyst for reduction of various carbonyl compounds using triethylsilane. *Appl. Organomet. Chem.* **2019**, *33*, 1–11. [CrossRef]
15. Zhang, Q.; Li, S.; Zhu, M.; Liu, Y.; He, H.; Cao, Y. Direct reductive amination of aldehydes with nitroarenes using bio-renewable formic acid as a hydrogen source. *Green Chem.* **2016**, *18*, 2507–2513. [CrossRef]
16. Li, H.; Guo, H.; Su, Y.; Hiraga, Y.; Fang, Z.; Watanabe, M.; Lee, R.; Smith, R.L., Jr.; Hensen, E.J.M. N-formyl-stabilizing quasi-catalytic species afford rapid and selective solvent-free amination of biomass-derived feedstocks. *Nat. Commun.* **2019**, *10*, 699. [CrossRef] [PubMed]
17. Guillena, G.; Ramo, D.J.; Yus, M. Hydrogen Autotransfer in the N-Alkylation of Amines and Related Compounds using Alcohols and Amines as Electrophiles. *Chem. Rev.* **2010**, *110*, 1611–1641. [CrossRef] [PubMed]
18. Irrgang, T.; Kempe, R. Transition-metal-catalyzed reductive amination employing hydrogen. *Chem. Rev.* **2020**, *120*, 9583–9674. [CrossRef] [PubMed]
19. Wang, Y.; Furukawa, S.; Fu, X.; Yan, N. Organonitrogen chemicals from oxygen-containing feedstock over heterogeneous catalysts. *ACS Catal.* **2020**, *10*, 311–335. [CrossRef]
20. Saberi, A.A. Recent advances in percolation theory and its applications. *Phys. Rep.* **2015**, *578*, 1–32. [CrossRef]
21. Chen, W.; Sun, Y.; Du, J.; Si, Z.; Tang, X.; Zeng, X.; Lin, L.; Liu, S.; Lei, T. Preparation of 5-(Aminomethyl)-2-furanmethanol by direct reductive amination of 5-Hydroxymethylfurfural with aqueous ammonia over the Ni/SBA-15 catalyst. *J. Chem. Technol. Biotechnol.* **2018**, *93*, 3028–3034. [CrossRef]
22. Nuzhdin, A.L.; Bukhtiyarova, M.V.; Bukhtiyarova, G.A. Cu-Al mixed oxide derived from layered double hydroxide as an efficient catalyst for continuous-flow reductive amination of aromatic aldehydes. *J. Chem. Technol. Biotechnol.* **2020**, *95*, 3292–3299. [CrossRef]
23. Nuzhdin, A.L.; Simonov, P.A.; Bukhtiyarova, G.A.; Eltsov, I.V.; Bukhtiyarov, V.I. Reductive amination of 5-acetoxymethylfurfural over Pt/Al_2O_3 catalyst in a flow reactor. *Mol. Catal.* **2021**, *499*, 111297. [CrossRef]
24. Galkin, K.I.; Ananikov, V.P. The Increasing Value of Biomass: Moving From C6 Carbohydrates to Multifunctionalized Building Blocks via 5-(hydroxymethyl)furfural. *ChemistryOpen* **2020**, *9*, 1135–1148. [CrossRef]
25. Lancien, A.; Wojcieszak, R.; Cuvelier, E.; Duban, M.; Dhulster, P.; Paul, S.; Dumeignil, F.; Froidevaux, R.; Heuson, E. Hybrid Conversion of 5-Hydroxymethylfurfural to 5-Aminomethyl-2-furancarboxylic acid: Toward New Bio-sourced Polymers. *ChemCatChem* **2021**, *13*, 247–259. [CrossRef]
26. Yang, Z.Y.; Hao, Y.C.; Hu, S.Q.; Zong, M.H.; Chen, Q.; Li, N. Direct Reductive Amination of Biobased Furans to N-Substituted Furfurylamines by Engineered Reductive Aminase. *Adv. Synth. Catal.* **2021**, *363*, 1033–1037. [CrossRef]
27. García-Ortiz, A.; Vidal, J.D.; Climent, M.J.; Concepción, P.; Corma, A.; Iborra, S. Chemicals from Biomass: Selective Synthesis of N-Substituted Furfuryl Amines by the One-Pot Direct Reductive Amination of Furanic Aldehydes. *ACS Sustain. Chem. Eng.* **2019**, *7*, 6243–6250. [CrossRef]
28. Wei, D.; Bruneau-Voisine, A.; Dubois, M.; Bastin, S.; Sortais, J.B. Manganese-Catalyzed Transfer Hydrogenation of Aldimines. *ChemCatChem* **2019**, *11*, 5256–5259. [CrossRef]

29. Tanaka, K.; Miki, T.; Murata, K.; Yamaguchi, A.; Kayaki, Y.; Kuwata, S.; Ikariya, T.; Watanabe, M. Reductive amination of ketonic compounds catalyzed by Cp*Ir(III) complexes bearing a picolinamidato ligand. *J. Org. Chem.* **2019**, *84*, 10962–10977. [CrossRef]
30. Yang, M.L.; Wu, Y.X.; Liu, Y.; Qiu, J.J.; Liu, C.M. A novel bio-based AB2 monomer for preparing hyperbranched polyamides derived from levulinic acid and furfurylamine. *Polym. Chem.* **2019**, *10*, 6217–6226. [CrossRef]
31. Chatterjee, M.; Ishizaka, T.; Kawanami, H. Reductive amination of furfural to furfurylamine using aqueous ammonia solution and molecular hydrogen: An environmentally friendly approach. *Green Chem.* **2016**, *18*, 487–496. [CrossRef]
32. Piccirilli, L.; Pinheiro, D.L.J.; Nielsen, M. Recent progress with pincer transition metal catalysts for sustainability. *Catalysts* **2020**, *10*, 773. [CrossRef]
33. Wang, D.; Astruc, D. The Golden Age of Transfer Hydrogenation. *Chem. Rev.* **2015**, *115*, 6621–6686. [CrossRef]
34. Wang, C.; Wu, X.; Xiao, J. Broader, greener, and more efficient: Recent advances in asymmetric transfer hydrogenation. *Chem. Asian J.* **2008**, *3*, 1750–1770. [CrossRef]
35. Farrar-tobar, R.A.; Dell'Acqua, A.; Tin, S.; de Vries, J.G. Metal-catalysed selective transfer hydrogenation of α,β-unsaturated carbonyl compounds to allylic alcohols. *Green Chem.* **2020**, *22*, 3323–3357. [CrossRef]
36. Clapham, S.E.; Hadzovic, A.; Morris, R.H. Mechanisms of the H2-hydrogenation and transfer hydrogenation of polar bonds catalyzed by ruthenium hydride complexes. *Coord. Chem. Rev.* **2004**, *248*, 2201–2237. [CrossRef]
37. Werkmeister, S.; Neumann, J.; Junge, K.; Beller, M. Pincer-Type Complexes for Catalytic (De)Hydrogenation and Transfer (De)Hydrogenation Reactions: Recent Progress. *Chem. Eur. J.* **2015**, *21*, 12226–12250. [CrossRef]
38. Farrar-Tobar, R.A.; Wozniak, B.; Savini, A.; Hinze, S.; Tin, S.; de Vries, J.G. Base-Free Iron Catalyzed Transfer Hydrogenation of Esters Using EtOH as Hydrogen Source. *Angew. Chem. Int. Ed.* **2019**, *58*, 1129–1133. [CrossRef] [PubMed]
39. Clarke, Z.E.; Maragh, P.T.; Dasgupta, T.P.; Gusev, D.G.; Lough, A.J.; Abdur-Rashid, K. A family of active iridium catalysts for transfer hydrogenation of ketones. *Organometallics* **2006**, *25*, 4113–4117. [CrossRef]
40. Castellanos-blanco, N.; Arévalo, A.; García, J.J. Nickel-catalyzed transfer hydrogenation of ketones using ethanol as a solvent and a hydrogen donor. *Dalt. Trans.* **2016**, *45*, 13604–13614. [CrossRef]
41. Aboo, A.H.; Begum, R.; Zhao, L.; Farooqi, Z.H.; Xiao, J. Methanol as hydrogen source: Chemoselective transfer hydrogena- tion of α,β -unsaturated ketones with a rhodacycle. *Chin. J. Catal.* **2019**, *40*, 1795–1799. [CrossRef]
42. Farrar-tobar, R.A.; Wei, Z.; Jiao, H.; Hinze, S.; Vries, J.G. De Selective Base-free Transfer Hydrogenation of α,β-Unsaturated Carbonyl Compounds using *i*PrOH or EtOH as Hydrogen Source. *Chem. Eur.J.* **2018**, *24*, 2725–2734. [CrossRef]
43. Padilla, R.; Koranchalil, S.; Nielsen, M. Efficient and selective catalytic hydrogenation of furanic aldehydes using well defined Ru and Ir pincer complexes. *Green Chem.* **2020**, *22*, 6767–6772. [CrossRef]
44. Padilla, R.; Nielsen, M.; Jørgensen, M.S.B. Efficient catalytic hydrogenation of alkyl levulinates to γ-valerolactone. *Green Chem.* **2019**, *21*, 5195–5200. [CrossRef]
45. Garbe, M.; Wei, Z.; Tannert, B.; Spannenberg, A.; Jiao, H.; Bachmann, S.; Scalone, M.; Junge, K.; Beller, M. Enantioselective Hydrogenation of Ketones using Different Metal Complexes with a Chiral PNP Pincer Ligand. *Adv. Synth. Catal.* **2019**, *361*, 1913–1920. [CrossRef]
46. Guan, C.; Pan, Y.; Ang, E.P.L.; Hu, J.; Yao, C.; Huang, M.H.; Li, H.; Lai, Z.; Huang, K.W. Conversion of CO_2 from air into formate using amines and phosphorus-nitrogen PN^3P-Ru(II) pincer complexes. *Green Chem.* **2018**, *20*, 4201–4205. [CrossRef]
47. Neumann, J.; Bornschein, C.; Jiao, H.; Junge, K.; Beller, M. Hydrogenation of Aliphatic and Aromatic Nitriles Using a Defined Ruthenium PNP Pincer Catalyst. *Eur. J. Org. Chem.* **2015**, *2015*, 5944–5948. [CrossRef]
48. Filonenko, G.A.; Van Putten, R.; Schulpen, E.N.; Hensen, E.J.M.; Pidko, E.A. Highly efficient reversible hydrogenation of carbon dioxide to formates using a ruthenium PNP-pincer catalyst. *ChemCatChem* **2014**, *6*, 1526–1530. [CrossRef]
49. Filonenko, G.A.; Hensen, E.J.M.; Pidko, E.A. Mechanism of CO_2 hydrogenation to formates by homogeneous Ru-PNP pincer catalyst: From a theoretical description to performance optimization. *Catal. Sci. Technol.* **2014**, *4*, 3474–3485. [CrossRef]
50. Oldenhuis, N.J.; Dong, V.M.; Guan, Z. Catalytic acceptorless dehydrogenations: Ru-Macho catalyzed construction of amides and imines. *Tetrahedron* **2014**, *70*, 4213–4218. [CrossRef]
51. Agapova, A.; Alberico, E.; Kammer, A.; Junge, H.; Beller, M. Catalytic Dehydrogenation of Formic Acid with Ruthenium-PNP-Pincer Complexes: Comparing N-Methylated and NH-Ligands. *ChemCatChem* **2019**, *11*, 1910–1914. [CrossRef]
52. Bertoli, M.; Choualeb, A.; Lough, A.J.; Moore, B.; Spasyuk, D.; Gusev, D.G. Osmium and ruthenium catalysts for dehydrogenation of alcohols. *Organometallics* **2011**, *30*, 3479–3482. [CrossRef]
53. Alberico, E.; Lennox, A.J.J.; Vogt, L.K.; Jiao, H.; Baumann, W.; Drexler, H.J.; Nielsen, M.; Spannenberg, A.; Checinski, M.P.; Junge, H.; et al. Unravelling the Mechanism of Basic Aqueous Methanol Dehydrogenation Catalyzed by Ru-PNP Pincer Complexes. *J. Am. Chem. Soc.* **2016**, *138*, 14890–14904. [CrossRef]
54. Nielsen, M.; Alberico, E.; Baumann, W.; Drexler, H.J.; Junge, H.; Gladiali, S.; Beller, M. Low-temperature aqueous-phase methanol dehydrogenation to hydrogen and carbon dioxide. *Nature* **2013**, *495*, 85–89. [CrossRef]
55. Sponholz, P.; Mellmann, D.; Cordes, C.; Alsabeh, P.G.; Li, B.; Li, Y.; Nielsen, M.; Junge, H.; Dixneuf, P.; Beller, M. Efficient and Selective Hydrogen Generation from Bioethanol using Ruthenium Pincer-type Complexes. *ChemSusChem* **2014**, *7*, 2419–2422. [CrossRef]
56. Li, Y.; Nielsen, M.; Li, B.; Dixneuf, P.H.; Junge, H.; Beller, M. Ruthenium-catalyzed hydrogen generation from glycerol and selective synthesis of lactic acid. *Green Chem.* **2015**, *17*, 193–198. [CrossRef]

57. Nielsen, M.; Junge, H.; Kammer, A.; Beller, M. Towards a green process for bulk-scale synthesis of ethyl acetate: Efficient acceptorless dehydrogenation of ethanol. *Angew. Chem. Int. Ed.* **2012**, *51*, 5711–5713. [CrossRef]
58. Dub, P.A.; Gordon, J.C. The role of the metal-bound N–H functionality in Noyori-type molecular catalysts. *Nat. Rev. Chem.* **2018**, *2*, 396–408. [CrossRef]
59. Kuriyama, W.; Matsumoto, T.; Ogata, O.; Ino, Y.; Aoki, K.; Tanaka, S.; Ishida, K.; Kobayashi, T.; Sayo, N.; Saito, T. Catalytic Hydrogenation of Esters. Development of an Efficient Catalyst and Processes for Synthesising (*R*)-1,2-Propanediol and 2-(*l*-Menthoxy)ethanol. *Org. Process Res. Dev.* **2012**, *16*, 166–171. [CrossRef]
60. Hu, L.; Lin, L.; Wu, Z.; Zhou, S.; Liu, S. Recent advances in catalytic transformation of biomass-derived 5-hydroxymethylfurfural into the innovative fuels and chemicals. *Renew. Sustain. Energy Rev.* **2017**, *74*, 230–257. [CrossRef]
61. Hou, Q.; Qi, X.; Zhen, M.; Qian, H.; Nie, Y.; Bai, C.; Zhang, S.; Bai, X.; Ju, M. Biorefinery roadmap based on catalytic production and upgrading 5-hydroxymethylfurfural. *Green Chem.* **2021**, *23*, 119–231. [CrossRef]
62. Xiao, J.; Jin, Q.; Yang, J.; Xiong, L.; Qiu, J.; Jiang, J.; Peng, Y.; Li, T.; Qiu, Z.; Yang, W. Catalytic Synthesis of N-(5-Methylfurfuryl)aniline from Bio-Derived Carbohydrates. *Asian J. Org. Chem.* **2019**, *8*, 328–334. [CrossRef]
63. Zubkov, F.I.; Nikitina, E.V.; Galeev, T.R.; Zaytsev, V.P.; Khrustalev, V.N.; Novikov, R.A.; Orlova, D.N.; Varlamov, A.V. General synthetic approach towards annelated 3a,6-epoxyisoindoles by tandem acylation/IMDAF reaction of furylazaheterocycles. Scope and limitations. *Tetrahedron* **2014**, *70*, 1659–1690. [CrossRef]
64. Wu, J.; Darcel, C. Iron-Catalyzed Hydrogen Transfer Reduction of Nitroarenes with Alcohols: Synthesis of Imines and Aza Heterocycles. *J. Org. Chem.* **2021**, *86*, 1023–1036. [CrossRef]
65. Ge, C.; Sang, X.; Yao, W.; Zhang, L.; Wang, D. Unsymmetrical indazolyl-pyridinyl-triazole ligand-promoted highly active iridium complexes supported on hydrotalcite and its catalytic application in water. *Green Chem.* **2018**, *20*, 1805–1812. [CrossRef]
66. Weickmann, D.; Frey, W.; Plietker, B. Synchronizing steric and electronic effects in {Ru^{II}(NNNN,P)} complexes: The catalytic dehydrative alkylation of anilines by using alcohols as a case study. *Chem. Eur. J.* **2013**, *19*, 2741–2748. [CrossRef]
67. Iovel, I.; Golomba, L.; Popelis, J.; Grinberga, S.; Lukevics, E. Synthesis and hydrosilylation of furan and thiophene N- methylene-fluoroanilines in the presence of Pd(I) complex. *Chem. Heterocycl. Compd.* **2005**, *41*, 1112–1118. [CrossRef]
68. Lim, C.H.; Kudisch, M.; Liu, B.; Miyake, G.M. C-N Cross-Coupling via Photoexcitation of Nickel-Amine Complexes. *J. Am. Chem. Soc.* **2018**, *140*, 7667–7673. [CrossRef]
69. Ware, R.W.; Hinkley, L.A.; Hardeman, K.P.; Jenks, M.G. Substituted Quinoline and Quinazoline Inhibitors of Quinone Reductase 2. U.S. Patent Application No. WO2006034235A3, 6 April 2006.
70. Nuzhdin, A.L.; Bukhtiyarova, M.V.; Bukhtiyarov, V.I. Two-Step One-Pot Reductive Amination of Furanic Aldehydes Using CuAlOx Catalyst in a Flow Reactor. *Molecules* **2020**, *25*, 4771. [CrossRef]

 catalysts

MDPI

Review

Recent Catalytic Advances in Hydrotreatment Processes of Pyrolysis Bio-Oil

Giuseppe Bagnato [1], Aimaro Sanna [2], Emilia Paone [3,*] and Enrico Catizzone [4]

1 School of Chemistry and Chemical Engineering, Queen's University Belfast, David Keir Building, 39-123 Stranmillis Rd, Belfast BT9 5AG, UK; G.Bagnato@qub.ac.uk
2 Advanced Biofuels Lab, Institute of Mechanical, Process and Energy Engineering, School of Engineering & Physical Sciences, Heriot-Watt University, Edinburgh EH14 4AS, UK; A.Sanna@hw.ac.uk
3 Dipartimento di Ingegneria Industriale (DIEF), Università degli Studi di Firenze, Via di S. Marta 3, I-50139 Firenze, Italy
4 ENEA—Italian National Agency for New Technologies, Energy and Sustainable Economic Development, Research Centre of Trisaia, I-75026 Rotondella, Italy; enrico.catizzone@enea.it
* Correspondence: emilia.paone@unifi.it; Tel.: +39-096-5169-2278

Abstract: Catalytic hydrotreatment (HT) is one of the most important refining steps in the actual petroleum-based refineries for the production of fuels and chemicals, and it will play also a crucial role for the development of biomass-based refineries. In fact, the utilization of HT processes for the upgrading of biomass and/or lignocellulosic residues aimed to the production of synthetic fuels and chemical intermediates represents a reliable strategy to reduce both carbon dioxide emissions and fossil fuels dependence. At this regard, the catalytic hydrotreatment of oils obtained from either thermochemical (e.g., pyrolysis) or physical (e.g., vegetable seeds pressing) processes allows to convert biomass-derived oils into a biofuel with properties very similar to conventional ones (so-called drop-in biofuels). Similarly, catalytic hydro-processing also may have a key role in the valorization of other biorefinery streams, such as lignocellulose, for the production of high-added value chemicals. This review is focused on recent hydrotreatment developments aimed to stabilizing the pyrolytic oil from biomasses. A particular emphasis is devoted on the catalyst formulation, reaction pathways, and technologies.

Keywords: pyrolysis oils; catalytic hydrotreatment; heterogeneous catalysis; hydrogenation; biorefinery; green chemistry

Citation: Bagnato, G.; Sanna, A.; Paone, E.; Catizzone, E. Recent Catalytic Advances in Hydrotreatment Processes of Pyrolysis Bio-Oil. *Catalysts* **2021**, *11*, 157. https://doi.org/10.3390/catal11020157

Received: 28 December 2020
Accepted: 21 January 2021
Published: 23 January 2021

1. Introduction

In a green and sustainable perspective, the world is moving from a strong fossil fuels' dependence to a consistent use of renewable feedstocks. In this view, Anastas and Green proposed in 1998 "the 12 principles of green chemistry" [1], where a particular attention was also given to (second and third generation) transportation biofuels, chemicals, commodities, and pharmaceuticals directly produced from biomass in modern biorefineries [2–6]. This transition is given not only by the matured awareness that fossil resources are running out, but it is mostly accelerated by the United Nation decision to adopt the 2030 Agenda for Sustainable Development, a program action of 17 ambitious goals (SDGs) and 169 targets aimed to eradicate the poverty, to protect the planet, and to ensure the prosperity for all [7]. Biomasses, that currently supply about 80% of global renewable energy and a low-emissions character, represent a unique sustainable pathway to successfully address SDGs [1,7,8]. Among several technologies that can use biomass waste as the feedstock to produce energy fuels, power, heat, and various high value-added chemicals [9–14], an interesting example is the use of lignocellulose (plant based biomasses mainly composed of cellulose, hemicellulose, and lignin) and microalgae (biomasses with high protein and carbohydrate content characterized by the absence of lignin) for the production of bio-oil that can be used as intermediate for the production of liquid bio-fuels [15]. Bio-oil is a dark brown-red colored liquid, with a characteristic smell of smoke and a chemical composition strictly related to the biomass feedstocks containing a wide number of unique compounds generated from the rapid quenching of pyrolytic fragments of lignocellulose [16]. Figure 1 shows the main compounds present in the bio-oil: an aqueous solution of several products derived from the fragmentation of cellulose and hemicellulose and from the depolymerization of lignin. The mixture consists of various organic compounds (20–30 wt%), water (19–20 wt%), water-soluble oligomers (WS, also known as pyrolityc humin), and water-insoluble oligomers (WIS, also known as pyrolityc lignin) (43–59 wt%) that can be efficiently used for several applications, such as drop-in fuel, production of chemicals, and various carbon-based materials [17–22].

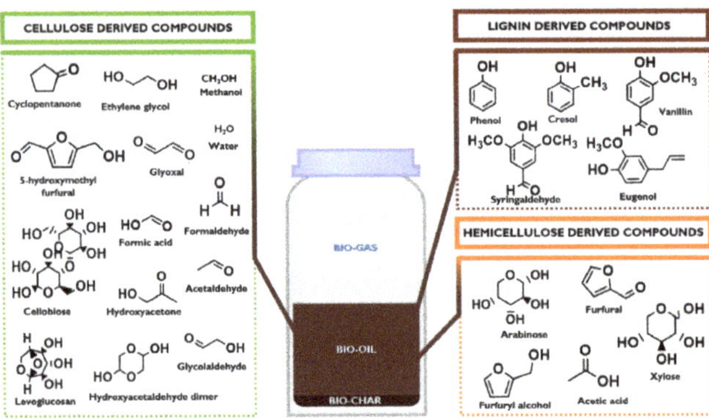

Figure 1. A simplified chemical composition of bio-oil: main lignocellulose-derived compounds.

Conventionally, bio-oil is produced by using a high energy demanding multistep process, such as pyrolysis (fast or slow, thermal or thermo-catalytic) and hydrothermal liquefaction (HTL). These thermochemical processes are conducted in the absence of oxygen and at high reaction temperature with the aim to allow the decomposition/depolymerisation of lignocellulose and microalgae into a bio-oil liquid (the major product), solid (bio-char), and gaseous products (CO_2, CO, CH_4, H_2) (bio-syngas [15,23,24]. Bio-char can access applications in several fields (e.g., soil amendment in agriculture, chemical sensing, adsorbent material in wastewater remediation) or combusted to recover energy for the pyrolysis stage [25], while bio-syngas may be directly utilized for many energy uses (e.g., electricity generation, fuel for transport, cooking fuel, feedstock for fuel cells) [26]. HTL processes were developed to improve the efficiency of direct thermal decomposition methods and differ from the pyrolysis for the adoption of lower reaction temperatures and for the presence of a homogeneous or heterogeneous catalyst by applying water and simple aliphatic (e.g., methanol, ethanol, and 2-propanol) alcohols used as such or in combination as reaction solvents. However, bio-oils arising from these two processes cannot be directly used as drop-in fuels in conventional engines due to problems related to the presence of a common limiting feature, the high oxygen content of biomass otherwise responsible of chemical unfavorable properties of bio-oil (high acidity, high viscosity, thermal and chemical instability) [16].

Thus, a biorefinery process in which biomass is first converted in bio-oil by pyrolytic or HTL step followed by an oxygen removal stage represents a most promising approach for the production of biofuels and chemicals.

To this regard, in order to mitigate the oxygen content and to improve the bio-oil properties for practical use, some catalytic approach (catalytic cracking, hydrodeoxygenation HDO, etc.), based on a thermal-catalytic treatment of biomass (hydrotreatment or hydrotreating process), come to help. Among them, one of the most promising strategies is the catalytic hydrodeoxygenation (HDO) that allows the reduction of oxygen content providing, at the same time, the highest C-atom efficiency.

Although this process allows to obtain bio-oil in a high yield, the formation of a variable quantity of coke remains a problem to be solved. In this context, it was reported that the presence of a suitable catalyst in a two-step biorefinery process can reduce the formation of coke by improving, at the same time, the bio-oil properties [27,28]. The first step (or stabilization step) permits the transformation of carbonyl and carboxyl functional groups into alcohols promoted by noble metals catalysts (Pt, Ru, and Pd) in a temperature range between 100 and 300 °C. The second step is conducted between 350 and 400 °C and is driven by sulphide conventional catalysts that allows to completely remove oxygen species.

This review aims to provide a brief overview on recent advances in the catalytic hydrogenation process of bio-oil arising from thermal treatment of lignocellulosic biomass and microalgae, highlighting progresses made in terms of enhancing catalyst efficient activity for upgrade bio-oil HDO.

2. Bio-Oil Proprieties

Bio-oil is the main product of biomass pyrolysis. Historical documents report that this process was already used in ancient Egypt to prepare sealants for boats and ointments. In the 18th century, wood distillation provided compounds such as soluble tar, pitch, creosote oil, as well as chemical and non-condensable gases. Interest in biomass pyrolysis was revived in the 1980s, when the process was perfected to have a high yield of liquid compound [29]. The pyrolysis process carried out with a temperature between 400 °C and 600 °C and varying the residence time and heating rate, the product distribution changes. To maximize the process in term of liquid yield, the fast pyrolysis at ~500 °C is usually preferred, advantageously producing a liquid yield up to three times larger than the conventional and slow pyrolysis [30].

As an example, Figure 2 shows the flow diagram of the BTG Bi-oliquids BV pyrolysis plant [31]. The first part of the plant consists in a drying unit where biomass from different origin (for example, wood, rice husk, bagasse, sludge, tobacco, energy crops, palm-oil residues, straw, olive stone residues, chicken manure) is dried to decrease the water content. The dry biomass, in presence of a hot carrier (sand), is then converted in a fluidized bed reactor into pyrolysis oil, gas, and char. After that, the products and the sand are separated from the vapor/gas phase by a series of cyclones. Then, the char and sand fraction is moved to a fluid bed combustor, where the char is used to heat the sand recycled in the fluidized bed. The vapor/gas phase is instead quenched by re-circulated oil to divide the bio-oil from the incondensable gases, where the latter are captured as high-pressure steam and utilized in a steam turbine system.

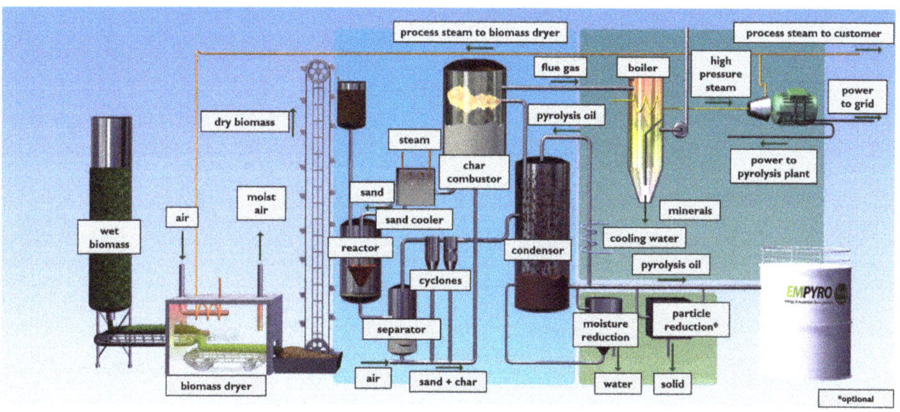

Figure 2. Pyrolysis plant. With permission from BTG Bioliquids BTG bioliquids BV [31].

Furthermore, recent researchers are focusing their attention to microalgae as feedstock for fast pyrolytic reaction [32]. Microalgae are classified as third-generation biofuel due to their fast growth cycle and high lipid content (~50%), easily converted in fuels. Moreover, microalgae do not require arable land and are adaptable at different water sources, including wastewater. The pyrolysis of microalgae is usually carried out in presence of a catalyst, such as zeolites, aluminosilicates, transitional metal-loaded zeolites, MOFs, silica gel [33–35]. The pyrolysis process for microalgae may be performed as (i) one-pot step process, where microalgae and catalyst are mixed together (ii) or a double-step process, the pyrolysis vapors from microalgae are swept over a catalyst at a specific temperature [36,37].

The fast pyrolysis of biomass produces hundreds of different compounds (Table 1), where their composition depends of the cellulose, hemicellulose, lignin, and extractive content in the respective feedstock. The influence of biomass composition on bio-oils composition can be appreciated from the variability of the bio-oils elemental composition reported in Table 2, where the C content can vary from 39% (pine sawdust) to about 60% (beech wood) under the same pyrolysis conditions. Furthermore, the operating conditions of the fast pyrolysis influence the bio-oil composition [38–40].

Table 1. Bio-oil composition.

Fraction/Chemical Groups	Compound Types	wt% (Wet Basis) [38]	wt% (Wet Basis) [39]	wt% [40]
	Water solubles 75–85%			
Acids alcohols	Small acids, small alcohols	5–10	6.5	8.5
Ether-solubles	Catechols, syringols, guaiacols, aldehydes, ketones, furans, and pyrans	5–15	15.4	20.3
Ether-insolubles	Sugars	30–40	34.4	45.3
Water	Water	20–30	23.9	-
	Water insoluble 15–25%			
Hexane-solubles	Extractives (High MW compounds with functional groups such as acids, alcohols)	2–6	4.35	5.7
DCM solubles	Stilbenes, Low MW lignin degraded compounds	5–10	13.4	17.7
DCM insolubles	High MW lignin degraded compounds	2–10	1.95	2.6

Table 2. Feedstock composition updated from [35].

Feedstock for Bio-Oil	C	H	O	N	S	Ref.
Beechwood	51.1	7.3	41.6			[41]
Pine wood	40.1	7.6	52.1	0.1		[42]
Rice husk	39.92	8.15	51.29	0.61	0.03	[43]
Beech wood	58.6	6.2	35.2			[44]
Pine sawdust	38.8	7.7	53.4	0.09	0.02	[45]
Eucalyptus	44.8	7.2	48.1	0.2		[46]
White spruce	49.6	6.4	43.1	0.2		[47]
Poplar	49.5	6.05	44.4	0.07		[47]
Sawdust	60.4	6.9	31.8	0.9		[48]
Microalgae	54.8	7.6	28.7	8.5	0.4	[49]
Scenedesmus	44.6	6.1	40.8	4.8	3.6	[50]
Nannochloropsis gaditana	40.3	5.97	14.5	6.3	0.37	[51]
Chlorella protothecoides	62.1	8.7	11.2	9.7	n/a	[52]
Spirulina	67.5	9.8	11.3	10.7	n/a	[53]
Nannochloropsis sp.	80.2	6.2	5.8	6.2	n/a	[54]

3. Catalytic Hydrogenation of BIO-Oil

In refineries, the hydrogenation reactions are common operations to limit the presence of oxygen, nitrogen, sulphur, olefins, and aromatics. The reaction is generally catalyzed by molybdenum together with Ni or Co supported by γAl_2O_3. The operating conditions depend on the type of feed: LHSV 0.2 to 8.0, H_2 circulation from 50 to 675 Nm^3/m^3, H_2 pressure between 14 and 138 bar, and temperatures between 290 and 470 °C [55]. Actually, there are not industrial processes for HDO of bio-oil, but several catalysts have been tested from noble metals to Ni and Co, in presence of acid supports such as Al_2O_3 and SiO_2, or C, in the temperature range 150–500 °C, pressure range between 2 and 200 bar [47]. In this section, recent advances on catalysts for HDO of bio-oil are summarized. Furthermore, technical aspects of emergent technologies (e.g., membrane reactors) for hydroprocessing are also discussed.

3.1. Catalysts

Hydro-processing is conventionally catalyzed in presence of metals from group VIII, such as nickel, palladium, and platinum [56]. Furthermore, group VIB metals (tungsten and molybdenum) have also been used for oxygen removal, since they are resistant to attack by oxygen, acids, and alkalis [57,58]. According to Masel [59], hydrogen is reactive in the surfaces of Co, Ni, Ru, Rh, Pd, Os, Ir, Pt as well as on Sc, Ti, V, Y, Zr, Nb, Mo, La, Hf, Ta, W, Cr, Mn, Fe, Tc, and Re. A slower uptake of hydrogen was observed with Cu [59]. Some authors increased the catalyst activity adding a second metal in order to promote an efficient adsorption of hydrogen at low temperature [60,61]. The most used supports were alumina-silica, carbon, titania (rutile), and zirconia (monoclinic form). Activated carbon is a well-known high-surface area (typically ~1000 m^2/g) support material, which has been shown to be stable in hot water processing environments; rutile titania and monoclinic zirconia have lesser surface area (typically 30–80 m^2/g) but have also demonstrated their utility as catalytic metal support and have been used in the hot water processing environment [62–64]. A possible pathway for upgrading bio-oils is represented by hydrogenation reactions in liquid phase, with the conversion of aldehydes, ketones, sugars, phenols, etc., in more stable alcohols. In order to improve the conversion of the bio-oils compounds and enhance the selectivity on desired products, several catalysts have been studied (Table 3). Interesting is the work of Wei et al. [65], where Pt over different ceria-zirconia supports were evaluated for the hydrogenation of cinnamaldehyde at 10 bar and 60 °C, obtaining a conversion in the range of 60–95%.

Table 3. Hydrogenation reaction.

Catalyst	Reactant	Pressure (bar)	Temperature (°C)	Time (h)	Conversion (%)	Note	Ref.
30% Ni/CNT	acetic acid	40	150	4	5.8	2 wt% cat	[66]
30% Cu/CNT	acetic acid	40	150	4	3.5	2 wt% cat	[66]
Ru/C	acetic acid	40	150	4	4.7	2 wt% cat	[66]
20% Mo/CNT	acetic acid	40	150	4	<2	2 wt% cat	[66]
10/10 wt% NiMo/CNT	acetic acid	40	150	4	14.8	2 wt% cat	[66]
3 wt% Ru/TiO_2	acetic acid	62	120	33 *	37.5	* time on stream	[67]
3 wt% Ru/TiO_2	Acetol	62	70	14 *	93.6	* time on stream	[67]

Table 3. *Cont.*

Catalyst	Reactant	Pressure (bar)	Temperature (°C)	Time (h)	Conversion (%)	Note	Ref.
3 wt% Ru/TiO_2	Bio-oil	62	120	21	27/38/79 **	** acetic acid/acetol/formic acid	[67]
3 wt% Ru/C	Bio-oil	52	120	6	33/99/97 **		[67]
Ru/Zr-MOFs	Furfural	5	20	5	20–95	TOF: 2–11 Selectivity to Furfuryl alcohol: 20–95	[68]
AuNCs/CNTs membrane	4-nitro-phenol				53/100	5/10 μmol Au/17 cm^2	[69]
Au/SiO_2	25 compounds	80	6	5–24	40–99	1 mmol of alkyne, 0.01 mmol of Au, and 1 mmol of piperazine	[70]
$Re–Pd/SiO_2$	Stearic acid	80	140	1	15	Re/Pd = 1/8	[71]
$Re–Pd/SiO_2$	Stearic acid	80	140	4	13		[71]
Ni/rutile	Crotonaldehyde	10	70		60		[72]
Pd-Cu/MgO	Furfural	6–8	80–130	0.5	100	98.7% selectivity of Furfuryl alcohol	[73]
Pt/MWNT	Furfural	20	150	5	75–100	Max Furfuryl alcohol selectivity: 79%	[74]
$ReOx–Pd/CeO_2$	16 compounds	80	140	4	1–60	substrate 0.5 g, 1,4 dioxane 4 g, Wcat = 150 mg (2 wt% Re, 0.3 wt% Pd)	[75]
Rh–$MoOx/SiO_2$+ CeO_2	cyclohexanecarboxamide	80	140	4	89		[76]

Liao et al. [77] used CeO as support with different metals (Ni, Co, and Cu) for the hydrogenation in liquid phase of maleic anhydride at 50 bar and 210 °C, converting all the reactant after 60, 180, and 420 min, for Ni, Co, and Cu, respectively. Elliott al. [78] elaborated a reactivity scale of hydrogenation of different organic compounds in presence of CoMo and NiMo sulphided catalysts (see Figure 3) based on literature work [79]. Olefins, aldehydes, and ketones were hydrogenated at low temperatures as low as 150–200 °C, while the alcohols at 250–300 °C. Carboxylic and phenolic ethers reacted at around 300 °C.

Recently, copper catalysts have attracted much attention for the conversion of glycerol to propylene glycol because of their intrinsic ability to selectively cleave the C-O bonds in glycerol rather than the C-C bonds. To increase the activity of Cu metal, Cu-based catalysts such as Cu-Cr, Cu-Al, and Cu-Mg have been developed to promote the hydrogenolysis reaction. Bienholz et al. prepared a highly dispersed silica-supported copper catalyst (Cu/SiO_2) using an ion-exchange method and achieved 100% glycerol conversion with 87% propylene glycol selectivity at optimum conditions of 5 mL/h of 40 wt% aqueous glycerol solution, 255 °C, and 300 mL/min of H_2 at 15 bar [80]. Liu's group studied the glycerol hydrogenolysis over Ru-Cu catalysts supported on different support materials including SiO_2, Al_2O_3, NaY zeolite, TiO_2, ZrO_2, and HY zeolite. The best activity was observed for $Ru-Cu/ZrO_2$ with 100% glycerol conversion and 78.5% propylene glycol selectivity. The high activity of this catalyst was attributed to the synergistic effect of Ru in the catalyst related to hydrogen spill-over, while the high selectivity was attributed mainly to the low acidity of the support and the Cu amount [81].The HDO of the Water soluble fraction of Bio-Oil (WBO) at different temperatures (220, 270, and 310 °C) at 190 bar, using 5 wt% Ru/C catalyst, was studied by de Miguel Mercader et al. [82], where the recovery of carbon in oil phase increased from 16.3 wt% to 38.5 wt%, when the temperature was increased from 220 to 310 °C. In another study, several lignin model compounds (phenol, m-cresol, anisole, guaiacol, and diphenyl ether) were tested for HDO reactions in presence of MoO_3 at atmospheric pressure and temperature between 150 and 250 °C [83]. The authors noted that, according to the bond dissociation energy, the highest catalytic reactivity was obtained with diphenyl ether, but important carburization phenomena have been noted onto the catalyst surface.

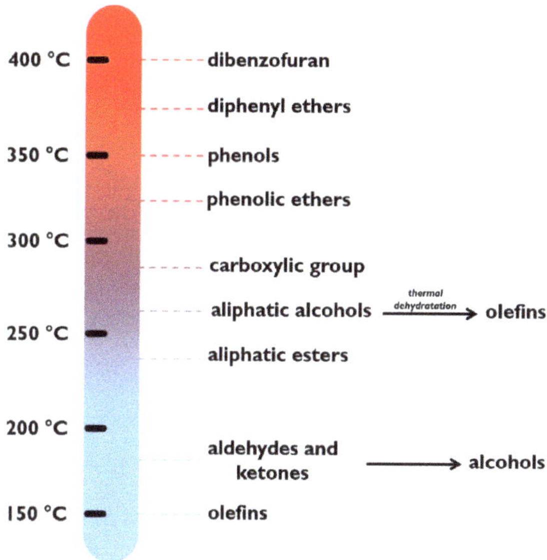

Figure 3. Reactivity scale of organic compounds under hydrotreatment conditions. Adapted with permission from [78].

Bagnato et al. [84] prepared by impregnation technique a series of monometallic and bimetallic metal catalysts in which the zirconia was doped with Pd and not noble metals (Cu and Fe), characterized, and their performances studied in term of conversion and selectivity for key bio-compounds. Vanillin was completely converted after 80 min at 100 °C and 50 bar, in presence of $PdFe/ZrO_2$. Meanwhile, the PdFe reached the conversion of 65.5% and 20% for furfural to furfuryl alcohol and glucose to sorbitol (74% selectivity), respectively.

The authors noted that the bimetallic catalyst was able to improve the conversion than the monometallic, mainly due the adsorption mechanism onto the catalytic surface: the not noble metal favoured the bonding to the aldehyde group, while the noble metal favoured the hydrogen molecule adsorption.

Bergem et al. [67] investigated the HDO of a model WBO using Ru/TiO_2 and Ru/C catalysts in a packet bed reactor (PBR) at a temperature between 100 and 140 °C, ~62 bar. A completed conversion was noted already a 100 °C for compounds such as acetone, acetaldehyde, propionaldehyde, 2-propen-1-ol, 1-hydroxy-2-butanone, 3-hydroxy-2-butanone, 2-hexanone, and 2-furanone. Other compounds such as furfural and hydroxyacetaldehyde required elevate temperature (>140 °C) for converting completely. Furthermore, the authors observed a decrement of catalyst activity, about 25% after 90 h, due at acid leaching. Sanna et al. [85] studied the HDO of a real WBO in presence of Ru/C and Pt/C catalysts in a two-stage continuous reactor. In the first stage, the reaction was carried out in presence of Ru/C catalyst at 125 °C, while in the second stage, it was carried out at a temperature between 200 and 250 °C with Pt/C, at 50 and 100 bar, and different weight hourly space velocities from 0.75 to 6 h^{-1}. During the first low temperature stage, the unstable bio-oil functionalities were stabilized into alcohols, where the main products were ethylene glycol, propylene glycol, and sorbitol, losing 7% of carbon as gas and solid phase. Furthermore, the catalyst showed a constant activity for about 80 h. In the second-high temperature stage, 45% of the carbon was converted in gasoline blend stocks and C_2 to C_6 diols.

3.2. Kinetic Mechanism

The reactions involved during the hydrotreating of bio-oil have been widely studied [86–88], as shown below:

$$\text{Hydrodeoxygenation (HDO)}: R-OH+H_2 \rightarrow R-H+H_2O; \quad (1)$$

$$\text{Hydrodesulphurisation (HDS)}: R-SH+H_2 \rightarrow R-H+H_2S; \quad (2)$$

$$\text{Hydrodenitrogenation (HDN)}: Pyridine+H_2 \rightarrow Pentane+NH_3; \quad (3)$$

$$\text{Hydrodealkylation}: R-C_6H_5+H_2 \rightarrow C_6H_6+R-H; \quad (4)$$

$$\text{Hydrocracking}: R_1-CH_2CH_2-R_2+H_2 \rightarrow R_1-CH_3+R_2-CH_3; \quad (5)$$

$$\text{Isomerisation of alkanes}: n-alkane \rightarrow i-alkane; \quad (6)$$

$$\text{Decarboxylation}: R-CO-OH \rightarrow R-H+CO_2; \quad (7)$$

$$\text{Decarbonilation}: R-CHO \rightarrow R-H+CO; \quad (8)$$

$$\text{Water gas shift reaction}: CO_2+H_2 \leftrightarrow CO+H_2O; \quad (9)$$

$$\text{Coke formation}: polyaromatic \rightarrow coke. \quad (10)$$

In the following section, the reaction mechanisms of some of the most representative bio-oil compounds will be discussed.

3.2.1. Phenol

The phenol hydrogenation has been widely studied [89–93]. The reaction pathways are shown in Figure 4, where hydrogen reacts with the phenol (PHE) attacking the hydroxyl group to produce benzene with subsequent production of cyclohexene (CHE) and cyclohexane (CHO). Another reaction pathway of the aromatic ring is the formation of cyclohexanol (CXO) with consecutive hydrogenation in cyclohexene and cyclohexane. A further reaction pathway is represented by the formation of cyclohexanone (COL) with subsequent cyclohexanol hydrogenation in cyclohexene and cyclohexane. Finally, methylcyclopentane (MCP) can be produced by isomerization reaction.

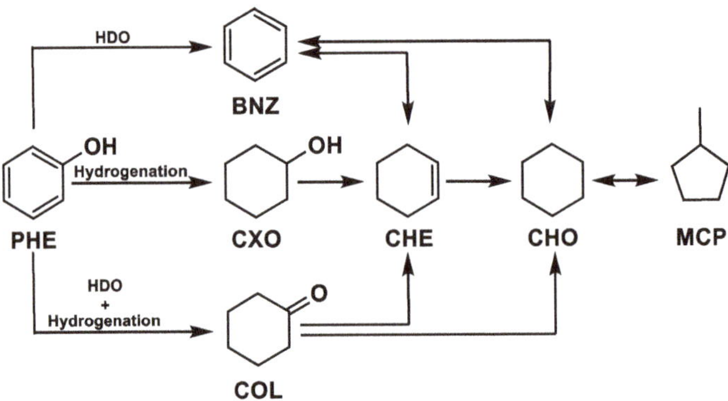

Figure 4. Phenol hydrogenation pathways.

3.2.2. Guaiacol

Another representative compound in bio-oil is guaiacol that reacts forming phenol [39–43] via two paths: (1) direct demethoxylation; and (2) indirect reaction through demethylation to catechol with subsequent hydrogenation of the latter compound. The undesired polymerisation of guaiacol (GCL) (Figure 5) leads to coke formation.

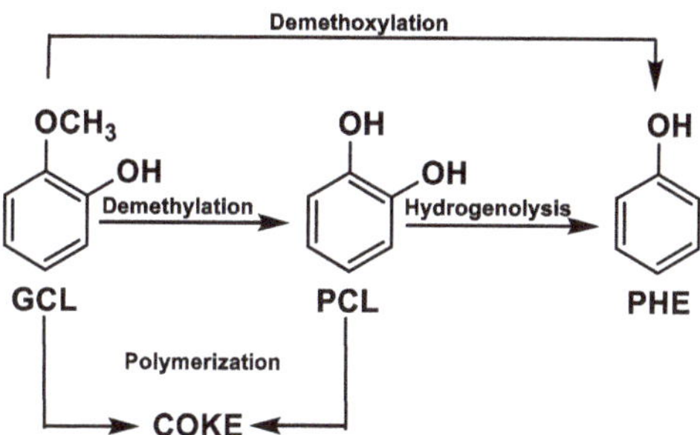

Figure 5. Guaiacol reaction path.

Bindwal et al. [94] proposed a kinetic rate for the hydrogenation of guaiacol in 1,2 cycloexanediol in presence of 5% Ru/C catalyst according to the Langmuir–Hinshe–wood–Hougen–Watson (LHHW) model. The authors, according to the experimental data obtained, identified the limitation step for the reaction taking place on the catalyst surface, assuming the dissociative adsorption of H_2. The reaction rate was described by the following equation:

$$r = \frac{k_{3,a} K_B C_B \sqrt{K_{H_2} C_{H_2}}}{\left(1 + \sqrt{K_{H_2} C_{H_2}} + K_B C_B\right)^2}, \quad (11)$$

where C_B C_{H2} are the molar concentration of guaiacol and hydrogen, respectively, $k_{3,a}$ the kinetic constant, K_B and K_{H2} are the adsorption constant of guaiacol and hydrogen.

3.2.3. Levoglucosan

The hydrolysis of levoglucosan has been studied in a solution of water and in the presence of Ru/C [95]. The path involves the production of glucose (hydrolysis reaction) with subsequent hydrogenation into sorbitol. Finally, ethylene glycol, 1,2-Propanediol, and 1,4-Butanediol are produced by the hydrogenation of sorbitol (Figure 6).

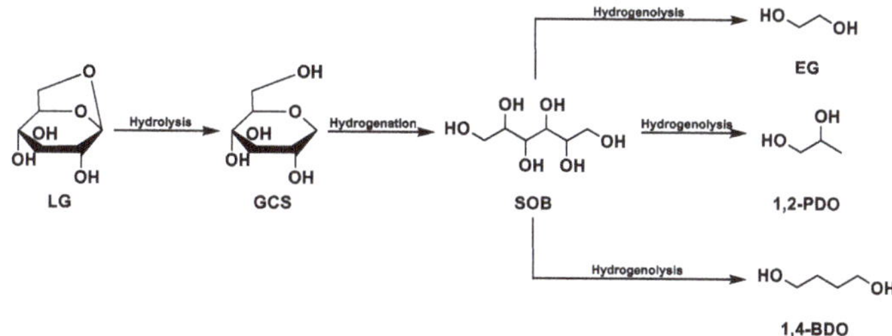

Figure 6. Levoglucosan reaction path.

Bindwal et al. [95] proposed a kinetic rate for the hydrogenation in presence of Ru/C, where the H_2 and levoglucosan (LG) chemisorbed and dissociated on the surface catalyst are as follows:

$$H_2 + X \leftrightarrow 2HX, \tag{12}$$

$$LG + X \leftrightarrow LGX, \tag{13}$$

$$2HX + LGX \leftrightarrow \ products, \tag{14}$$

represented by the following equation:

$$r = \frac{k_3 K_{H2} K_{LG} C_{H2} C_{LG}}{\left(1 + \ \sqrt{K_{H2} C_{H2}} + K_{LG} C_{LG}\right)^3}. \tag{15}$$

3.2.4. Other Compounds

Bindwal et al. [94] studied the kinetics rate of other compounds using 5% Ru/C catalyst to convert hydroxycetone, hydroxyacetaldehyde and 2-furanone in 1,2 propanediol, ethylene glycol and γ-butyrolactone, according to the reactions in Figure 7.

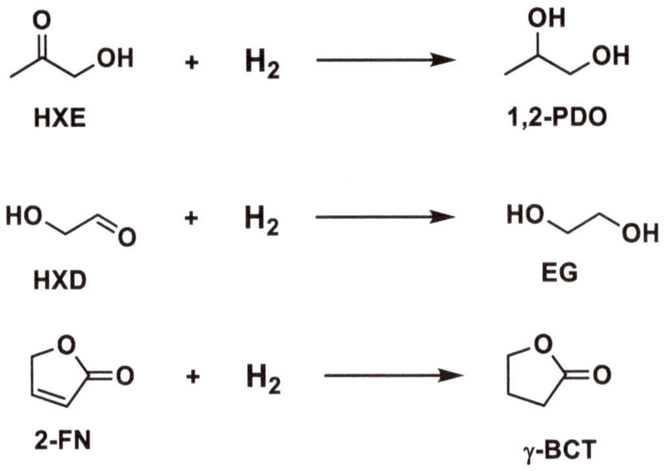

Figure 7. Hydrogenation of hydroxyacetone, hydroxyacetaldehyde, and 2-furanone.

The authors proposed different kinetics rates varying the limitation step and the possibility to have an atomic or molecular H_2 adsorption.

The kinetics rates hypothesized were validated experimentally confirming that the reactions are surface-reaction limited in presence of dissociative adsorption of H_2. The equation for the kinetics rate were

$$r = \frac{k_3 \sqrt{K_{H_2} C_{H_2}} K_B C_B}{\left(1 + \sqrt{K_{H_2} C_{H_2}} + K_B C_B\right)^2}, \tag{16}$$

$$r = \frac{k_3 K_{H_2} K_B C_{H_2} C_B}{\left(1 + \sqrt{K_{H_2} C_{H_2}} + K_B C_B\right)^3}. \tag{17}$$

Zhang et al. [96,97] described the reaction kinetics by dividing the products as Light oil ranged from 36 °C to 250 °C, heavy oil from 250 °C to 450 °C, vapors, water, and coke. They assumed a series of parallel reactions with a first-order kinetics in presence of $CoMo/\gamma Al_2O_3$ catalyst.

$$Bio-oil \underset{k_1}{\rightarrow} light\ fraction\ oil, \tag{18}$$

$$Bio-oil \underset{k_2}{\rightarrow} heavy\ fraction\ oil \underset{k_4}{\rightarrow} light\ fraction\ oil, \tag{19}$$

$$Bio-oil \underset{k_2}{\rightarrow} heavy\ fraction\ oil \underset{k_5}{\rightarrow} gas + water + char, \tag{20}$$

$$Bio-oil \underset{k_3}{\rightarrow} gas + water + char. \tag{21}$$

Furthermore, Sheu et al. [98] divided the bio-oil into six groups (heavy non-volatiles, light non-volatile, phenols, aromatics, alkanes, Coke + H_2O + Outlet Gases) and used three different catalysts ($Pt/Al_2O_3/SiO_2$, $CoMo/\gamma$-Al_2O_3, and Ni-W/γ-Al_2O_3) to study the hydrogenation of bio-oil. Moreover, the authors proposed a reaction pathway by series-parallels of first-order reactions.

$$heavy\ nonvolatiles \underset{k_1}{\rightarrow} light\ nonvolatiles \underset{k_3}{\rightarrow} phenols, \tag{22}$$

$$heavy\ nonvolatiles \underset{k_2}{\rightarrow} aromatics + alkanes, \tag{23}$$

$$phenols \underset{k_4}{\rightarrow} aromatics + alkanes \underset{k_5}{\rightarrow} Coke + H_2O + outlet\ gases. \tag{24}$$

k_i is a kinetic constant and depend of the temperature and pressure by

$$k_i = k_{i0} P^{n_i} \exp\left(-\frac{Ea}{RT}\right), \tag{25}$$

where k_{i0} and n_i are the parameters of the reaction and the catalysts used.

3.3. Reactor Technologies

3.3.1. Conventional Reactors

The hydrogenation reaction is largely used in refinery to convert the heavy oil fraction into light hydrocarbons. The existing process have been based on the following reactors: fixed beds (FBRs), moving beds (MBRs), and expanded or ebullated beds (EBRs). The main difference among the reactors involves the transport phenomena and some technical details.

The FBRs are the main reactor systems used commercially and used for hydrogenating light hydrocarbon mixture such as naphtha and middle distillate. The FBRs are designed for operating in an adiabatic condition. The reactor is divided into three catalytic zones separated to an inert material (ceramic balls), the liquid and gas stream through the first catalytic bed. The output fed exchange heat by the inert bed and subsequently quenched adding fresh gas reactant and then fed inlet of the second catalytic bed. The output of the second reactor is cooled again by the inert bed and by quenching. EBR reactors have been also used to hydrogenate feeds such as vacuum residue.

The EBRs are used for heavy feeds with a large amount of metals and asphaltenes, where the liquid and gas streams are fed from bottom expanding and mixing the catalyst bed, reducing the pressure drop effect. In the output of the catalytic bed the hydrogen not reacted is recycled, while the liquid products are recovered by a flash unit.

3.3.2. Membrane Reactor

The main disadvantage of hydrogenation reaction is represented by mass transport limitation, because the reaction takes place in contact with the gaseous, liquid, and solid phase. The system has to have a high operating pressure, improving the gas solubility into the liquid system and high temperature to advantage the kinetic, but at the same time, the H_2 solubility decreases under those conditions. A membrane reactor (MR) is an operation unit to produce new species by chemical reaction and separation process in a single equipment [99]. The use of MRs can reduce the process footprint, since the plant will be more compact and can result in lower investment costs, improving the economics of the process [100,101].

In a MBR, the fresh catalyst is fed from the top and trough the reactor, while the reactant stream is fed from the bottom. Afterthought, the products leave the MBR and the deactivated catalyst is sent to the regenerator reactor, where the coke deposition is burned and the activated catalyst returns to MBR.

One of the features of the MRs is to act as a contactor between the three phases during HDO reaction. Furthermore, the membrane can have catalytic activity chancing the product distribution as reported by Liu et al. [102], who compared packed bed MR and catalytic MR for the hydrogenation of nitrobenzene in presence of Pd/γ-Al_2O_3 catalyst. The CMMR showed best performance in term of conversion and catalytic stability (~85% for 10 h) as shown in Figure 8.

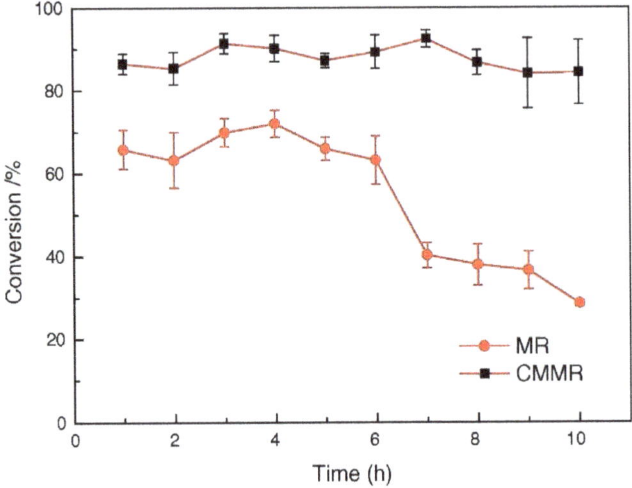

Figure 8. Comparison between catalytic membrane reactor (CMR) and packed bed membrane reactor (PBMR) [102].

Liu et al. [103] studied the selectivity hydrogenation of butadiene in butene at 40 °C and 10 bar by catalytic membrane reactors (CMRs), obtaining a butene selectivity higher than 99% and butadiene concentration in the output stream lower than 10 ppm. Another example of hydrogenation reaction is reported in Table 4. Despite the increasing interest in catalytic membranes, the HDO of bio-oil in MR is a novelty, since in literature there is only one article [104] available on the topic, where the authors used a MR for the hydrogenation of levulinic acid (compound present in bio-oils) by a porous expanded polytetrafluoroethylene (ePTFE) membrane with Ru catalyst particles. Moreover, the same membranes were coated only in one side with a dense Matrimid layer, which was used to control the hydrogen flux through the membrane. The reaction was studied in a temperature and reaction pressure between 40 and 90 °C and 0.7 and 5.6 bar, respectively. Furthermore, the authors compared the result obtained with a PBR as shown in Figure 9, where the kinetic rate is presented as ratio of gamma-valerolactone product (g/h) over grams of Ru. In particular, the MR without the Matrimid layer obtained the best performance (four times more than PBR) with a conversion of 0.0065%, while the MR with the control layer (Matrimid) showed a kinetic rate two times less than the PBR.

Recent studies have emphasized the functionality of MR to be able to achieve a TOF equal to 48,000 h^{-1} for the partial hydrogenation of furfural in presence of Ru–polyethersulfone (PES) catalytic membrane at 70 °C and 7 bar [105].

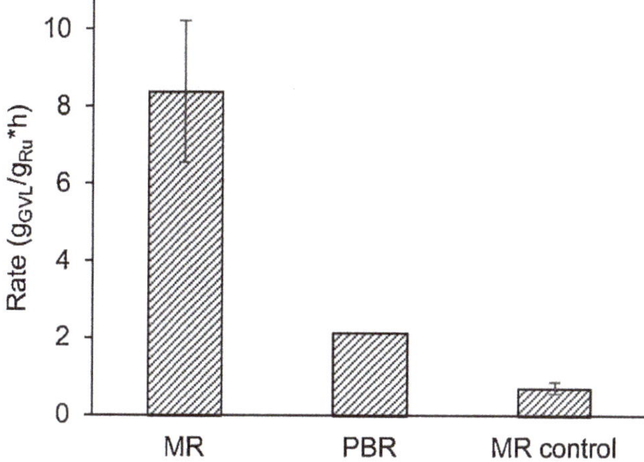

Figure 9. Kinetic rate of hydrogenation of levulonic acid using membrane reactors (MRs) and a packet bed reactor (PBR) [104].

Table 4. MR for hydrogenation reaction.

Hydrogenation of	Catalyst	Support		Pressure (bar)	Temperature (°C)	Ref.
3-hexyn-1-ol	Pd nanoparticles (4.6 nm)	zirconia/polyvinyl alcohol	Batch	5–10	25	[106]
Nitrite	Pd	γ-Al_2O_3	Continuous	1	25	[107]
Methylenecyclohexane (and isomerization)	Pt, Pd, Ru in γ-Al_2O_3	macroporous α-Al_2O_3	Continuous	1.5 liquid 2 gas	15–70	[108]
Methylenecyclohexane	Pd-PVDS PVP	macroporous α-Al_2O_3	Continuous		25–50	[109] *
Edible oil	Pd, Pt	porous polyamideimide (PAI)	Continuous	4	100	[110]
Nitrobenzene	Pd	zirconia/polyvinyl alcohol	Continuous	1–2	25	[111]
Nitrobenzene	Pd/γ-Al_2O_3	PDMS	Continuous	1–2	20	[102]
Butadiene	PVP-Pd, PVP-Pd, EC-Pd, AR-Pd, AR-Pd, PVP-Pd, PVP-Pd-0.5 $Co(OAc)_2$, PVP-Pd-0.5 $Co(OAc)_2$	CA, PSF, CA, CA, PSF, CA, CA, CA	Continuous	10	40	[103] **
Furfural	Ru	PES	Continous	7	70	[105]

* Ceramic membrane showed a higher selectivity toward the hydrogenated product than the polymeric membrane but exhibited a lower TON (= converted moles in a second per gram of Pd) value. ** PVP-Pd-0.5 $Co(OAc)_2$ showed best performance. The presence of Co inhibited isomerization reaction.

4. Concluding Remarks and Future Outlook

The valorization of biomass and residues for the production of liquid fuels by both thermochemical (e.g., pyrolysis) or physical (e.g., pressing) methods has attracted a great attention from both scientific and technological point of view. In fact, the utilization of vegetable raw materials for the production of synthetic chemical intermediates and hydrocarbons is considered one of the most investigated strategies aimed at reducing both the carbon dioxide emissions and the dependence on fossils fuels.

In this review, we summarize the main aspects related to pyrolysis and to the properties of the obtained bio-oils, focusing great attention to the hydrotreatment process alternatives for converting the pyrolysis bio-oil into drop-in fuel.

Research studies usually are focused on the pyrolysis of a well-defined biomass, while a limited number of papers are devoted to biomass residues with variable composition. The former approach is useful for understanding the complex mechanism involved during the pyrolysis and the effect of process parameters on reaction pathways, while the latter is of paramount importance for developing the technology at a pilot and demonstrative scale. In fact, the compositional variations in the feedstock modify both yield and composition of bio-oil, and this aspect has a significant impact on the viability of the process. As previously mentioned, biomass availability at a low cost is one of the biggest challenges of biorefinery. Therefore, more effort should be put on the experimental investigation at a pilot and demonstrative scale on the production of bio-oil from biomass residues with a large compositional variability. The research is also focused on the role of biomass pre-treatment on bio-oil quality. In particular, physical, chemical, and thermal methods may be adopted. As an example, the modification of size and shape of biomass particles has an effect on heat transfer with an impact on bio-oil quality. Whereas, the reduction of hemicellulose by a thermal method, such as torrefaction, decreases the amount of organic acids, acetals, and water in the bio-oil, with a positive impact on bio-oil stability, but with a higher inorganics content. The amount of inorganics may be reduced by physical pre-treatment such as biomass washing with water or acids. Unfortunately, there is a lack of information and knowledge about the economic feasibility of the biomass pre-treatment methods. Another important aspect that should be investigated in more detail is the stability of produced bio-oil. In fact, the bio-oil is a complex mixture containing water and both polar and nonpolar organics that cause several reactions, e.g., oligomerization condensation and dehydration, with aging of the bio-oil and formation of a more complex multiphase systems. The addition of alcohols, such as methanol, usually improves stability, homogeneity, and viscosity of bio-oil. Further research on bio-oil stabilization is needed to address technical issues during bio-oil storage and processing.

This review aims to summarize recent advances on the conversion to pyrolysis bio-oil into drop-in fuels by catalytic hydrogenation. In this regard, the research efforts should be better focused on (i) catalytic assessment of novel catalysts, and (ii) experimental investigation at pilot and demonstrative scale of hydrotreatment of real bio-oil. Concerning the first point, several metals and metals supported over moderately acid solids have been investigated. Ni-Mo or Co-Mo bimetallic systems supported over gamma-alumina are the most investigated catalysts for hydrotreatment, since they are well-known catalysts for hydroprocessing oil-derived streams, i.e., hydrodesulphurization. In these systems, Mo represents the active phase for the removal of heteroatoms, while Ni or Co acts as promoters for the hydrogenation step. Several alternative catalysts have been studied mostly for the hydrogenation of model compounds, whose catalytic behavior is in part discussed in this review. For instance, different metals and different supports have been studied, while a less attention was paid to the design of innovative hybrid systems, where the catalytic functionalities requested by the process, e.g., redox, acids, are carefully tuned with the aim to improve catalyst effectiveness. In this regard, research should be also devoted to the study of reaction mechanism as a function of surface properties of the catalysts. This approach has brought advances in other fields, such as hydrogenation of carbon dioxide to synthetic fuels, and it may be useful for a better understanding of catalysis of hydrogenation of bio-oils.

Nevertheless, bio-oil strongly differs from typical crude oil derived streams, due to the presence of a large amount of oxygenated compounds, e.g., carboxylic acids, phenols, aldehydes, ketones, sugars, and water. For this reason, the physic-chemical features of the catalyst for hydrotreatment of bio-oil should be carefully tuned as a function of bio-oil composition. On the contrary, a limited number of studies were carried out on the hydrotreatment of real pyrolysis bio-oil. In that case, the number of variables and the issues strongly increase. As an example, the presence of unsaturated oligomers in real bio-oil may lead to the formation of coke with deactivation of the catalyst. Therefore, future focus should be on the separation of bio-oil fractions in order to assess the most suitable bio-oil cut for hydroprocessing. In fact, the presence of oligomers inevitably causes a large amount of coke formation with catalyst deactivation. Of course, the presence of an additional step between pyrolysis and hydrotreatment has a significant effect on the process costs. The experimental investigation of hydrotreatment of bio-oil fractions at an either pilot or demonstrative scale may push towards more research in the field of bio-oil pre-treatments, as well as address also challenges, such as hydrogen consumption. In fact, most of the papers on hydrotreatment of bio-oil or bio-oil models are focused on product yield and quality, but it is difficult to find quantitative information on hydrogen consumption, which is usually used in large excess. As in the case of pyrolysis step, investigations at scales larger than laboratory of hydrogenation steps may surely provide quantitative data useful for viability studies on the production of drop-in fuels from biomass via pyrolysis.

Author Contributions: Conceptualization and organization: G.B., E.P. and E.C.; Writing—original draft preparation: G.B., A.S., E.P. and E.C.; Writing—review and editing: E.P.; All authors have read and agreed to the published version of the manuscript.

Funding: This research received no external funding.

Conflicts of Interest: The authors declare no conflict of interest.

Abbreviations

BNZ	Benzene
CHE	Cyclohexene
CHO	Cyclohexanol
CMR	Catalytic membrane reactor
COL	Cyclohexanone
CXO	Cyclohexanol
Ea	Activation energy
EBR	Ebullated bed reactor
EG	Ethylene glycol
γ-BCT	γ-Butyrolactone
GCL	Guaiacol
GCS	Glucose
HDO	Hydrodeoxygenation reaction
HD	Hydrotreating
HTL	Hydrothermal liquefaction
HXD	Hydroxyacetaldehyde
HXE	Hydroxyacetone
k	Kinetic rate
k0	Pre-exponential number
LG	Levoglucosan
LHHW	Langmuir–Hinshelwood–Hougen–Watson
LHSV	Liquid hourly space velocity
MR	Membrane reactor
MBR	Moving bed reactor
MCP	Methylcyclopentane
n	Kinetic order
PBR	Packet bed reactor
PBMR	Packed bed membrane reactor
PCL	Pyrocatechol
PHE	Phenol
SOB	Sorbitol
TEA	Techno-economical assessment
TOF	Turnover of frequency
WBO	Water soluble bio-oil fraction
1,2-PDO	1,2-Propanediol
1,4-BDO	1,4-Butanediol
2-FN	2-Furanone

References

1. Anastas, P.; Eghbali, N. Green Chemistry: Principles and Practice. *Chem. Soc. Rev.* **2010**, *39*, 301–312. [CrossRef] [PubMed]
2. Ragauskas, A.J.; Williams, C.K.; Davison, B.H.; Britovsek, G.; Cairney, J.; Eckert, C.A.; Frederick, W.J.; Hallett, J.P.; Leak, D.J.; Liotta, C.L.; et al. The Path Forward for Biofuels and Biomaterials. *Science* **2006**, *311*, 484–489. [CrossRef] [PubMed]
3. Paone, E.; Tabanelli, T.; Mauriello, F. The rise of lignin biorefinery. *Curr. Opin. Green Sustain. Chem.* **2020**, *24*, 1–6. [CrossRef]
4. Paone, E.; Beneduci, A.; Corrente, G.A.; Malara, A.; Mauriello, F. Hydrogenolysis of aromatic ethers under lignin-first conditions. *Mol. Catal.* **2020**, *497*, 111228. [CrossRef]
5. Malara, A.; Paone, E.; Frontera, P.; Bonaccorsi, L.; Panzera, G.; Mauriello, F. Sustainable exploitation of coffee silverskin in water remediation. *Sustainability* **2020**, *10*, 3547. [CrossRef]
6. Xu, C.; Paone, E.; Rodríguez-Padrón, D.; Luque, R.; Mauriello, F. Recent catalytic routes for the preparation and the upgrading of biomass derived furfural and 5-hydroxymethylfurfural. *ChemSocRev* **2020**, *49*, 4273–4306. [CrossRef]
7. Transforming Our World: The 2030 Agenda for Sustainable Development. Available online: https://sustainabledevelopment.un.org/post2015/transformingourworld (accessed on 11 January 2021).
8. Carlos, R.M.; Khang, D.B. Characterization of biomass energy projects in Southeast Asia. *Biomass Bioenergy* **2008**, *32*, 525–532. [CrossRef]
9. Paone, E.; Espro, C.; Pietropaolo, R.; Mauriello, F. Selective arene production from transfer hydrogenolysis of benzyl phenyl ether promoted by a co-precipitated Pd/Fe_3O_4 catalyst. *Catal. Sci. Technol.* **2016**, *6*, 7937–7941. [CrossRef]
10. Mauriello, F.; Paone, E.; Pietropaolo, R.; Balu, A.M.; Luque, R. Catalytic transfer hydrogenolysis of lignin-derived aromatic ethers promoted by bimetallic Pd/Ni systems. *ACS Sustain. Chem. Eng.* **2018**, *6*, 9269–9276. [CrossRef]
11. Espro, C.; Gumina, B.; Paone, E.; Mauriello, F. Upgrading lignocellulosic biomasses: Hydrogenolysis of platform derived molecules promoted by heterogeneous Pd-Fe catalysts. *Catalysts* **2017**, *7*, 78. [CrossRef]
12. Espro, C.; Gumina, B.; Szumelda, T.; Paone, E.; Mauriello, F. Catalytic transfer hydrogenolysis as an effective tool for the reductive upgrading of cellulose, hemicellulose, lignin, and their derived molecules. *Catalysts* **2018**, *8*, 313. [CrossRef]
13. Mauriello, F.; Ariga-Miwa, H.; Paone, E.; Pietropaolo, R.; Takakusagi, S.; Asakura, K. Transfer hydrogenolysis of aromatic ethers promoted by the bimetallic Pd/Co catalyst. *Catal. Today* **2020**, *357*, 511–517. [CrossRef]
14. Malara, A.; Paone, E.; Bonaccorsi, L.; Mauriello, F.; Macario, A.; Frontera, P. Pd/Fe_3O_4 Nanofibers for the Catalytic Conversion of Lignin-Derived Benzyl Phenyl Ether under Transfer Hydrogenolysis Conditions. *Catalysts* **2020**, *10*, 20. [CrossRef]
15. Srifa, A.; Chaiwat, W.; Pitakjakpipop, P.; Anutrasakda, W.; Faungnawakij, K. Chapter 6—Advances in bio-oil production and upgrading technologies. In *Sustainable Bioenergy*; Elsevier: Amsterdam, The Netherlands, 2019; pp. 167–198.
16. Han, Y.; Gholizadeh, M.; Tran, C.-C.; Kaliaguine, S.; Li, C.-Z.; Olarte, M.; Garcia-Perez, M. Hydrotreatment of pyrolysis bio-oil: A review. *Fuel Process. Technol.* **2019**, *195*, 106140. [CrossRef]
17. Tabanelli, T.; Paone, E.; Blair Vásquez, P.; Pietropaolo, R.; Cavani, F.; Mauriello, F. Transfer Hydrogenation of Methyl and Ethyl Levulinate Promoted by a ZrO_2 Catalyst: Comparison of Batch vs Continuous Gas-Flow Conditions. *ACS Sustain. Chem. Eng.* **2019**, *7*, 9937–9947. [CrossRef]
18. Fasolini, A.; Cucciniello, R.; Paone, E.; Mauriello, F.; Tabanelli, T. Short Overview on the Hydrogen Production Via Aqueous Phase Reforming (APR) of Cellulose, C6-C5 Sugars and Polyols. *Catalysts* **2019**, *9*, 917. [CrossRef]
19. Xu, C.; Paone, E.; Rodríguez-Padrón, D.; Luque, R.; Mauriello, F. Reductive catalytic routes towards sustainable production of hydrogen, fuels and chemicals from biomass derived polyols. *Renew. Sustain. Energy Rev.* **2020**, *127*, 109852. [CrossRef]
20. Weldekidan, H.; Strezov, V.; Town, G. Review of solar energy for biofuel extraction. *Renew. Sustain. Energy Rev.* **2018**, *88*, 184–192. [CrossRef]
21. Pistone, A.; Espro, C. Current trends on turning biomass wastes into carbon materials for electrochemical sensing and rechargeable battery applications. *Curr. Opin. Green Sustain. Chem.* **2020**, *26*, 100374. [CrossRef]
22. Hu, X.; Gholizadeh, M. Progress of the applications of bio-oil. *Renew. Sustain. Energy Rev.* **2020**, *134*, 110124. [CrossRef]
23. Bagnato, G.; Boulet, F.; Sanna, A. Effect of Li-LSX zeolite, NiCe/Al_2O_3 and NiCe/ZrO_2 on the production of drop-in bio-fuels by pyrolysis and hydrotreating of Nannochloropsis and isochrysis microalgae. *Energy* **2019**, *179*, 199–213. [CrossRef]
24. Kumar, R.; Strezova, V.; Weldekidan, H.; He, J.; Singh, S.; Kan, T.; Dastjerdi, B. Lignocellulose biomass pyrolysis for bio-oil production: A review of biomass pre-treatment methods for production of drop-in fuels. *Renew. Sustain. Energy Rev.* **2020**, *123*, 109763. [CrossRef]
25. Fang, J.; Zhan, L.; Ok, Y.S.; Gao, B. Minireview of potential applications of hydrochar derived from hydrothermal carbonization of biomass. *J. Ind. Eng. Chem.* **2018**, *57*, 15–21. [CrossRef]
26. Ullah Khan, I.; Othman, M.H.D.; Hashim, H.; Matsuura, T.; Ismail, A.F.; Rezaei-DashtArzhandi, M.; Wan Azelee, I. Biogas as a renewable energy fuel—A review of biogas upgrading, utilisation and storage. *Energy Convers. Manag.* **2017**, *150*, 277–294. [CrossRef]
27. Baker, E.G.; Elliott, D.C. Catalytic Hydrotreating of Biomass-Derived Oils, in Pyrolysis Oils from Biomass. *ACS Symp. Ser.* **1988**, *21*, 228–240.
28. Bagnato, G.S.; Sanna, A. Process and Techno-Economic Analysis for Fuel and Chemical Production by Hydrodeoxygenation of Bio-Oil. *Catalysts* **2019**, *9*, 1021. [CrossRef]
29. Lima, S.; Fernandes, A.; Antunes, M.M.; Pillinger, M.; Ribeiro, F.; Valente, A.A. Dehydration of Xylose into Furfural in the Presence of Crystalline Microporous Silicoaluminophosphates. *Catal. Lett.* **2010**, *135*, 41–47. [CrossRef]

30. Demirbas, A. Thermochemical Processes. In *Biorefineries: For Biomass Upgrading Facilities. Green Energy and Technology*; Springer: London, UK, 2010; pp. 135–192.
31. BTG Bioliquids BV. 2020. Available online: https://www.btg-bioliquids.com (accessed on 11 January 2021).
32. Li, F.; Srivatsa, S.C.; Bhattacharya, S. A review on catalytic pyrolysis of microalgae to high-quality bio-oil with low oxygeneous and nitrogenous compounds. *Renew. Sustain. Energy Rev.* **2019**, *108*, 481–497. [CrossRef]
33. Furimsky, E. Catalytic hydrodeoxygenation. *Appl. Catal. A Gen.* **2000**, *199*, 147–190. [CrossRef]
34. Fisk, C.A.; Morgan, T.; Ji, Y.; Crocker, M.; Crofcheck, C.; Lewis, S.A. Bio-oil upgrading over platinum catalysts using in situ generated hydrogen. *Appl. Catal. A Gen.* **2009**, *358*, 150–156. [CrossRef]
35. Li, P.; Chen, X.; Wang, X.; Shao, J.; Lin, G.; Yang, H.; Yang, Q.; Chen, H. Catalytic Upgrading of Fast Pyrolysis Products with Fe-, Zr-, and Co-Modified Zeolites Based on Pyrolyzer–GC/MS Analysis. *Energy Fuels* **2017**, *31*, 3979–3986. [CrossRef]
36. Dong, T.; Wang, J.; Miao, C.; Zheng, Y.; Chen, S. Two-step in situ biodiesel production from microalgae with high free fatty acid content. *Bioresour. Technol.* **2013**, *136*, 8–15. [CrossRef] [PubMed]
37. Zainan, N.H.; Srivatsa, S.C.; Bhattacharya, S. Catalytic pyrolysis of microalgae Tetraselmis suecica and characterization study using in situ Synchrotron-based Infrared Microscopy. *Fuel* **2015**, *161*, 345–354. [CrossRef]
38. Demirbas, A. The influence of temperature on the yields of compounds existing in bio-oils obtained from biomass samples via pyrolysis. *Fuel Process. Technol.* **2007**, *88*, 591–597. [CrossRef]
39. Venderbosch, R.H.; Ardiyanti, A.R.; Wildschut, J.; Oasmaa, A.; Heeres, H.J. Stabilization of biomass-derived pyrolysis oils. *J. Chem. Technol. Biotechnol.* **2010**, *85*, 674–686. [CrossRef]
40. Oasmaa, A.; Solantausta, Y.; Arpiainen, V.; Kuoppala, E.; Sipilä, K. Fast pyrolysis bio-oils from wood and agricultural residues. *Energy Fuels* **2010**, *2*, 1380–1388. [CrossRef]
41. Wildschut, J.; Mahfud, F.H.; Venderbosch, R.H.; Heeres, H.J. Hydrotreatment of Fast Pyrolysis Oil Using Heterogeneous Noble-Metal Catalysts. *Ind. Eng. Chem. Res.* **2009**, *48*, 10324–10334. [CrossRef]
42. App, A.R.; Khromova, S.A.; Venderbosch, R.H.; Yakovlev, V.A.; Heeres, H.J. Catalytic hydrotreatment of fast-pyrolysis oil using non-sulfided bimetallic Ni-Cu catalysts on a δ-Al_2O_3 support. *Appl. Catal. B* **2012**, *117–118*, 105–117.
43. Lu, Q.; Yang, X.-L.; Zhu, X.-F. Analysis on chemical and physical properties of bio-oil pyrolyzed from rice husk. *J. Anal. Appl. Pyrolysis* **2008**, *82*, 191–198. [CrossRef]
44. Wildschut, J.; Iqbal, M.; Mahfud, F.H.; Cabrera, I.M.; Venderbosch, R.H.; Heeres, H.J. Insights in the hydrotreatment of fast pyrolysis oil using a ruthenium on carbon catalyst. *Energy Environ. Sci.* **2010**, *3*, 962–970. [CrossRef]
45. Elliott, D.C.; Hart, T.R.; Neuenschwande, G.G.; Rotness, L.J.; Olarte, M.V.; Zacher, A.H.; Solantausta, Y. Catalytic Hydroprocessing of Fast Pyrolysis Bio-oil from Pine Sawdust. *Energy Fuels* **2012**, *26*, 3891–3896. [CrossRef]
46. Elliott, D.C.; Neuenschwander, G.G. Liquid Fuels by Low-Severity Hydrotreating of Biocrude. In *Developments in Thermochemical Biomass Conversion: Volume 1/Volume 2*; Bridgwater, A.V., Boocock, D.G.B., Eds.; Springer: Dordrecht, The Netherlands, 1997; pp. 611–621.
47. Patel, M.; Kumar, A. Production of renewable diesel through the hydroprocessing of lignocellulosic biomass-derived bio-oil: A review. *Renew. Sustain. Energy Rev.* **2016**, *58*, 1293–1307. [CrossRef]
48. Zhang, S.; Yan, Y.; Li, T.; Ren, Z. Upgrading of liquid fuel from the pyrolysis of biomass. *Bioresour. Technol.* **2005**, *96*, 545–550. [CrossRef] [PubMed]
49. López, R.; Fernández, C.; Gómez, X.; Martínez, O.; Sánchez, M.E. Thermogravimetric analysis of lignocellulosic and microalgae biomasses and their blends during combustion. *J. Therm. Anal.* **2013**, *114*, 295–305. [CrossRef]
50. Liu, H.; Chen, Y.; Yang, H.; Gentili, F.G.; Söderlind, U.; Wang, X.; Zhang, W.; Chen, H. Hydrothermal carbonization of natural microalgae containing a high ash content. *Fuel* **2019**, *249*, 441–448. [CrossRef]
51. Plis, A.; Lasek, J.; Skawińska, A. Kinetic analysis of the combustion process of Nannochloropsis gaditana microalgae based on thermogravimetric studies. *J. Anal. Appl. Pyrolysis* **2017**, *127*, 109–119. [CrossRef]
52. Miao, X.; Wu, Q. High yield bio-oil production from fast pyrolysis by metabolic controlling of Chlorella protothecoides. *J. Biotechnol.* **2004**, *110*, 85–93. [CrossRef]
53. Jena, U.; Das, K.C. Comparative Evaluation of Thermochemical Liquefaction and Pyrolysis for Bio-Oil Production from Microalgae. *Energy Fuels* **2011**, *25*, 5472–5482. [CrossRef]
54. Guo, Q.; Wu, M.; Wang, K.; Zhang, L.; Xu, X. Catalytic Hydrodeoxygenation of Algae Bio-oil over Bimetallic Ni–Cu/ZrO_2 Catalysts. *Ind. Eng. Chem. Res.* **2015**, *54*, 890–899. [CrossRef]
55. Gruia, A. Hydrotreating. In *Handbook of Petroleum Processing*; Jones, D.S.J.S., Pujado, P.R., Eds.; Springer: Dordrecht, The Netherlands, 2006; pp. 321–354.
56. Jacobson, K.; Maheria, K.C.; Kumar Dalai, A. Bio-oil valorization: A review. *Renew. Sustain. Energy Rev.* **2013**, *23*, 91–106. [CrossRef]
57. Mauchausse, C.; Kural, E.; Trimm, D.L.; Cant, N.W. Optimization of tungsten-based catalysts for the hydrotreatment of coal-derived liquids. *Fuel* **1992**, *71*, 203–209. [CrossRef]
58. Bulushev, D.A.; Ross, J.R.H. Catalysis for conversion of biomass to fuels via pyrolysis and gasification: A review. *Catal. Today* **2011**, *171*, 1–13. [CrossRef]
59. Masel, R.I. *Principles of Adsorption and Reaction on Solid Surfaces*; Wiley: New York, NY, USA; Chichester, UK, 1996.
60. Ptushinskiĭ, Y.G. Low-temperature adsorption of gases on metal surfaces (Review). *Low Temp. Phys.* **2004**, *30*, 1–26. [CrossRef]

61. Ferrin, P.; Kandoi, S.; Nilekar, A.U.; Mavrikakis, M. Hydrogen adsorption, absorption and diffusion on and in transition metal surfaces: A DFT study. *Surf. Sci.* **2012**, *606*, 679–689. [CrossRef]
62. Elliott, D.C.; Hu, J.; Hart, T.R.; Neuenschwander, G.G. Palladium Catalyzed Hydrogenation of Bio-Oils and Organic Compounds. U.S. Patent 7,425,657, 16 September 2008.
63. Mortensen, P.M.; Grunwaldt, J.D.; Jensen, P.A.; Jensen, A.D. Screening of Catalysts for Hydrodeoxygenation of Phenol as a Model Compound for Bio-oil. *ACS Catal.* **2013**, *3*, 1774–1785. [CrossRef]
64. Ardiyanti, A.R.; Khromova, S.A.; Venderbosch, R.H.; Yakovlev, V.A.; Melián-Cabrera, I.V.; Heeres, H.J. Catalytic hydrotreatment of fast pyrolysis oil using bimetallic Ni–Cu catalysts on various supports. *Appl. Catal. A* **2012**, *449*, 121–130. [CrossRef]
65. Wei, S.; Zhao, Y.; Fan, G.; Yang, L.; Li, F. Structure-dependent selective hydrogenation of cinnamaldehyde over high-surface-area CeO_2-ZrO_2 composites supported Pt nanoparticles. *Chem. Eng. J.* **2017**, *322*, 234–245. [CrossRef]
66. Zhou, M.; Liu, P.; Wang, K.; Xu, J.; Jiang, J. Catalytic hydrogenation and one step hydrogenation-esterification to remove acetic acid for bio-oil upgrading: Model reaction study. *Catal. Sci. Technol.* **2016**, *6*, 7783–7792. [CrossRef]
67. Bergem, H.; Xu, R.; Brown, R.C.; Huber, G.W. Low temperature aqueous phase hydrogenation of the light oxygenate fraction of bio-oil over supported ruthenium catalysts. *Green Chem.* **2017**, *19*, 3252–3262. [CrossRef]
68. Yuan, Q.; Zhang, D.; van Haandel, L.; Ye, F.; Xue, T.; Hensen, E.J.M.; Guan, Y. Selective liquid phase hydrogenation of furfural to furfuryl alcohol by Ru/Zr-MOFs. *J. Mol. Catal. A Chem.* **2015**, *406*, 58–64. [CrossRef]
69. Liu, Y.; Zheng, Y.; Du, B.; Nasaruddin, R.R.; Chen, T.; Xie, J. Golden Carbon Nanotube Membrane for Continuous Flow Catalysis. *Ind. Eng. Chem. Res.* **2017**, *56*, 2999–3007. [CrossRef]
70. Fiorio, J.L.; López, N.; Rossi, L.M. Gold–Ligand-Catalyzed Selective Hydrogenation of Alkynes into cis-Alkenes via H_2 Heterolytic Activation by Frustrated Lewis Pairs. *ACS Catal.* **2017**, *7*, 2973–2980. [CrossRef]
71. Takeda, Y.; Tamura, M.; Nakagawa, Y.; Okumura, K.; Tomishige, K. Characterization of Re–Pd/SiO_2 Catalysts for Hydrogenation of Stearic Acid. *ACS Catal.* **2015**, *5*, 7034–7047. [CrossRef]
72. Raj, K.J.A.; Prakash, M.G.; Mahalakshmy, R.; Elangovan, T.; Viswanathan, B. Liquid phase hydrogenation of crotanaldehyde over nickel supported on titania. *J. Mol. Catal. A Chem.* **2013**, *366*, 92–98.
73. Fulajtárova, K.; Soták, T.; Hronec, M.; Vávra, I.; Dobročka, E.; Omastová, M. Aqueous phase hydrogenation of furfural to furfuryl alcohol over Pd–Cu catalysts. *Appl. Catal. A* **2015**, *502*, 78–85. [CrossRef]
74. Wang, C.; Guo, Z.; Yang, Y.; Chang, J.; Borgna, A. Hydrogenation of Furfural as Model Reaction of Bio-Oil Stabilization under Mild Conditions Using Multiwalled Carbon Nanotube (MWNT)-Supported Pt Catalysts. *Ind. Eng. Chem. Res.* **2014**, *53*, 11284–11291. [CrossRef]
75. Ota, N.; Tamura, M.; Nakagawa, Y.; Okumura, K.; Tomishige, K. Performance, Structure, and Mechanism of ReOx–Pd/CeO_2 Catalyst for Simultaneous Removal of Vicinal OH Groups with H_2. *ACS Catal.* **2016**, *6*, 3213–3226. [CrossRef]
76. Nakagawa, Y.; Tamura, R.; Tamura, M.; Tomishige, K. Combination of supported bimetallic rhodium–molybdenum catalyst and cerium oxide for hydrogenation of amide. *Sci. Technol. Adv. Mater.* **2015**, *16*, 014901. [CrossRef]
77. Liao, X.; Zhang, Y.; Guo, J.; Zhao, L.; Hill, M.; Jiang, Z.; Zhao, Y. The Catalytic Hydrogenation of Maleic Anhydride on $CeO_{2-\delta}$-Supported Transition Metal Catalysts. *Catalysts* **2017**, *7*, 272. [CrossRef]
78. Elliott, D.C. Historical developments in hydroprocessing bio-oils. *Energy Fuels* **2007**, *21*, 1792–1815. [CrossRef]
79. Laurent, E.; Delmon, B. Study of the hydrodeoxygenation of carbonyl, car□ylic and guaiacyl groups over sulfided CoMo/γ-Al_2O_3 and NiMo/γ-Al_2O_3 catalysts: I. Catalytic reaction schemes. *Appl. Catal. A* **1994**, *109*, 77–96. [CrossRef]
80. Bienholz, A.; Schwab, F.; Claus, P. Hydrogenolysis of glycerol over a highly active CuO/ZnO catalyst prepared by an oxalate gel method: Influence of solvent and reaction temperature on catalyst deactivation. *Green Chem.* **2010**, *12*, 290–295. [CrossRef]
81. Liu, H.; Liang, S.; Jiang, T.; Han, B.; Zhou, Y. Hydrogenolysis of Glycerol to 1,2-Propanediol over Ru–Cu Bimetals Supported on Different Supports. *CLEAN Soil Air Water* **2012**, *40*, 318–324. [CrossRef]
82. de Miguel Mercader, F.; Groeneveld, M.J.; Kersten, S.R.A.; Geantet, C.; Toussaint, G.; Way, N.W.J.; Schaverien, C.J.; Hogendoorn, K.J.A. Hydrodeoxygenation of pyrolysis oil fractions: Process understanding and quality assessment through co-processing in refinery units. *Energy Environ. Sci.* **2011**, *4*, 985–997. [CrossRef]
83. Prasomsri, T.; Shetty, M.; Murugappan, K.; Román-Leshkov, Y. Insights into the catalytic activity and surface modification of MoO_3 during the hydrodeoxygenation of lignin-derived model compounds into aromatic hydrocarbons under low hydrogen pressures. *Energy Environ. Sci.* **2014**, *7*, 2660–2669. [CrossRef]
84. Bagnato, G.; Signoretto, M.; Pizzolitto, C.; Menegazzo, F.; Xi, X.; ten Brink, G.H.; Kooi, B.J.; Heeres, H.J.; Sanna, A. Hydrogenation of Biobased Aldehydes to Monoalcohols Using Bimetallic Catalysts. *ACS Sustain. Chem. Eng.* **2020**, *8*, 11994–12004. [CrossRef]
85. Sanna, A.; Vispute, T.P.; Huber, G.W. Hydrodeoxygenation of the aqueous fraction of bio-oil with Ru/C and Pt/C catalysts. *Appl. Catal. B* **2015**, *165*, 446–456. [CrossRef]
86. Oasmaa, A.; Kuoppala, E.; Solantausta, Y. Fast Pyrolysis of Forestry Residue. 2. Physicochemical Composition of Product Liquid. *Energy Fuels* **2003**, *17*, 433–443. [CrossRef]
87. Oasmaa, A.; Kuoppala, E.; Gust, S.; Solantausta, Y. Fast Pyrolysis of Forestry Residue. 1. Effect of Extractives on Phase Separation of Pyrolysis Liquids. *Energy Fuels* **2003**, *17*, 1–12. [CrossRef]
88. Adjaye, J.D.; Bakhshi, N.N. Production of hydrocarbons by catalytic upgrading of a fast pyrolysis bio-oil. Part II: Comparative catalyst performance and reaction pathways. *Fuel Process. Technol.* **1995**, *45*, 185–202. [CrossRef]

89. De la Puente, G.; Gil, A.; Pis, J.J.; Grange, P. Effects of Support Surface Chemistry in Hydrodeoxygenation Reactions over CoMo/Activated Carbon Sulfided Catalysts. *Langmuir* **1999**, *15*, 5800–5806. [CrossRef]
90. Gutierrez, A.; Kaila, R.K.; Honkela, M.L.; Slioor, R.; Krause, A.O.I. Hydrodeoxygenation of guaiacol on noble metal catalysts. *Catal. Today* **2009**, *147*, 239–246. [CrossRef]
91. Bui, V.N.; Laurenti, D.; Afanasiev, P.; Geantet, C. Hydrodeoxygenation of guaiacol with CoMo catalysts. Part I: Promoting effect of cobalt on HDO selectivity and activity. *Appl. Catal. B* **2011**, *101*, 239–245. [CrossRef]
92. Centeno, A.; Laurent, E.; Delmon, B. Influence of the Support of CoMo Sulfide Catalysts and of the Addition of Potassium and Platinum on the Catalytic Performances for the Hydrodeoxygenation of Carbonyl, Carboxyl, and Guaiacol-Type Molecules. *J. Catal.* **1995**, *154*, 288–298. [CrossRef]
93. Ferrari, M.; Delmon, B.; Grange, P. Influence of the active phase loading in carbon supported molybdenum–cobalt catalysts for hydrodeoxygenation reactions. *Microporous Mesoporous Mater.* **2002**, *56*, 279–290. [CrossRef]
94. Bindwal, A.B.; Bari, A.H.; Vaidya, P.D. Kinetics of low temperature aqueous-phase hydrogenation of model bio-oil compounds. *Chem. Eng. J.* **2012**, *207–208*, 725–733. [CrossRef]
95. Bindwal, A.B.; Vaidya, P.D. Kinetics of Aqueous-Phase Hydrogenation of Levoglucosan over Ru/C Catalyst. *Ind. Eng. Chem. Res.* **2013**, *52*, 17781–17789. [CrossRef]
96. Zhang, S.; Yan, Y.; Li, T.; Ren, Z. Lumping Kinetic Model for Hydrotreating of Bio-oil from the Fast Pyrolysis of Biomass. *Energy Sources Part A* **2009**, *31*, 639–645. [CrossRef]
97. Zhang, S.-P. Study of Hydrodeoxygenation of Bio-Oil from the Fast Pyrolysis of Biomass. *Energy Sources* **2003**, *25*, 57–65.
98. Sheu, Y.-H.E.; Anthony, R.G.; Soltes, E.J. Kinetic studies of upgrading pine pyrolytic oil by hydrotreatment. *Fuel Process. Technol.* **1988**, *19*, 31–50. [CrossRef]
99. Bagnato, G.; Iulianelli, A.; Sanna, A.; Basile, A. Glycerol Production and Transformation: A Critical Review with Particular Emphasis on Glycerol Reforming Reaction for Producing Hydrogen in Conventional and Membrane Reactors. *Membranes* **2017**, *7*, 17. [CrossRef] [PubMed]
100. Iulianelli, A.; Longo, T.; Basile, A. Methanol steam reforming reaction in a Pd–Ag membrane reactor for CO-free hydrogen production. *Int. J. Hydrogen Energy* **2008**, *33*, 5583–5588. [CrossRef]
101. Iulianelli, A.; Longo, T.; Basile, A. CO-free hydrogen production by steam reforming of acetic acid carried out in a Pd–Ag membrane reactor: The effect of co-current and counter-current mode. *Int. J. Hydrogen Energy* **2008**, *33*, 4091–4096. [CrossRef]
102. Liu, M.; Zhu, X.; Chen, R.; Liao, Q.; Feng, H.; Li, L. Catalytic membrane microreactor with Pd/γ-Al_2O_3 coated PDMS film modified by dopamine for hydrogenation of nitrobenzene. *Chem. Eng. J.* **2016**, *301*, 35–41. [CrossRef]
103. Liu, C.; Xu, Y.; Liao, S.; Yu, D. Mono- and bimetallic catalytic hollow- fiber reactors for the selective hydrogenation of butadiene in 1-butene. *Appl. Catal. A Gen.* **1998**, *172*, 23–29. [CrossRef]
104. Stanford, J.P.; Soto, M.C.; Pfromm, P.H.; Rezac, M.E. Aqueous phase hydrogenation of levulinic acid using a porous catalytic membrane reactor. *Catal. Today* **2016**, *268*, 19–28. [CrossRef]
105. Bagnato, G.; Figoli, A.; Ursino, C.; Galiano, F.; Sanna, A. A novel Ru–polyethersulfone (PES) catalytic membrane for highly efficient and selective hydrogenation of furfural to furfuryl alcohol. *J. Mater. Chem. A* **2018**, *6*, 4955–4965. [CrossRef]
106. Liguori, F.; Barbaro, P.; Giordano, C.; Sawa, H. Partial hydrogenation reactions over Pd-containing hybrid inorganic/polymeric catalytic membranes. *Appl. Catal. A Gen.* **2013**, *459*, 81–88. [CrossRef]
107. Aran, H.C.; Klooster, H.; Jani, J.M.; Wessling, M.; Lefferts, L.; Lammertink, R.G.H. Influence of geometrical and operational parameters on the performance of porous catalytic membrane reactors. *Chem. Eng. J.* **2012**, *207–208*, 814–821. [CrossRef]
108. Bottino, A.; Capannelli, G.; Comite, A.; Del Borghi, A.; Di Felice, R. Catalytic ceramic membrane in a three-phase reactor for the competitive hydrogenation–isomerisation of methylenecyclohexane. *Sep. Purif. Technol.* **2004**, *34*, 239–245. [CrossRef]
109. Bottino, A.; Capannelli, G.; Comite, A.; Di Felice, R. Polymeric and ceramic membranes in three-phase catalytic membrane reactors for the hydrogenation of methylenecyclohexane. *Desalination* **2002**, *144*, 411–416. [CrossRef]
110. Bengtson, G.; Fritsch, D. Catalytic membrane reactor for the selective hydrogenation of edible oil: Platinum versus palladium catalyst. *Desalination* **2006**, *200*, 666–667. [CrossRef]
111. Liguori, F.; Barbaro, P.; Sawa, H. Continuous flow hydrogenation reactions by Pd catalysts onto hybrid ZrO_2/PVA materials. *Appl. Catal. A Gen.* **2014**, *488*, 58–65. [CrossRef]

Article

Syntheses, Characterization, and Application of Tridentate Phenoxyimino-Phenoxy Aluminum Complexes for the Coupling of Terminal Epoxide with CO_2: From Binary System to Single Component Catalyst

Zhichao Zhang [1,*], Tianming Wang [1], Peng Xiang [2], Qinqin Du [3] and Shuang Han [1]

1 School of Science, Shenyang University of Chemical Technology, Shenyang 110142, China; wtm1333@gmail.com (T.W.); unihanshuang@syuct.edu.cn (S.H.)
2 PolyAnalytik Inc., 700 Collip Circle, Suite 202, London, ON N6G 4X8, Canada; peng.xiang@polyanalytik.com
3 DUT Chemistry Analysis & Research Centre, Dalian University of Technology, Dalian 116024, China; duqinqin@dlut.edu.cn
* Correspondence: zzhang@syuct.edu.cn; Tel.: +86-24-8938-3297

Abstract: A series of binuclear aluminum complexes **1–3** supported by tridentate phenoxyimino-phenoxy ligands was synthesized and used as catalysts for the coupling reaction of terminal epoxide with carbon dioxide. The aluminum complex 1, which is catalytically inactive toward the coupling of epoxide with CO_2 by itself, shows moderate activity in the presence of excess nucleophiles or organic bases at high temperature. In sharp contrast to complex **1**, bifunctional complexes **2** and **3**, which incorporate tertiary amine groups as the built-in nucleophile, are able to efficiently transform terminal epoxide with CO_2 to corresponding cyclic carbonates as a sole product by themselves at 100 °C. The number of amine groups on the ligand skeleton and the reaction temperature exert a great influence on the catalytic activity. The bifunctional complexes **2** and **3** are also active at low carbon dioxide pressure such as 2 atm or atmospheric CO_2 pressure. Kinetic studies of the coupling reactions of chloropropylene oxide/CO_2 and styrene oxide/CO_2 using bifunctional catalysts under atmospheric pressure of CO_2 demonstrate that the coupling reaction has a first-order dependence on the concentration of the epoxide.

Keywords: aluminum complex; CO_2 fixation; bifunctional catalyst; cyclic carbonate

Citation: Zhang, Z.; Wang, T.; Xiang, P.; Du, Q.; Han, S. Syntheses, Characterization, and Application of Tridentate Phenoxyimino-Phenoxy Aluminum Complexes for the Coupling of Terminal Epoxide with CO_2: From Binary System to Single Component Catalyst. *Catalysts* **2021**, *11*, 145. https://doi.org/10.3390/catal11020145

Received: 30 December 2020
Accepted: 15 January 2021
Published: 20 January 2021

1. Introduction

With the rapid development of the industry as well as the ever-growing human activity, a huge amount of CO_2 has been released to the air, which has caused severe environmental problems [1–4]. The utilization of CO_2 as feedstock to produce valuable chemicals is a promising way to solve this problem, but this conversion is limited due to the inertness of CO_2 [5]. The catalytic coupling reaction with epoxides represents one of the promising processes employing CO_2 to produce organic cyclic carbonate, which has found widespread applications in many respects [6–10]. Generally, the catalyst capable of achieving the transformation is comprised of nucleophile and Lewis acid. Thus, organic compounds and metal-based complexes have been employed as the Lewis acid for the coupling reaction, with the aid of nucleophiles to constitute the binary catalytic system. Usually, metal-based complexes display higher reactivity than organocatalysts for the coupling reaction. As a result of the diversity of the metal across the periodic table and the organic ligands of different structures, miscellaneous metal complexes based on aluminum [11–32], magnesium [33–36], chromium [37–42], cobalt [43–48], iron [49–53], and rare earth metals [54,55] have been innovated for the coupling reaction of CO_2 with epoxides. Of many metal complexes, aluminum complexes have been drawing attention due to their low price, easy availability, and the superior selectivity for the cyclic carbonate over polymeric products [56]. It is well established that the electronic effects and steric hindrance around the aluminum center that can be finely tuned are crucial for the catalytic activity and the product selectivity. For example, optically active carbonate can be produced from the coupling of racemic epoxide with CO_2 when the chiral complexes are employed [44,45,57].

In addition to binary catalytic system, there is a tendency to employ a bifunctional catalyst in which nucleophile and Lewis metal complexes are constructed into one molecule as the one-component catalyst for the coupling of CO_2 with epoxides in the past decade [58–68]. Quaternary ammonium halide as a nucleophile is usually integrated into the metal complex to construct an ionic bifunctional catalyst. The pre-association of the nucleophilic halide to the metal complex via electrostatic attraction would effectively reduce the translational entropy penalty at the ring-opening step and the cyclization step as compared with the binary catalytic system [69]. Thus, the bifunctional catalyst loading for the coupling reaction is significantly reduced, and usually, a higher catalytic reactivity is achieved than that with the corresponding binary catalytic system. However, the incorporation of the quaternary onium salts into the metal complex impairs the solubility of the bifunctional catalyst. Another drawback is the thermostability of the ammonium halide-functionalized catalyst at elevated temperature [59], which undergoes ammonium salt decomposition pathways, including Zaitsev and Hoffman type eliminations [70–72] and retro-Menschutkin reactions [73–76]. Hence, it is important and necessary to develop bifunctional catalyst of high efficiency and stability at elevated temperature.

In this work, we present the syntheses of a series of aluminum complexes supported by novel tridentate phenoxyimino-phenoxy ligands and their catalytic behaviors for the coupling of terminal epoxides with CO_2. Compared with tetradentate salen-type ligand stabilized metal complexes, which are extensively employed for the coupling of epoxide with CO_2, aluminum complexes supported by tridentate ligands for this transformation are rare. As a result of the low coordination number of tridentate ligands, the resultant complexes are expected to have a more acidic metal center, which is supposed to benefit the activation of the epoxide ring and to facilitate the ring-opening. Moreover, an amine group was incorporated onto the ligand framework as the built-in nucleophile to construct neutral bifunctional complexes **2** and **3**, in order to avoid the decomposition of ammonium halide moiety under elevated temperature. Systematic investigation of the catalytic behaviors of the aluminum complexes **2** and **3** reveals that the neutral and ionic bifunctional complexes are highly active toward the coupling of epoxide to CO_2, whereas the unfunctionalized aluminum complex 1 is catalytically inactive by itself.

2. Results and Discussion

2.1. *Syntheses and Characterization of Complexes* **1–3**

The tridentate ligand precursors H_2L^1–H_2L^3 (see Scheme 1) were readily synthesized by reacting the substituted salicylaldehyde with substituted aminophenol in 1:1 molar ratio (see the experimental section in the supporting material). The three ligand precursors were identified by proton nuclear magnetic resonance (^{1}H NMR) and ^{13}C-nuclear magnetic resonance (^{13}C NMR) (Figures S1–S3 in the supporting material).

Et$_2$AlCl/Hexane, THF, −30 °C

H_2L^1: R^1 = *t*-Bu, R^2 = H
H_2L^2: R^1 = CH_2NMe_2, R^2 = H
H_2L^3: R^1 = CH_2NMe_2, R^2 = CH_2NMe_2

1: R^1 = *t*-Bu, R^2 = H
2: R^1 = CH_2NMe_2, R^2 = H
3: R^1 = CH_2NMe_2, R^2 = CH_2NMe_2

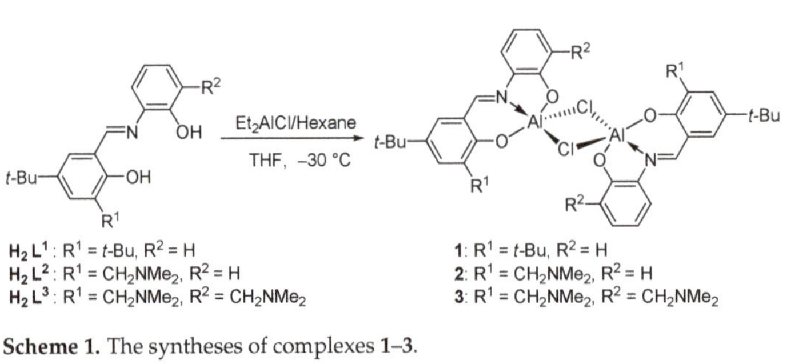

Scheme 1. The syntheses of complexes **1–3**.

Subsequent syntheses of complexes **1–3** were conducted by the alkyl-elimination reaction of Et_2AlCl with the corresponding ligand precursors at low temperature, as illustrated in Scheme 1. In a typical procedure, a tetrahydrofuran (THF) solution of H_2L^1 was added dropwise to a stirred hexane solution of $AlEt_2Cl$ at −30 °C to afford a yellow solution. The reaction was slowly warmed up to room temperature and stirred for another 4 h. After the removal of all the volatiles under reduced pressure, yellow solids were separated, which was followed by washing with cold hexane three times and drying to constant weight under vacuum. The structure of the isolated solids was characterized by the NMR technique with $CDCl_3$ as the solvent. The ^1H NMR spectrum shows no resonances of hydroxy groups or ethyl groups bound to aluminum. In the low field, the resonance of 8.06 ppm assignable to the imino moiety was observed, which slightly shifted to the low field when compared with that in free ligand precursor H_2L^1 (see Figure S4 in the supporting material). Two distinct resonances of 1.32 ppm and 0.84 ppm for the tBu groups were found (1.49 and 1.35 ppm in the H_2L^1), indicating chelation of the tridentate ligand to aluminum. Moreover, the upfield shift of the imino group in the H_2L^1 from 8.72 to 8.06 ppm in the isolated complex suggested the coordination of nitrogen of the imino group to the aluminum center. Remarkably, no solvent molecules were detected in the ^1H NMR spectrum, suggesting that the isolated complex is THF-free.

In addition to the characterization by NMR, mass spectrometry (MS) was employed to characterize the isolated complex. A very weak peak at m/z = 350.1566 was detected, which corresponds to the cationic species **1a** having the formula of $C_{21}H_{25}O_2NAl$, as shown in Scheme 2. Interestingly, a strong signal was observed at m/z = 414.2219, which was assigned to the cationic species 1b with the formula of $C_{23}H_{33}O_4NAl$. We supposed that the species **1b** was generated by the coordination of two methanol molecules to the cationic species **1a**. Thus, it is reasonable to interpret the weak signal of **1a** and the strong peak of **1b** on the mass spectrum. In the light of the strong signal of **1b** on the MS, no solvent detection on the ^1H NMR spectrum, and the coordination number of 5 usually adopted by aluminum complex, we expected that complex **1** should be a binuclear structure, with each aluminum center being surrounded by a phenoxyimino-phenoxy ligand and two chlorides, as presented in Scheme 2.

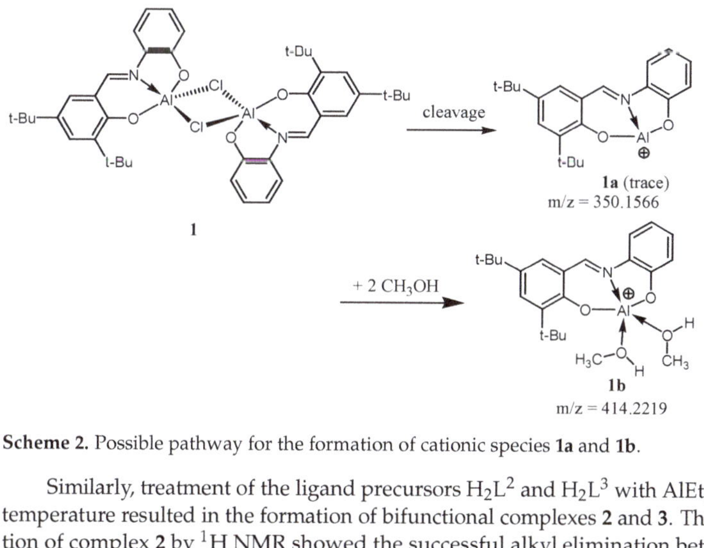

Scheme 2. Possible pathway for the formation of cationic species **1a** and **1b**.

Similarly, treatment of the ligand precursors H_2L^2 and H_2L^3 with $AlEt_2Cl$ under low temperature resulted in the formation of bifunctional complexes **2** and **3**. The characterization of complex **2** by ^1H NMR showed the successful alkyl elimination between $AlEt_2Cl$ and H_2L^2. Moreover, there are no THF molecules detected on the ^1H NMR spectrum, suggesting that complex **2** is THF-free. The MS spectrum of complex **2** shows a m/z peak positioning at 397.1085. We tentatively ascribed this signal to the structure **2a**, which probably is an ethanol coordinated species, as illustrated in Figure 1. Meanwhile, a peak at m/z = 783.3226 was detected, which is ascribed to the cationic species **2b**. Although a tertiary amine group is introduced into the molecule, the inner coordination of the amine group to the Al center is impossible because of the orientation of the tertiary amine group. Taking into account the coordination saturation of the aluminum, it is reasonable to assume the complex **2** adopts binuclear structure with dangling tertiary amine groups. Similar results were also observed for complex **3**, as indicated by the ^{1}H NMR and MS, which should also adopt a binuclear structure with dangling amine groups via chloride bridges.

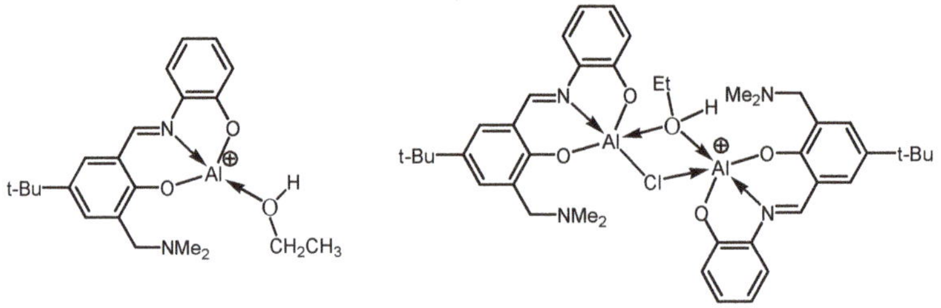

2a: $C_{22}H_{30}AlN_2O_3^+$
m/z: 397.1085

2b: $C_{42}H_{54}Al_2ClN_4O_5^+$
m/z: 783.3226

Figure 1. Possible structures of cationic species of **2a** and **2b** found in the MS of complex **2**.

2.2. Catalytic Performance of Complexes **1–3** *in the Coupling Reactions of Terminal Epoxide with* CO_2

To begin with, we examined the catalytic performance of complex **1** for the coupling of propylene oxide (PO) with CO_2 (see Scheme 3). Unfortunately, complex **1** was catalytically inactive by itself, which was in agreement with the previous literature reports [21,22]. When it was paired with a nucleophile or organic base, complex **1** showed moderate activity for the PO/CO_2 coupling (see Tables S1–S4 in the supporting information). For instance, complex **1** in combination with tetrabutylammonium bromide (TBAB, 40 eq) catalyzed the PO/CO_2 coupling to produce propylene carbonate (PC) as the sole product at 80 °C under 20 atm CO_2 in 10 h with 91% PO conversion (see entry 8 in Table S1). There was no detection of any polymeric products as evidenced by the ^{1}H NMR spectrum of the reaction mixtures. It was also interesting to find that the complex **1**/TBAB (40 eq) system displayed a zeroth-order dependence on the PO throughout the coupling (see Table S2 and Figure S7, in the supporting material). This result is analogous to Han's report in which polymeric ionic liquid was employed as the catalyst for the PO/CO_2 coupling reaction [77]. When other organobases such as 4-dimethylaminopyridine (DMAP) or 1-methylimidazole (1-MeIm) were employed as the cocatalysts, high cocatalyst loading was required for effective transformation (see Tables S3 and S4 in the supporting material). For instance, the binary system composed of complex **1**/DMAP (1/20) displayed lower catalytic activity than complex **1**/TBAB (1/20) at 60 °C.

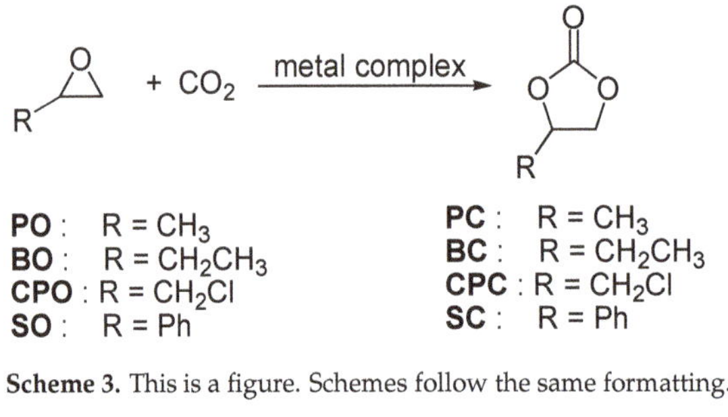

+ CO_2 —metal complex→

PO : R = CH_3
BO : R = CH_2CH_3
CPO : R = CH_2Cl
SO : R = Ph

PC : R = CH_3
BC : R = CH_2CH_3
CPC : R = CH_2Cl
SC : R = Ph

Scheme 3. This is a figure. Schemes follow the same formatting.

It has been reported by North and Pasquale that quaternary ammonium halide would decompose at high temperature to generate tertiary amine and halohydrocarbon [78]. The in situ generated tertiary amine, together with the remaining ammonium halide salts, assists the metal complex to convert propylene oxide and CO_2 to cyclic carbonate. Recently, Kim and his coworkers reported the coupling of epoxide with CO_2 catalyzed by a series of tertiary amines with moderate catalytic activity [79]. These results enlightened us to design amine-functionalized complex **2** as the single-component catalyst for the coupling of epoxide with CO_2. The catalytic results were compiled in Table 1. It was satisfying that the bifunctional complex **2** by itself succeeded in transforming the CO_2 and PO to PC. Analysis of the reaction mixture showed that PC was exclusively produced without the formation of any polymeric products as evidenced by ^{1}H NMR spectrum. Although complex **2** displayed very low activity at 80 °C (entries 1 and 2, in Table 1), however, the catalytic activity was considerably enhanced when the reaction was conducted at 100 °C. For instance, 68% of PO was converted to PC under 10 atm CO_2 in 10 h (entry 3, Table 1). In addition to the reaction temperature, increasing the CO_2 pressure accelerates the coupling reaction. Then, 94% of PO was transformed to PC at 20 atm CO_2 pressure within 10 h (entry 4, Table 1), suggesting that high CO_2 pressure, namely, the high CO_2 concentration facilitated the coupling reaction.

Table 1. The coupling reaction of terminal epoxide with CO_2 by the sole complex **2** [1].

Entry	Epoxide	P(CO_2) (atm)	Temperature (°C)	Time (h)	Conversion [2] (%)
1	PO	10	80	10	7
2	PO	20	80	10	8
3	PO	10	100	10	68
4	PO	20	100	10	94
5	BO	10	80	10	7
6	BO	20	100	10	49
7	CPO	10	80	10	12
8	CPO	20	80	10	26
9	CPO	10	100	5	36
10	CPO	10	100	10	52
11	CPO	10	100	24	90
12	CPO	20	100	5	40
13	CPO	20	100	10	62
14	SO	10	100	5	5
15	SO	10	100	10	11
16	SO	10	100	18	32
17	SO	10	100	24	75
18	SO	20	100	5	7
19	SO	20	100	10	15

[1] Conditions: Complex **2** (38.7 mg, 50 µmol), $n_{(2)}/n_{(epoxide)} = 1/1000$. [2] Determined by ^{1}H NMR; No polymeric products were discovered as evidenced by ^{1}H NMR.

Subsequent examination of the catalytic performance of complex **2** in other terminal epoxides/CO_2 coupling reactions confirmed again the significance of the reaction temperature and CO_2 pressure. The 1,2-butylene oxide (BO) conversion of 7% at 80 °C under 10 atm CO_2 was improved to 49% when the reaction was performed at 100 °C under 20 atm of CO_2 pressure (entries 5 and 6, Table 1). Similar trends were also observed for the coupling of CO_2 with epichlorohydrin (or chloropropylene oxide, abbreviated as CPO) or styrene oxide (SO). The CPO conversion of 12% at 80 °C under 10 atm of CO_2 in 10 h was enhanced to 36%, 52%, and 90% within 5, 10, and 24 h, respectively, when the reaction temperature was simply increased to 100 °C (entries 7, 9–11, Table 1). As the CO_2 pressure was increased to 20 atm, the conversion of CPO rose to 40% in 5 h and 62% in 10 h under 20 atm CO_2 (entries 12 and 13, Table 1). As for the SO/CO_2 coupling, the reaction was quite sluggish (entries 14–19, Table 1). High SO conversion can only be achieved after a long reaction time (entry 17, Table 1).

Complex **2** was capable of transforming CPO/CO_2 and SO/CO_2 to cyclic carbonate under atmospheric CO_2 pressure. The results were collected in Table 2, and the kinetic curves (conversion vs. reaction time) were depicted in Figure 2. As expected, the rate of the coupling of CO_2 with either CPO or SO under 1 atm CO_2 was extremely slow, which was partly due to the low CO_2 pressure. As can be seen from Figure 2, there is a surge in the CPO conversion after 36 h (the blue curve). This might be interpreted by the poor solubility of complex **2** in CPO. Therefore, the metal complex **2** was not well dispersed in the epoxide, leading to a heterogeneous system. Thus, the effective concentration of complex **2** was lower than the theoretical value at the beginning. With the proceeding of the coupling reaction, highly polar cyclic carbonate was generated, which in turn acted as solvent to dissolve complex **2**. Thus, the effective concentration of complex **2** was continuously increased in the course of the reaction until the full dissolution of complex **2**. This would accelerate the coupling, as is shown in Figure 2 (the blue curve). A similar situation took place in the case of SO/CO_2 coupling as well, where the reaction rate gradually increased with the reaction time, as indicated by the kinetic curve (the red curves in Figure 2).

Table 2. Complex **2** catalyzed chloropropylene oxide (CPO)/CO_2 and styrene oxide (SO)/CO_2 coupling under atmospheric CO_2 [1].

Entry	Epoxide	Time (h)	Conversion (%)
1	CPO	5	6
2	CPO	10	11
3	CPO	24	28
4	CPO	36	35
5	CPO	48	68
6	SO	5	0.7
7	SO	10	2
8	SO	24	9
9	SO	36	17
10	SO	48	30
11	SO	60	42

[1] Conditions: Complex **2** (38.7 mg, 50 μmol), $n_{(2)}/n_{(epoxide)}$ = 1/1000, P(CO_2) = 1 atm, temperature is 100 °C. Other notes are the same as those in Table 1.

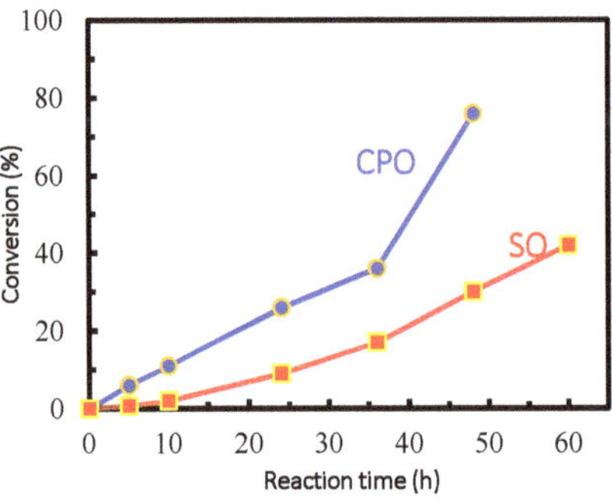

Figure 2. Kinetic curves (substrate conversion vs reaction time) of the coupling of CPO/CO_2 and SO/CO_2 by complex **2** under atmospheric CO_2 pressure at 100 °C.

The success of complex **2** as a single-component catalyst motivated us to examine the catalytic performance of complex **3**, which has more tertiary amine groups on the ligand skeleton for the coupling of epoxides toward CO_2. The catalytic results were collected in Table 3. As anticipated, complex **3** displayed higher activity than complex **2** did under the same conditions, which was tentatively attributed to the incorporation of one more amine group. Similar to complex **2**, complex **3** exhibited low activity at 80 °C. For example, 14% of PO was transformed to PC under 20 atm CO_2 in 10 h (entry 1, Table 3). However, the catalytic activity was dramatically improved by increasing the temperature to 100 °C, with the 64% conversion of PO in 5 h, and 99% in 10 h (entries 2 and 3, Table 3). The significance of reaction temperature was also observed in the BO/CO_2 coupling. The BO conversion of 5% at 80 °C was dramatically increased to 76% when the reaction temperature was increased to 100 °C at 10 atm CO_2 pressure (entries 4 and 5, Table 3). Similarly, enhancing the CO_2 pressure to 20 atm gave rise to 95% of BO conversion in 10 h (entry 6, Table 3), which was much higher than that of the coupling reaction catalyzed by complex 2 (entry 6, Table 1).

Table 3. The coupling reaction of terminal epoxide with CO_2 by the sole complex **3** [1].

Entry	Epoxide	P(CO_2) (atm)	Temperature (°C)	Time (h)	Conversion (%)
1	PO	20	80	10	14
2	PO	20	100	5	64
3	PO	20	100	10	99
4	BO	10	80	10	5
5	BO	10	100	10	76
6	BO	20	100	10	95
7	CPO	20	100	5	92
8	CPO	20	100	10	98
9	SO	20	100	10	50
10	SO	20	100	24	99

[1] Conditions: Complex **3** (44.4 mg, 50 µmol), $n_{(3)}/n_{(epoxide)}$ = 1/1000. Other notes are the same as those in Table 1.

Complex **3** displayed high catalytic activity for the CPO/CO_2 and SO/CO_2 coupling. Taking the CPO/CO_2 coupling as an example, complex **3** converted 92% of CPO in 5 h at 100 °C under 20 atm of CO_2 to produce the corresponding carbonate (entry 7, Table 3), and a higher conversion rate of 98% was achieved in 10 h (entry 8, Table 3). When SO was used as the substrate, 50% of the SO was converted to the corresponding cyclic carbonate in 10 h at 100 °C under 20 atm of CO_2 (entry 9, Table 3). Nearly complete conversion of SO was achieved by extension of the reaction time to 24 h (entry 10, Table 3).

Complex **3** also displayed high activity toward CPO/CO_2 coupling under low CO_2 pressure. For example, 85% of CPO was converted to cyclic carbonate at 100 °C within 5 h under constant 2 atm CO_2 pressure (entry 1, Table 4). Prolonging the reaction time to 10 h gave rise to the nearly complete conversion of CPO (entry 2, Table 4). Complex **3** was still active under atmospheric CO_2 pressure, but the activity dropped sharply. For example, the conversion of CPO was only 28% in 5 h and 55% in 10 h (entries 1 and 2, Table 4), owing to the low concentration of CO_2 dissolved in CPO. By further extending the reaction time, high conversion of CPO was achieved (entries 3–6, Table 4). However, for SO/CO_2 coupling, complex **3** displayed very low activity. The conversion of SO was only 26% in 10 h when the reaction was performed at 100 °C, maintaining the CO_2 pressure at constant 2 atm (entry 7, Table 4). Lower SO conversion was expected, as the reaction was carried out at atmospheric CO_2 pressure. Analogous to CPO/CO_2 coupling, a high conversion of SO can be achieved by extending the reaction time (entries 8–14, Table 4).

Table 4. The coupling of epoxide/CO_2 under low CO_2 pressure by complex **3** [1].

Entry	Epoxide	P(CO_2) (atm)	Time (h)	Conversion (%)
1	CPO	2	5	85
2	CPO	2	10	98
3	CPO	1	5	28
4	CPO	1	10	55
5	CPO	1	24	83
6	CPO	1	36	91
7	SO	2	10	26
8	SO	1	5	10
9	SO	1	10	19
10	SO	1	24	50
11	SO	1	36	82
12	SO	1	48	92
13	SO	1	60	96

[1] Conditions: Complex **3** (44.4 mg, 50 μmol), $n_{(3)}/n_{(epoxide)}$ = 1/1000, the temperature is 100 °C. Other notes are the same as those in Table 1.

Kinetic studies of the coupling reactions of CPO/CO_2 and SO/CO_2 at atmospheric CO_2 pressure by complex **3** are plotted in Figure 3. It is clear that the CPO/CO_2 coupling is faster than the SO/CO_2 coupling under the same conditions (the blue curves in Figure 3). It is worth noting that the coupling of CPO/CO_2 has a first-order dependence on CPO concentration throughout the coupling reaction, as illustrated in Figure 3 (the red dash line, -ln(1-Conversion) vs. time). This may indicate a better solubility of complex **3** in CPO than that of complex **2**, which was tentatively ascribed to the incorporation of more amine groups on the ligand framework. Differing from CPO/CO_2 coupling, the kinetic curve (-ln(1-Conversion) vs. time) of SO/CO_2 coupling shows an increase in the slope during the early stage of the reaction (the red dot curve in Figure 3), reflecting the heterogeneous nature of the system owing to the poor solubility of complex **3** in SO. As the reaction proceeded, more SC was produced, which in turn helped to dissolve complex **3**, increasing the catalyst concentration and speeding up the coupling reaction. After some time, complex **3** was totally dissolved, forming a homogeneous system and maintaining constant concentration. Thus, a linear relationship was observed at the late stage of the coupling reaction, demonstrating a first-order dependence on SO concentration. The linear relation at the late stage also suggested the high stability of complex **3** under elevated temperature such as 100 °C.

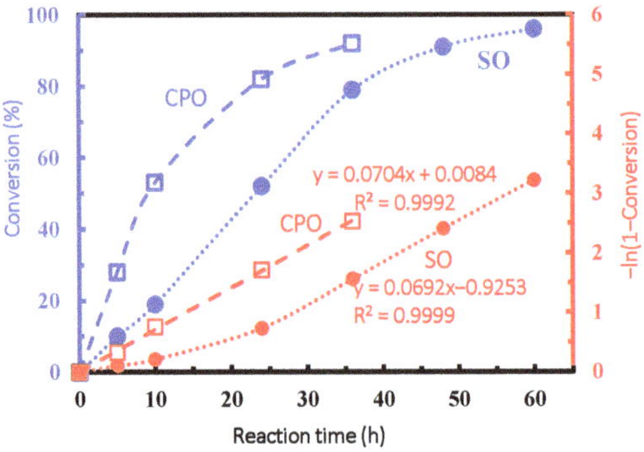

Figure 3. Kinetic curves of CPO/CO_2 and SO/SO_2 coupling catalyzed by complex **3** at 100 °C at atmospheric CO_2 pressure.

3. Materials and Methods

2-aminophenol (99%), 4-tertbutylphenol, aqueous dimethylamine solution (40 wt%), 3,5-ditertbutylsalicyaldehyde (98%), and diethyl aluminum chloride (1 mol/L, in hexane) were purchased from Energy Chemical (Anqing, China) and used as received. Propylene oxide, 1,2-butylene oxide, epichlorohydrin, and styrene oxide obtained from Energy Chemical (Anqing, China) were distilled over CaH_2 before use. THF and hexane were pre-dried over 4 Å molecular sieves before distillation over sodium with benzophenone as the indicator under an argon atmosphere and stored over freshly cut sodium in a glovebox. CO_2 (99.999% purity) was purchased from Shenyang Hongsheng Gas Limited Corporation, Suzhou, China. Ethanol and formic acid were used as received without further handling.

General procedures: All reactions sensitive to air and moisture were carried out in a glovebox filled with dry argon. Proligands H_2L^1-H_2L^3 were synthesized according to the literature methods and structurally identified by ^{1}H NMR and ^{13}C NMR [24]. The aluminum complexes **1–3** were prepared by equimolar reaction of the proligands with $AlEt_2Cl$ in a glovebox filled with argon. The structures of complexes **1–3** were characterized by ^{1}H NMR, ^{13}C NMR, and MS. The coupling of epoxide with CO_2 was carried in a stainless-steel autoclave equipped with a magnetic stirring bar. The reaction mixture was analyzed by the ^{1}H NMR spectra with $CDCl_3$ as a solvent. The ^{1}H and ^{13}C NMR spectra were recorded using Bruker AVANCE III 500 MHz spectrometer (Billerica, MA, USA). Mass spectra of complexes **1–3** were obtained in the electrospray positive mode (ESI+) on Thermo LTQ Orbitrap XL spectrometer (Waltham, MA, USA), samples were diluted in methanol or ethanol, at DUT Chemistry Analysis & Research Centre, Dalian University of Technology (Dalian, China).

3.1. Coupling Reaction of Epoxide with CO_2 by Aluminum Complex **1**

The typical procedure for the coupling of PO with CO_2 by complex **1** under elevated pressure is as follows. Complex **1** and cocatalyst were dissolved in PO in a Schlenk tube. Then, the solution was transferred via syringe into the pre-dried autoclave under CO_2 atmosphere. Then, the autoclave was pressurized with CO_2 and heated. After the designated time, the autoclave was cooled in an ice bath. The excess of the CO_2 was vented out. The conversion of PO was determined by GC analysis and ^{1}H NMR of the reaction mixture, and the yield was calculated based on isolated propylene carbonate. The results showed that the GC result was consistent with that by weight analysis of the PC.

3.2. Coupling Reaction of Epoxide with CO_2 by Bifunctional Aluminum Complexes

The typical procedure for the coupling of PO with CO_2 by complex **2** or **3** under elevated pressure is as follows. Complex **2** (38.7 mg, 50 μmol) was first placed in the pre-dried autoclave. After the internal atmosphere of the autoclave was displaced by CO_2 three times, PO was injected, which was followed by charging CO_2 to 20 atm. Then, the autoclave was heated for the designated time. The analysis was the same as the above method.

The typical procedure for the coupling of CO_2 with CPO by complex **2** or **3** at atmospheric CO_2 pressure is as follows. Complex **2** (38.7 mg, 50 μmol) was placed in a 10 mL round-bottom flask in a glovebox. The flask was taken out, which was equipped with a balloon. The flask was vacuumed and recharged with CO_2. After the injection of a prescribed amount of CPO, the flask was heated and maintained at 100 °C. The samples were taken out periodically and analyzed by Agilent GC to determine the conversion of CPO.

4. Conclusions

We have demonstrated the synthesis of a series of aluminum complexes **1–3**. Structural characterization by NMR and MS suggested the formation of binuclear structures via the chloride. Although complex **1** can be activated by either a nucleophile or an organic base to enable the coupling of terminal epoxide with CO_2, the excess use of cocatalyst unambiguously hampers its practical application. In contrast, the bifunctional complexes **2** and **3** containing tertiary amine as the built-in nucleophile are highly active catalysts by their own for the coupling of CO_2 with terminal epoxides. For bifunctional complexes **2** and **3**, high temperature facilitates the rapid transformation. It is found that the incorporation of more amine groups in the metal complex greatly enhances the catalytic activity. The bifunctional complexes are even active under low CO_2 pressure at the expense of catalytic activity to some extent. This work may shed light on the design of metal-based catalyst of high activity for the coupling of terminal epoxide with CO_2.

Supplementary Materials: The following are available online at https://www.mdpi.com/2073-4344/11/2/145/s1, Figure S1: ^1H NMR spectrum of proligand H_2L^1 in $CDCl_3$, Figure S1: ^1H NMR spectrum of proligand H_2L^2 in DMSO-d_6, Figure S3: ^1H NMR spectrum of proligand H_2L^3 in DMSO-d_6, Figure S4: ^1H NMR spectrum of the complex **1** in $CDCl_3$, Figure S5: ^1H NMR spectrum of the complex **2** in DMSO-d_6, Figure S6: ^1H NMR spectrum of the complex **3** in DMSO-d_6, Figure S7: Kinetic study of coupling of PO/CO_2 mediated by **1**/TBAB (1:40 molar ratio), PO conversion vs. reaction time, Table S1: Cycloaddition of CO_2 to PO mediated by complex **1** and TBAB, Table S2: Effect of reaction time on the PO conversion, Table S3: Coupling of CO_2/PO catalyzed by **1**/DMAP, Table S4: Coupling of CO_2/PO catalyzed by **1**/1-MeIm.

Author Contributions: Conceptualization, Z.Z. and S.H.; validation, S.H., Q.D., T.W. and Z.Z.; investigation, Z.Z., Q.D. and T.W.; writing—original draft preparation, Z.Z.; writing—review and editing, Q.D., P.X. and S.H.; funding acquisition, Z.Z. and S.H. All authors have read and agreed to the published version of the manuscript.

Funding: This research was funded by "National Natural Science Foundation of China, grant number 21674066", and "Ministry of Education of Liaoning province, China, grant number LJ2020009, LQ2020028".

Institutional Review Board Statement: Not applicable.

Informed Consent Statement: Not applicable.

Data Availability Statement: Data are contained within the article or supplementary material.

Acknowledgments: We thank Meiheng Lv for her constructive discussions about the identification of bifunctional complexes **2** and **3**.

Conflicts of Interest: The authors declare no conflict of interest.

References

1. Dibenedetto, A.; Angelini, A.; Stufano, P. Use of carbon dioxide as feedstock for chemicals and fuels: Homogeneous and heterogeneous catalysis. *J. Chem. Technol. Biotechnol.* **2014**, *89*, 334–353. [CrossRef]
2. Yu, B.; Diao, Z.-F.; Guo, C.-X.; He, L.-N. Carboxylation of olefins/alkynes with CO_2 to industrially relevant acrylic acid derivatives. *J. CO2 Util.* **2013**, *1*, 60–68. [CrossRef]
3. Drees, M.; Cokoja, M.; Kühn, F.E. Recycling CO_2? Computational Considerations of the Activation of CO_2 with Homogeneous Transition Metal Catalysts. *ChemCatChem* **2012**, *4*, 1703–1712. [CrossRef]
4. Cokoja, M.; Bruckmeier, C.; Rieger, B.; Herrmann, W.A.; Kuehn, F.E. Transformation of Carbon Dioxide with Homogeneous Transition-Metal Catalysts: A Molecular Solution to a Global Challenge? *Angew. Chem. Int. Ed.* **2011**, *50*, 8510–8537. [CrossRef] [PubMed]
5. Omae, I. Aspects of carbon dioxide utilization. *Catal. Today* **2006**, *115*, 33–52. [CrossRef]
6. Cokoja, M.; Wilhelm, M.E.; Anthofer, M.H.; Herrmann, W.A.; Kuehn, F.E. Synthesis of Cyclic Carbonates from Epoxides and Carbon Dioxide by Using Organocatalysts. *ChemSusChem* **2015**, *8*, 2436–2454. [CrossRef] [PubMed]
7. Maeda, C.; Miyazaki, Y.; Ema, T. Recent progress in catalytic conversions of carbon dioxide. *Catal. Sci. Technol.* **2014**, *4*, 1482–1497. [CrossRef]
8. Xu, K. Nonaqueous liquid electrolytes for lithium-based rechargeable batteries. *Chem. Rev.* **2004**, *104*, 4303–4417. [CrossRef]
9. Schaeffner, B.; Schaeffner, F.; Verevkin, S.P.; Boerner, A. Organic Carbonates as Solvents in Synthesis and Catalysis. *Chem. Rev.* **2010**, *110*, 4554–4581. [CrossRef]
10. Shaikh, A.A.G.; Sivaram, S. Organic carbonates. *Chem. Rev.* **1996**, *96*, 951–976. [CrossRef]
11. Lu, X.B.; Feng, X.J.; He, R. Catalytic formation of ethylene carbonate from supercritical carbon dioxide/ethylene oxide mixture with tetradentate Schiff-base complexes as catalyst. *Appl. Catal. Gen.* **2002**, *234*, 25–33. [CrossRef]
12. Fuchs, M.A.; Altesleben, C.; Zevaco, T.A.; Dinjus, E. An Efficient Homogeneous Chloro-Aluminum- N_2O_2 Catalyst for the Coupling of Epoxides with Carbon Dioxide. *Eur. J. Inorg. Chem.* **2013**, *2013*, 4541–4545. [CrossRef]

13. Clegg, W.; Harrington, R.W.; North, M.; Pasquale, R. Cyclic Carbonate Synthesis Catalysed by Bimetallic Aluminium-Salen Complexes. *Chem. Eur. J.* **2010**, *16*, 6828–6843. [CrossRef] [PubMed]
14. Kasuga, K.; Nagao, S.; Fukumoto, T.; Handa, M. Cycloaddition of carbon dioxide to propylene oxide catalysed by tetra-t-butylphthalocyaninatoaluminium(III) chloride. *Polyhedron* **1996**, *15*, 69–72. [CrossRef]
15. Woo, W.H.; Hyun, K.; Kim, Y.; Ryu, J.Y.; Lee, J.; Kim, M.; Park, M.H.; Kim, Y. Highly Active Salen-Based Aluminum Catalyst for the Coupling of Carbon Dioxide with Epoxides at Ambient Temperature. *Eur. J. Inorg. Chem.* **2017**, 5372–5378. [CrossRef]
16. Wu, X.; North, M. A Bimetallic Aluminium(Salphen) Complex for the Synthesis of Cyclic Carbonates from Epoxides and Carbon Dioxide. *ChemSusChem* **2017**, *10*, 74–78. [CrossRef] [PubMed]
17. Li, C.-Y.; Liu, D.-C.; Ko, B.-T. Synthesis, characterization and reactivity of single-site aluminium amides bearing benzotriazole phenoxide ligands: Catalysis for ring-opening polymerization of lactide and carbon dioxide/propylene oxide coupling. *Dalton Trans.* **2013**, *42*, 11488–11496. [CrossRef]
18. Melendez, J.; North, M.; Pasquale, R. Synthesis of cyclic carbonates from atmospheric pressure carbon dioxide using exceptionally active aluminium(salen) complexes as catalysts. *Eur. J. Inorg. Chem.* **2007**, 3323–3326. [CrossRef]
19. Qin, Y.; Guo, H.; Sheng, X.; Wang, X.; Wang, F. An aluminum porphyrin complex with high activity and selectivity for cyclic carbonate synthesis. *Green Chem.* **2015**, *17*, 2853–2858. [CrossRef]
20. Aida, T.; Inoue, S. Activation of carbon dioxide with aluminum porphyrin and reaction with epoxide. Studies on (tetraphenylporphinato)aluminum alkoxide having a long oxyalkylene chain as the alkoxide group. *J. Am. Chem. Soc.* **1983**, *105*, 1304–1309. [CrossRef]
21. Lu, X.B.; Zhang, Y.J.; Jin, K.; Luo, L.M.; Wang, H. Highly active electrophile-nucleophile catalyst system for the cycloaddition of CO_2 to epoxides at ambient temperature. *J. Catal.* **2004**, *227*, 537–541. [CrossRef]
22. Lu, X.B.; Zhang, Y.J.; Liang, B.; Li, X.; Wang, H. Chemical fixation of carbon dioxide to cyclic carbonates under extremely mild conditions with highly active bifunctional catalysts. *J. Mol. Catal. Chem.* **2004**, *210*, 31–34. [CrossRef]
23. Rintjema, J.; Kleij, A.W. Aluminum-Mediated Formation of Cyclic Carbonates: Benchmarking Catalytic Performance Metrics. *ChemSusChem* **2017**, *10*, 1274–1282. [CrossRef] [PubMed]
24. Gao, P.; Zhao, Z.; Chen, L.; Yuan, D.; Yao, Y. Dinuclear Aluminum Poly(phenolate) Complexes as Efficient Catalysts for Cyclic Carbonate Synthesis. *Organometallics* **2016**, *35*, 1707–1712. [CrossRef]
25. Cozzolino, M.; Press, K.; Mazzeo, M.; Lamberti, M. Carbon Dioxide/Epoxide Reactions Catalyzed by Bimetallic Salalen Aluminum Complexes. *ChemCatChem* **2016**, *8*, 455–460. [CrossRef]
26. Castro-Osma, J.A.; Lara-Sanchez, A.; North, M.; Otero, A.; Villuendas, P. Synthesis of cyclic carbonates using monometallic, and helical bimetallic, aluminium complexes. *Catal. Sci. Technol.* **2012**, *2*, 1021–1026. [CrossRef]
27. Rios Yepes, Y.; Quintero, C.; Osorio Melendez, D.; Daniliuc, C.G.; Martinez, J.; Rojas, R.S. Cyclic Carbonates from CO_2 and Epoxides Catalyzed by Tetra- and Pentacoordinate Amidinate Aluminum Complexes. *Organometallics* **2019**, *38*, 469–478. [CrossRef]
28. Whiteoak, C.J.; Kielland, N.; Laserna, V.; Castro-Gomez, F.; Martin, E.; Escudero-Adan, E.C.; Bo, C.; Kleij, A.W. Highly Active Aluminium Catalysts for the Formation of Organic Carbonates from CO_2 and Oxiranes. *Chem. Eur. J.* **2014**, *20*, 2264–2275. [CrossRef]
29. Castro-Osma, J.A.; North, M.; Wu, X. Development of a Halide-Free Aluminium-Based Catalyst for the Synthesis of Cyclic Carbonates from Epoxides and Carbon Dioxide. *Chem. Eur. J.* **2014**, *20*, 15005–15008. [CrossRef]
30. Supasitmongkol, S.; Styring, P. A single centre aluminium(III) catalyst and TBAB as an ionic organo-catalyst for the homogeneous catalytic synthesis of styrene carbonate. *Catal. Sci. Technol.* **2014**, *4*, 1622–1630. [CrossRef]
31. Kim, S.H.; Han, S.Y.; Kim, J.H.; Kang, Y.Y.; Lee, J.; Kim, Y. Monomeric or Dimeric Aluminum Complexes as Catalysts for Cycloaddition between CO_2 and Epoxides. *Eur. J. Inorg. Chem.* **2015**, *2015*, 2323–2329. [CrossRef]
32. Whiteoak, C.J.; Kielland, N.; Laserna, V.; Escudero-Adan, E.C.; Martin, E.; Kleij, A.W. A Powerful Aluminum Catalyst for the Synthesis of Highly Functional Organic Carbonates. *J. Am. Chem. Soc.* **2013**, *135*, 1228–1231. [CrossRef] [PubMed]
33. Ema, T.; Miyazaki, Y.; Koyama, S.; Yano, Y.; Sakai, T. A bifunctional catalyst for carbon dioxide fixation: Cooperative double activation of epoxides for the synthesis of cyclic carbonates. *Chem. Commun.* **2012**, *48*, 4489–4491. [CrossRef] [PubMed]
34. Bai, D.; Duan, S.; Hai, L.; Jing, H. Carbon Dioxide Fixation by Cycloaddition with Epoxides, Catalyzed by Biomimetic Metalloporphyrins. *ChemCatChem* **2012**, *4*, 1752–1758. [CrossRef]
35. Raghavendra, B.; Shashank, P.V.S.; Pandey, M.K.; Reddy, N.D. CO_2/Epoxide Coupling and the ROP of epsilon-Caprolactone: Mg and Al Complexes of gamma-Phosphino-ketiminates as Dual-Purpose Catalysts. *Organometallics* **2018**, *37*, 1656–1664. [CrossRef]
36. Ema, T.; Miyazaki, Y.; Shimonishi, J.; Maeda, C.; Hasegawa, J.-Y. Bifunctional Porphyrin Catalysts for the Synthesis of Cyclic Carbonates from Epoxides and CO_2: Structural Optimization and Mechanistic Study. *J. Am. Chem. Soc.* **2014**, *136*, 15270–15279. [CrossRef]
37. Darensbourg, D.J.; Fang, C.C.; Rodgers, J.L. Catalytic coupling of carbon dioxide and 2,3-epoxy-1,2,3,4-tetrahydronaphthalene in the presence of a (Salen)(CrCl)-Cl-III derivative. *Organometallics* **2004**, *23*, 924–927. [CrossRef]
38. Castro-Osma, J.A.; North, M.; Wu, X. Synthesis of Cyclic Carbonates Catalysed by Chromium and Aluminium Salphen Complexes. *Chem. Eur. J.* **2016**, *22*, 2100–2107. [CrossRef]

39. Ramidi, P.; Sullivan, S.Z.; Gartia, Y.; Munshi, P.; Griffin, W.O.; Darsey, J.A.; Biswas, A.; Shaikh, A.U.; Ghosh, A. Catalytic Cyclic Carbonate Synthesis Using Epoxide and Carbon Dioxide: Combined Catalytic Effect of Both Cation and Anion of an Ionic Cr-v(O) Amido Macrocyclic Complex. *Ind. Eng. Chem. Res.* **2011**, *50*, 7800–7807. [CrossRef]
40. Cuesta-Aluja, L.; Djoufak, M.; Aghmiz, A.; Rivas, R.; Christ, L.; Masdeu-Bulto, A.M. Novel chromium (III) complexes with N-4-donor ligands as catalysts for the coupling of CO_2 and epoxides in supercritical CO_2. *J. Mol. Catal. Chem.* **2014**, *381*, 161–170. [CrossRef]
41. Adolph, M.; Zevaco, T.A.; Altesleben, C.; Walter, O.; Dinjus, E. New cobalt, iron and chromium catalysts based on easy-to-handle N-4-chelating ligands for the coupling reaction of epoxides with CO_2. *Dalton Trans.* **2014**, *43*, 3285–3296. [CrossRef] [PubMed]
42. Iksi, S.; Aghmiz, A.; Rivas, R.; Dolores Gonzalez, M.; Cuesta-Aluja, L.; Castilla, J.; Orejon, A.; El Guemmout, F.; Masdeu-Bulto, A.M. Chromium complexes with tridentate NN′O Schiff base ligands as catalysts for the coupling of CO_2 and epoxides. *J. Mol. Catal. Chem.* **2014**, *383*, 143–152. [CrossRef]
43. Ambrose, K.; Robertson, K.N.; Kozak, C.M. Cobalt amino-bis(phenolate) complexes for coupling and copolymerization of epoxides with carbon dioxide. *Dalton Trans.* **2019**, *48*, 6248–6260. [CrossRef]
44. Lu, X.B.; Liang, B.; Zhang, Y.J.; Tian, Y.Z.; Wang, Y.M.; Bai, C.X.; Wang, H.; Zhang, R. Asymmetric catalysis with CO_2: Direct synthesis of optically active propylene carbonate from racemic epoxides. *J. Am. Chem. Soc.* **2004**, *126*, 3732–3733. [CrossRef] [PubMed]
45. Chang, T.; Jin, L.; Jing, H. Bifunctional Chiral Catalyst for the Synthesis of Chiral Cyclic Carbonates from Carbon Dioxide and Epoxides. *ChemCatChem* **2009**, *1*, 379–383. [CrossRef]
46. Reiter, M.; Altenbuchner, P.T.; Kissling, S.; Herdtweck, E.; Rieger, B. Amine-bis(phenolato)cobalt(II) Catalysts for the Formation of Organic Carbonates from Carbon Dioxide and Epoxides. *Eur. J. Inorg. Chem.* **2015**, *2015*, 1766–1774. [CrossRef]
47. Yu, C.-Y.; Chuang, H.-J.; Ko, B.-T. Bimetallic bis(benzotriazole iminophenolate) cobalt, nickel and zinc complexes as versatile catalysts for coupling of carbon dioxide with epoxides and copolymerization of phthalic anhydride with cyclohexene oxide. *Catal. Sci. Technol.* **2016**, *6*, 1779–1791. [CrossRef]
48. Adolph, M.; Zevaco, T.A.; Altesleben, C.; Staudt, S.; Walter, O.; Dinjus, E. New ionic cobalt(III) complexes based on the N, N-bis(2-pyrazinecarboxamide)-1,2-benzene ligand: Application to the formation of organic carbonates from epoxides and CO_2. *New J. Chem.* **2015**, *39*, 9858–9865. [CrossRef]
49. Buonerba, A.; De Nisi, A.; Grassi, A.; Milione, S.; Capacchione, C.; Vagin, S.; Rieger, B. Novel iron(III) catalyst for the efficient and selective coupling of carbon dioxide and epoxides to form cyclic carbonates. *Catal. Sci. Technol.* **2015**, *5*, 118–123. [CrossRef]
50. Buchard, A.; Kember, M.R.; Sandeman, K.G.; Williams, C.K. A bimetallic iron(III) catalyst for CO_2/epoxide coupling. *Chem. Commun.* **2011**, *47*, 212–214. [CrossRef]
51. Whiteoak, C.J.; Martin, E.; Martinez Belmonte, M.; Benet-Buchholz, J.; Kleij, A.W. An Efficient Iron Catalyst for the Synthesis of Five- and Six-Membered Organic Carbonates under Mild Conditions. *Adv. Synth. Catal.* **2012**, *354*, 469–476. [CrossRef]
52. Della Monica, F.; Buonerba, A.; Capacchione, C. Homogeneous Iron Catalysts in the Reaction of Epoxides with Carbon Dioxide. *Adv. Synth. Catal.* **2019**, *361*, 265–282. [CrossRef]
53. Della Monica, F.; Vummaleti, S.V.C.; Buonerba, A.; De Nisi, A.; Monari, M.; Milione, S.; Grassi, A.; Cavallo, L.; Capacchione, C. Coupling of Carbon Dioxide with Epoxides Efficiently Catalyzed by Thioether-Triphenolate Bimetallic Iron(III) Complexes: Catalyst Structure-Reactivity Relationship and Mechanistic DFT Study. *Adv. Synth. Catal.* **2016**, *358*, 3231–3243. [CrossRef]
54. Zhao, Z.; Qin, J.; Zhang, C.; Wang, Y.; Yuan, D.; Yao, Y. Recyclable Single-Component Rare-Earth Metal Catalysts for Cycloaddition of CO_2 and Epoxides at Atmospheric Pressure. *Inorg. Chem.* **2017**, *56*, 4568–4575. [CrossRef] [PubMed]
55. Wang, C.; Liu, X.; Dai, Z.; Sun, Y.; Tang, N.; Wu, J. Yttrium complex supported by a sterically encumbering N-anchored tris-arylphenoxide ligand: Heteroselective ROP of rac-lactide and CO_2/epoxide coupling. *Inorg. Chem. Commun.* **2015**, *56*, 69–72. [CrossRef]
56. Luinstra, G.A.; Haas, G.R.; Molnar, F.; Bernhart, V.; Eberhardt, R.; Rieger, B. On the formation of aliphatic polycarbonates from epoxides with chromium(III) and aluminum(III) metal-salen complexes. *Chem. Eur. J.* **2005**, *11*, 6298–6314. [CrossRef]
57. North, M.; Quek, S.C.Z.; Pridmore, N.E.; Whitwood, A.C.; Wu, X. Aluminum(salen) Complexes as Catalysts for of Terminal Epoxides via CO_2 Coupling. *ACS Catal.* **2015**, *5*, 3398–3402. [CrossRef]
58. North, M.; Young, C. Bimetallic aluminium(acen) complexes as catalysts for the synthesis of cyclic carbonates from carbon dioxide and epoxides. *Catal. Sci. Technol.* **2011**, *1*, 93–99. [CrossRef]
59. Rulev, Y.A.; Gugkaeva, Z.; Maleev, V.I.; North, M.; Belokon, Y.N. Robust bifunctional aluminium-salen catalysts for the preparation of cyclic carbonates from carbon dioxide and epoxides. *Beilstein. J. Org. Chem.* **2015**, *11*, 1614–1623. [CrossRef]
60. Martinez, J.; Castro-Osma, J.A.; Alonso-Moreno, C.; Rodriguez-Dieguez, A.; North, M.; Otero, A.; Lara-Sanchez, A. One-Component Aluminum(heteroscorpionate) Catalysts for the Formation of Cyclic Carbonates from Epoxides and Carbon Dioxide. *ChemSusChem* **2017**, *10*, 1175–1185. [CrossRef]
61. de la Cruz-Martinez, F.; Martinez, J.; Gaona, M.A.; Fernandez-Baeza, J.; Sanchez-Barba, L.F.; Rodriguez, A.M.; Castro-Osma, J.A.; Otero, A.; Lara-Sanchez, A. Bifunctional Aluminum Catalysts for the Chemical Fixation of Carbon Dioxide into Cyclic Carbonates. *ACS Sustain. Chem. Eng.* **2018**, *6*, 5322–5332. [CrossRef]
62. Melendez, J.; North, M.; Villuendas, P. One-component catalysts for cyclic carbonate synthesis. *Chem. Commun.* **2009**, 2577–2579. [CrossRef] [PubMed]

63. Luo, R.; Zhou, X.; Chen, S.; Li, Y.; Zhou, L.; Ji, H. Highly efficient synthesis of cyclic carbonates from epoxides catalyzed by salen aluminum complexes with built-in "CO_2 capture" capability under mild conditions. *Green Chem.* **2014**, *16*, 1496–1506. [CrossRef]
64. Martinez, J.; de la Cruz-Martinez, F.; Gaona, M.A.; Pinilla-Penalver, E.; Fernandez-Baeza, J.; Rodriguez, A.M.; Castro-Osma, J.A.; Otero, A.; Lara-Sanchez, A. Influence of the Counterion on the Synthesis of Cyclic Carbonates Catalyzed by Bifunctional Aluminum Complexes. *Inorg. Chem.* **2019**, *58*, 3396–3408. [CrossRef]
65. Ren, W.-M.; Liu, Y.; Lu, X.-B. Bifunctional Aluminum Catalyst for CO_2 Fixation: Regioselective Ring Opening of Three-Membered Heterocyclic Compounds. *J. Org. Chem.* **2014**, *79*, 9771–9777. [CrossRef]
66. Tian, D.; Liu, B.; Zhang, L.; Wang, X.; Zhang, W.; Han, L.; Park, D.-W. Coupling reaction of carbon dioxide and epoxides efficiently catalyzed by one-component aluminum-salen complex under solvent-free conditions. *J. Ind. Eng. Chem.* **2012**, *18*, 1332–1338. [CrossRef]
67. Tian, D.; Liu, B.; Gan, Q.; Li, H.; Darensbourg, D.J. Formation of Cyclic Carbonates from Carbon Dioxide and Epoxides Coupling Reactions Efficiently Catalyzed by Robust, Recyclable One-Component Aluminum-Salen Complexes. *ACS Catal.* **2012**, *2*, 2029–2035. [CrossRef]
68. Ren, Y.; Jiang, O.; Zeng, H.; Mao, Q.; Jiang, H. Lewis acid-base bifunctional aluminum-salen catalysts: Synthesis of cyclic carbonates from carbon dioxide and epoxides. *RSC Adv.* **2016**, *6*, 3243–3249. [CrossRef]
69. Hong, M.; Kim, Y.; Kim, H.; Cho, H.J.; Baik, M.-H.; Kim, Y. Scorpionate Catalysts for Coupling CO_2 and Epoxides to Cyclic Carbonates: A Rational Design Approach for Organocatalysts. *J. Org. Chem.* **2018**, *83*, 9370–9380. [CrossRef]
70. Hanhart, W.; Ingold, C.K. CXXXIX.—The nature of the alternating effect in carbon chains. Part XVIII. Mechanism of exhaustive methylation and its relation to anomalous hydrolysis. *J. Chem. Soc.* **1927**, 997–1020. [CrossRef]
71. Hughes, E.D.; Ingold, C.K.; Patel, C.S. 135. Influence of poles and polar linkings on the course pursued by elimination reactions. Part XVI. Mechanism of the thermal decomposition of quaternary ammonium compounds. *J. Chem. Soc.* **1933**, 526–530. [CrossRef]
72. De la Zerda, J.; Neumann, R.; Sasson, Y. Hofmann decomposition of quaternary ammonium salts under phase-transfer catalytic conditions. *J. Chem. Soc. Perkin Trans. 2* **1986**, 823–826. [CrossRef]
73. Hofman, A.W., IX. Contributions towards the history of the monamines.—No. III. Compound ammonias by inverse substitution. *Proc. R. Soc. Lond.* **1859**, *10*, 594–596.
74. Collie, N.; Schryver, S.B. LIII. The action of heat on the chlorides and hydroxides of mixed quaternary ammonium compounds. *J. Chem. Soc. Trans.* **1890**, *57*, 767–782. [CrossRef]
75. Zaki, A.; Fahim, H. Some quaternary ammonium salts and their decomposition products. *J. Chem. Soc.* **1942**, 270–272. [CrossRef]
76. Gordon, J.E. Fused Organic Salts. III.1a Chemical Stability of Molten Tetra-n-alkylammonium Salts. Medium Effects on Thermal R4N+X- Decomposition. RBr + I- = RI + Br- Equilibrium Constant in Fused Salt Medium. *J. Org. Chem.* **1965**, *30*, 2760–2763. [CrossRef]
77. Xie, Y.; Zhang, Z.; Jiang, T.; He, J.; Han, B.; Wu, T.; Ding, K. CO_2 cycloaddition reactions catalyzed by an ionic liquid grafted onto a highly cross-linked polymer matrix. *Angew. Chem. Int. Ed.* **2007**, *46*, 7255–7258. [CrossRef]
78. North, M.; Pasquale, R. Mechanism of Cyclic Carbonate Synthesis from Epoxides and CO_2. *Angew. Chem. Int. Ed.* **2009**, *48*, 2946–2948. [CrossRef]
79. Cho, W., Shin, M.S., Hwang, S., Kim, H., Kim, M., Kim, J.G., Kim, Y. Tertiary amines. A new class of highly efficient organocatalysts for CO_2 fixations. *J. Ind. Eng. Chem.* **2016**, *44*, 210–215. [CrossRef]

MDPI

Article

CO Oxidation Efficiency and Hysteresis Behavior over Mesoporous Pd/SiO_2 Catalyst

Rola Mohammad Al Soubaihi [1], Khaled Mohammad Saoud [2], Myo Tay Zar Myint [3], Mats A. Göthelid [4] and Joydeep Dutta [1,*]

1 Functional Materials, Applied Physics Department, School of Engineering Sciences, KTH Royal Institute of Technology, AlbaNova Universitetscentrum, 106 91 Stockholm, Sweden; rolaas@kth.se
2 Liberal Arts and Sciences Program, Virginia Commonwealth University in Qatar, Doha, Qatar; s2kmsaou@vcu.edu
3 Department of Physics, College of Science, Sultan Qaboos University, P.O. Box 36, Muscat PC 123, Oman; myotayzar.myint@gmail.com
4 Materialfysik, SCI, Albano, Hannes Alfvéns väg 12, KTH Royal Institute of Technology, 114 19 Stockholm, Sweden; gothelid@kth.se
* Correspondence: joydeep@kth.se

Abstract: Carbon monoxide (CO) oxidation is considered an important reaction in heterogeneous industrial catalysis and has been extensively studied. Pd supported on SiO_2 aerogel catalysts exhibit good catalytic activity toward this reaction owing to their CO bond activation capability and thermal stability. Pd/SiO_2 catalysts were investigated using carbon monoxide (CO) oxidation as a model reaction. The catalyst becomes active, and the conversion increases after the temperature reaches the ignition temperature (T_{ig}). A normal hysteresis in carbon monoxide (CO) oxidation has been observed, where the catalysts continue to exhibit high catalytic activity (CO conversion remains at 100%) during the extinction even at temperatures lower than T_{ig}. The catalyst was characterized using BET, TEM, XPS, TGA-DSC, and FTIR. In this work, the influence of pretreatment conditions and stability of the active sites on the catalytic activity and hysteresis is presented. The CO oxidation on the Pd/SiO_2 catalyst has been attributed to the dissociative adsorption of molecular oxygen and the activation of the C-O bond, followed by diffusion of adsorbates at T_{ig} to form CO_2. Whereas, the hysteresis has been explained by the enhanced stability of the active site caused by thermal effects, pretreatment conditions, Pd-SiO_2 support interaction, and PdO formation and decomposition.

Keywords: CO oxidation; hysteresis; thermal stability; pretreatment; structure-activity

Citation: Al Soubaihi, R.M.; Saoud, K.M.; Myint, M.T.Z.; Göthelid, M.A.; Dutta, J. CO Oxidation Efficiency and Hysteresis Behavior over Mesoporous Pd/SiO_2 Catalyst. *Catalysts* **2021**, *11*, 131. https://doi.org/10.3390/catal11010131

Received: 16 December 2020
Accepted: 13 January 2021
Published: 16 January 2021

1. Introduction

Low-temperature carbon monoxide (CO) oxidation is considered a prototype reaction for heterogeneous catalysis. It has garnered attention in recent years due to its interesting catalytic behavior and the screening of new heterogeneous catalysts [1]. In recent years, "metal oxide supported palladium catalysts" have been studied in detail due to their high intrinsic activity, lower cost, metal-support interaction, and other non-linear dynamic behaviors, which are beneficial for CO oxidation [2,3]. The properties of Pd catalysts are affected by types of support, preparation conditions, and the dispersion of the Pd particles [4]. Support materials are a crucial factor in the catalytic behavior of Pd toward CO oxidation where the synergistic effect between Pd and the support depends on the nature of support (i.e., reducible vs. non-reducible) [5]. Stabilization of the catalyst can be attained by anchoring Pd particles on the surface of the support to resist sintering at high temperatures and dispersion in metal oxides, such as silica (SiO_2) [6]. Silica materials have been widely explored as catalyst support owing to their unique morphologies, narrow pore sizes, large surface areas, and thermal stability [6]. Despite being inert and irreducible, SiO_2 can exhibit a metal–support interaction with Pd, that affects the morphology, wetting, and interdiffusion in the Pd/SiO_2 catalyst, which is known to improve catalytic properties and stability [7]. The heat treatment under oxidation and reduction conditions is crucial for the preparation of supported Pd/SiO_2 catalysts. Such treatment can induce morphological changes and affect the dispersion of Pd particles resulting from the sintering of the Pd particles. Therefore, it is crucial to study the optimal preparation and pretreatment conditions and activation of the Pd catalysts. Morphological changes due to sintering have been reported wherein encapsulation, inter-diffusion, and alloy formation are found to be highly dependent on heating conditions [8]. The thermal treatment can cause deactivation of catalysts impacting the support, oxidation state, particle size, and the surface area of the catalyst due to sintering at high temperatures, which can directly influence the catalytic activity [2]. Palladium nanoparticles dispersion within a narrow pore size distribution and high surface area mesoporous silica (SiO_2) aerogel increases the catalytic activity of supported Pd catalysts. Moreover, the formation of Pd intermediate has been reported to influence the catalytic behavior and stability during ignition/extinction cycles [9]. The Pd catalyst has a wide operating temperature range during CO oxidation and can exist in two thermodynamically stable phases depending on the partial pressure of oxygen and the reaction temperature, either as palladium oxide (PdO) or in its metallic form (Pd) [10]. Therefore, the Pd catalyst exhibits nonlinear dynamics such as hysteresis effects and self-sustained oscillations due to the PdO decomposition and re-formation. This behavior could be crucial for the future development of heterogeneous catalysts for a range of reactions in addition to CO oxidation reactions [11].

Hysteresis is a complex phenomenon attributed to many factors such as surface coverage, multiplicity, thermal inertia, exotherm, and catalyst oxidation states. The effects of CO conversion hysteresis over supported and unsupported palladium catalysts were investigated and reported by researchers, including our group [2,12]. Hysteresis loops arise from the difference between the activity during the heating and cooling processes [13]. The reversible oxidation of Pd due to the dissociative adsorption of oxygen on the subsurface layer of the catalyst to form a surface layer of PdO leads to the hysteresis and self-sustained oscillations in CO oxidation [12]. This phenomenon was observed in many exothermic oxidation reactions in addition to CO oxidation and has been attributed to the heat released at the surface of the catalyst during the exothermic oxidation, wherein the surface temperature exceeds the reactor temperature [14]. "Normal" hysteresis observed in CO oxidation when the catalytic activity during CO ignition takes place at a higher temperature than the temperature during extinction, while inverse hysteresis is observed for some gas mixtures, where the catalytic activity during ignition exceeds the activity during extinction [15]. This behavior has been realized and plays a crucial role in many applications, including long-life carbon dioxide (CO_2) lasers, partial oxidation in chemical synthesis, and removal of pollutants in catalytic converters during prototype exothermic reactions including CO oxidation [16]. Most of the reports in the literature focus on the CO hysteresis as a function of inlet temperatures [15]. Although reasons for the hysteresis phenomena are still unclear, some researchers associate its rise with the multiplicity of steady states, oxidation of the catalyst, and temperature fluctuations. Due to the coexistence of PdO or Pd states in the palladium catalyst and the strong interdependence of catalytic activity and hysteresis on the support materials, palladium catalysts exhibit normal hysteresis during CO oxidation due to the decomposition and reformation of PdO.

In our previous work, we have reported the synthesis and the catalytic activity of the Pd/SiO_2 catalyst. In continuation of this work, we provide detailed information on the influence of pretreatment conditions on the catalyst microstructure and dispersion of Pd, thermal stability, and catalyst performance. The mesoporous Pd/SiO_2 aerogel catalyst with well-dispersed Pd nanoparticles was evaluated for CO oxidation under ignition/extinction conditions, wherein the possibility of using the hysteresis effect to attain high conversions at lower reaction temperatures were explored. The Pd/SiO_2 catalyst is a CO oxidation catalyst with high catalytic activity at a temperature range of (150 to 250 °C). It has excellent thermal stability, which could stabilize the active phase for CO oxidation by increasing the metal-support interaction and dispersion of Pd species. This study reports the experimental data on the catalytic activity, hysteresis, and thermal stability during the CO oxidation over the Pd/SiO_2 catalyst, which is found to be intensely dependent on the preparation, pretreatment conditions, and external parameters. The preparation and treatment conditions' effect under dynamic reaction conditions provide a correlation between the catalyst structure and activity, as well as the role of thermal effects on the observed hysteresis behavior.

2. Results and Discussion

2.1. Catalytic Activity and Hysteresis Behavior of Pd/SiO_2 Catalysts

2.1.1. CO Conversion Efficiency and Light-Off Testing

Catalytic CO oxidation was performed under constant gas composition where the concentration of CO is 3.5%, with a surplus of oxygen (20% concentration) balanced in helium to allow for the complete CO conversion. The result was obtained under controlled heating (heating rate is 10 °C/min), and subsequent normal cooling conditions (ignition/extinction) as a function of the reaction (catalyst) temperature and not the feed gas inlet temperature, as reported by other researchers in this field [17]. In this context, we defined the ignition or light-off temperature (T_{ig}) as the temperature at which the CO conversion efficiency reaches 3% during heating and extinction or light-out temperature (T_{ext}) as the temperature at which the CO conversion efficiency reaches 3% during cooling. Furthermore, we define the hysteresis as the difference between the temperatures at which the CO conversion efficiency is 50% during ignition and extinction.

Figure 1a presents the conversion efficiency and catalyst temperature during the catalytic CO conversion test as a function of time, while Figure 1b shows the conversion efficiency as a function of the catalyst temperature for two consecutive light-off cycles. The figures show clearly the exothermic evolution as a function of time and the hysteresis effect for oxidation reaction. During heating (ignition), three active conversion zones can be distinguished in which the reaction takes place at varying catalyst temperatures. In Zone I, when the catalyst temperature is below 214 °C, the reaction rate is slow, leading to low CO conversion as it is kinetically controlled. In Zone II, intermediate conversion zone when the temperature is above 214 °C, the reaction rate is slightly higher, leading to better CO conversions as it is controlled by internal diffusion and mass transfer limitations in the SiO_2 porous support [18]. In Zone III, the full conversion zone when the temperature increased to above 276 °C, at which the CO conversion efficiency reaches 50%, the reaction rate is high and controlled by external (gas phase) diffusion without any mass transfer limitations.

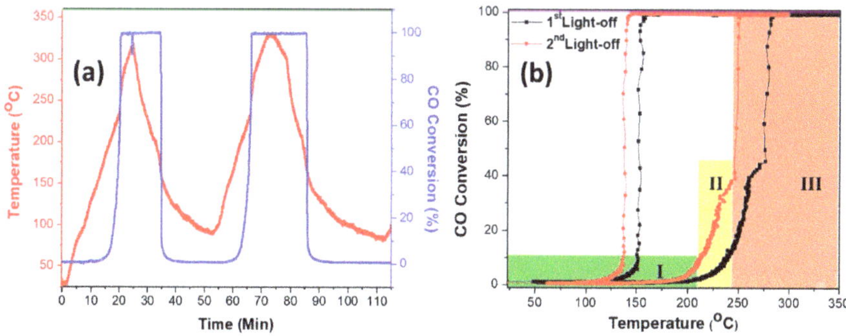

Figure 1. (**a**) Reaction temperature and carbon monoxide (CO) conversion efficiency during typical light-off testing in a synthetic gas reactor as a function of time, (**b**) CO conversion efficiency, and hysteresis effect as a function of the reaction (catalyst) temperature.

On the other hand, during cooling (extinction), the reaction kinetics and diffusion of gas molecules play a role due to temperature differences between the high exothermic reaction heat generated at the catalyst surface and the reduced gas temperature at the reactor inlet, when the reactor is below the ignition temperature (below 214 °C). This leads to kinetic bi-stability, where the surface nanoparticles alternate between active (PdO) and in-active (Pd) states [19]. The self-sustained oscillations and hysteresis appear during cooling of the catalyst under the flow of the reaction gas mixture. The cycling experiments for the first and second light-off were performed, where the first ignition was obtained at 88.6 °C. These results confirm that the hysteresis effect could help attain and maintain high CO conversions while oscillating between higher and lower temperatures. Upon heating/cooling, a significant hysteresis was observed at 121–296 °C. This effect can help in reducing the temperature for the second light-off for CO conversions over a Pd/SiO_2 catalyst by lowering the temperature along the extinction leg to a point below the ignition temperature as a result of the convective heat transfer associated with the temperature of the incoming gas, the CO oxidation reaction exotherm, and conduction of heat along the catalyst. Table 1 summarizes and compares the catalytic ignition and extinction profile and hysteresis width of the Pd/SiO_2 aerogel catalyst during the first light-off (fresh catalyst) and the second light-off (heat-treated catalyst in the CO/O_2 mixture) cycles. The second light-off demonstrated higher activity compared to the first light-off cycle, as shown in Figure 1b. The temperature of the second light-off and hysteresis width were shifted to lower values compared to the first light-off cycle. Reportedly, Pd^0 metal particles are preferably formed under a reducing CO atmosphere, whereas, under oxidizing O_2 atmosphere, Pd^{2+} or PdO particles should be formed [20]. The heating in the CO/O_2 mixture caused the CO conversion ignition temperature (T_{ig}) of Pd/SiO_2 to decrease from 214 to 195 °C, and the full CO conversion was achieved at (T_{100}) of 272 °C rather than 296 °C, while the hysteresis width decreased from 121 to 108 °C. The results might be attributed to the removal of moisture and hydrocarbons from the surface of the catalyst, increasing the metal-support interaction, as observed in XPS and XRD [2], the formation of palladium silicide, reduction of metal hydroxide, and oxidation of metallic Pd lead to improving the active site for the CO oxidation reaction.

Table 1. Comparison between the catalytic ignition and extinction profile, and hysteresis width of Pd/SiO_2 aerogel catalysts of the first and the second light-off (ignition) cycles.

Light-Off Cycle	T_{ig} (°C)	T_{50} (°C)	T_{100} (°C)	T_{ex} (°C)	Hysteresis Width (°C)
First light-off	214	276	296	138	121
Second light-off	195	247	272	123	108

Where T_{ig} is the ignition temperature at 3% conversion in °C, T_{50} is the temperature at 50% conversion in °C, T_{100} is the temperature at 100% conversion in °C, T_{ex} is the extinction temperature at 3% conversion in °C, and hysteresis width is at 50% conversion in °C.

Furthermore, the removal of silanol groups increases the metal dispersion and the catalytic activity of the Pd/SiO_2 catalyst [21]. Note that the conversion curve remained the same even after four cycles. The effect of heating rate on the catalytic activity, hysteresis, and ignition/extinction profile of the Pd/SiO_2 aerogel catalyst will be reported in a future manuscript.

To further investigate the catalytic activity and hysteresis behavior of the catalyst during the first and second light-off (or before and after catalytic conversion experiments), changes in the catalyst's porosity, surface area, Pd particle size, the oxidation state of Pd, crystallography, and pore size were investigated before and after the CO oxidation reaction.

The porosity of the catalyst assumes a role in the catalytic activity and the hysteresis activity during the CO oxidation reaction. The N_2 adsorption-desorption characteristics of fresh (as prepared) Pd/SiO_2 aerogel isotherms and pore size distribution before the CO oxidation reaction was performed. The isotherms of the fresh sample with the N_2 uptake consistent with type IV (according to the IUPAC classification), and a narrow H_2 type hysteresis loop at $p/p_0 > 0.75$ are shown in Figure S1, suggesting mesoporous and microporous characteristics due to capillary condensation in silica [22]. The increment of adsorption at $p/p_0 = 1.0$ was caused by larger mesopores, typical in mesoporous materials [23]. This wider hysteresis loop is known to occur when the distributions of the pore radius are wide [24]. The effect of heat treatment in the CO/O_2 atmosphere (after the first light-off) on the morphologies, surface area, and pore volume distributions were investigated to clarify the higher catalytic activity in the second light-off observed for CO oxidation on the Pd/SiO_2 catalyst. The structural properties of both samples were shown in Table 2 and illustrated in Figure S1. The fresh Pd/SiO_2 aerogel catalyst shows continuous pore volume distribution with diameters between 2 and 80 nm. However, N_2 adsorption-desorption characteristics of the Pd/SiO_2 aerogel pore size distribution after CO oxidation (after the first light-off) show a pore volume with continuous distribution of pore diameters (between 2 and 60 nm). The higher temperature during the CO oxidation does not affect the integrity of the catalyst. However, the quantity of large mesopores and macropores is eliminated due to the collapse following the heat treatment, leading to a higher ratio of micropores and small mesopores, as reported by Gage et al. [25]. BET results show that the average pore diameter of the Pd/SiO_2 aerogel slightly reduced to 15.6 nm and the pore volume slightly increased to 0.06 cm^3/g, although the surface area reduced to ~940.9 m^2/g. The decrease in the surface area could be attributed to the sintering of the large excess palladium particles outside the pores as found in electron microscopy, which will be discussed later.

Table 2. Structural properties of fresh Pd/SiO_2 aerogel catalysts and after the first light-off.

Catalyst Pd/SiO_2	BET Surface Area (gm^{-1})	Pore Diameter [a] (nm)	Pore Volume [a] (cm^3g^{-1})
Fresh	1113.65	15.7	0.05
After the first light-off	940.86	15.6	0.06

[a] Calculated by the Barrett, Joyner, and Halenda (BJH) method from the desorption isotherm.

To study the effect of the particles size of the catalyst on the catalytic activity and hysteresis width during the CO oxidation reaction, the TEM of the catalyst before (fresh) (Figure 2a) and after CO oxidation (first light-off) (Figure 2b) was performed. The presence of a large number of smaller Pd particles (2–5 nm) in the framework (marked by yellow arrows) located inside the mesoporous framework of SiO_2 aerogels homogeneously dispersed within the SiO_2 network and a few large surface particles (20–40 nm) (marked by magenta arrows). This unique texture is believed to occur due in the samples prepared by the sol-gel synthesis resulting in a better catalytic activity and effectively hindering the sintering of Pd particles [26]. The average size estimated from over 200 particles shows a mean size of the Pd particles in as-synthesized particles as 6.1 nm. Figure 2d,e shows the TEM micrographs of typical samples after CO oxidation. The mean size of Pd particles following the completion of the reaction was found to grow to ~7.8 nm due to the sintering, as also observed from the X-ray diffraction (XRD) analysis, and reported in a previous work [2].

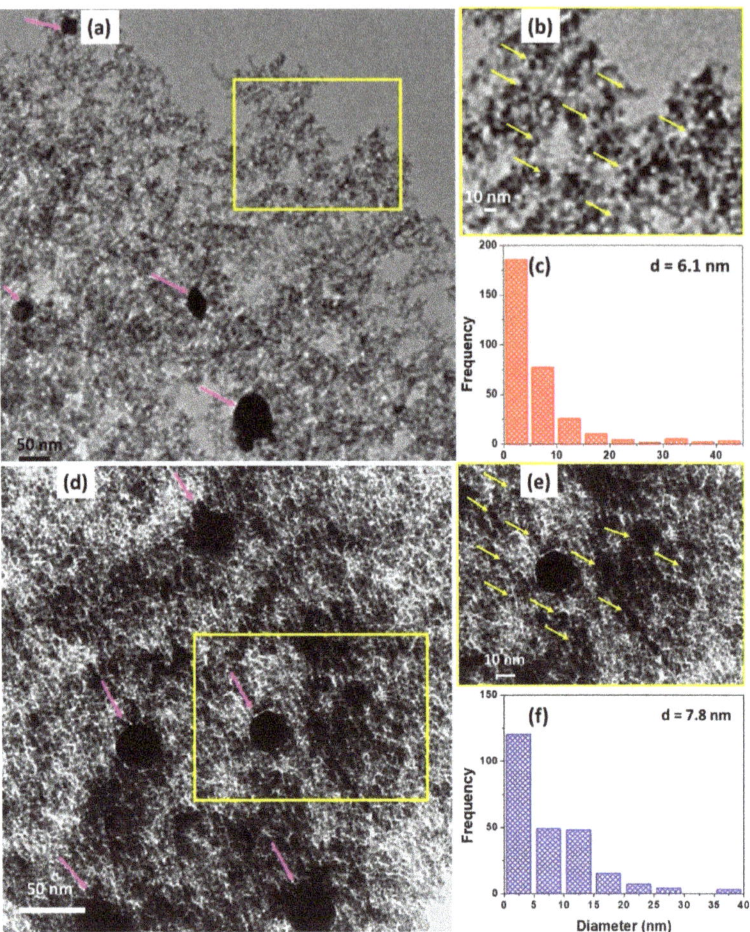

Figure 2. TEM micrographs and the corresponding estimated particle size distribution of Pd/SiO_2 aerogel catalysts (**a–c**) before, and (**d–f**) after the catalytic CO oxidation.

The XPS spectrum of the core level Pd 3d peaks were obtained from the samples at different points and verified by the NIST Standard XPS Database for PdO_x/Pd and Pd/SiO_x. The deconvolution of the spectrum of Pd 3d shows two spin-orbital states, $3d_{5/2}$ and $3d_{3/2}$ [2]. The Pd 3d and O 1s XPS peaks located in the near-surface region of freshly prepared Pd/SiO_2 samples, at 250 °C (full CO conversion), and after catalytic CO oxidation were investigated to understand the chemical environment, palladium oxidation state, and active species before and after the CO oxidation reaction. The Pd 3d showed peaks for Pd $3d_{5/2}$ and Pd $3d_{3/2}$ and fitted with the mixed Pd(0) and PdO combined spectrum. Table 3 summarizes the binding energies (BEs) of palladium Pd $3d_{5/2}$, Pd $3d_{3/2}$, and the binding energy (BE) of the corresponding O 1s peak [27]. The O 1s peak was fitted with three components due to the overlap of O 1s and Pd 3P3/2 peaks following Zemlyanove et al. [28]. Figure 3 shows Pd 3d and O 1s spectra of the reduced Pd/SiO_2 catalysts along with the deconvoluted peaks. Figure 3a shows Pd 3d spectra for the fresh (or as prepared) catalyst with two peaks at binding energies (BE) at 334.06 and 339.3 eV assigned to the metallic palladium Pd(0) and two peaks observed at 335.8 and 341 eV assigned to Pd^{2+} or PdO, respectively. This clearly indicates that most of Pd exists in the form of Pd metal with only a small fraction in the Pd^{2+} form. Upon increasing the temperature to 250 °C (Figure 3b), the obtained spectrum of the sample shows that the low energy doublet of $3d_{5/2}$ is shifted to 335.1 eV which could be assigned to the photoemission of electrons from Pd^{2+} and the high energy doublet is shifted to 336.1 eV due to the oxidation of smaller Pd particles (2–3 nm) or Pd^{2+} cations within the catalyst structure [29]. The increase in BE of Pd 3d for the catalyst annealed at 250 °C affirms the increase in formation of PdO, indicating that the surface of Pd is highly oxidized. After the CO oxidation reaction (Figure 3c,f) the deconvolution of the Pd 3d peaks show Pd(0) and Pd^{2+} (PdO) peaks with higher concentration of PdO compared to freshly prepared samples.

Table 3. List of the binding energies of palladium species (Pd and PdO) and the binding energy of the corresponding O 1s peak located in the near-surface region of Pd/SiO_2 catalysts.

Catalyst	Binding Energy, eV		
	Pd $3d_{5/2}$	Pd $3d_{3/2}$	O 1s
Fresh	334.1	339.3	531.5
At 250 °C	335.13	340.23	531.6
After CO oxidation	334.8	339.9	531.8

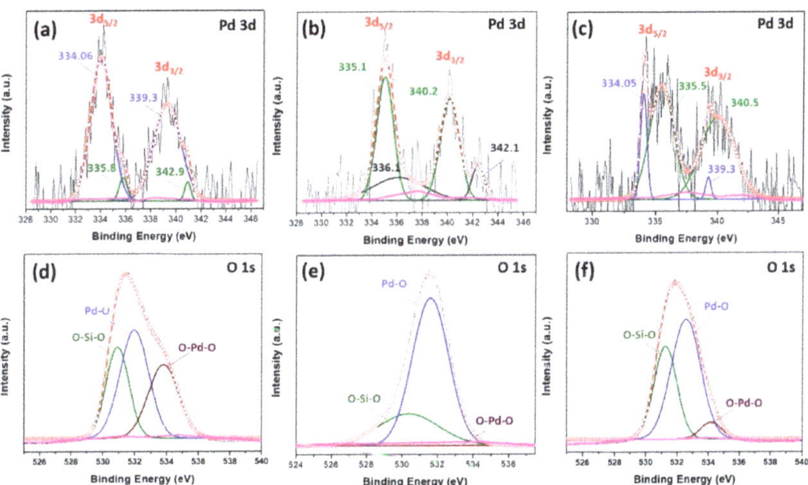

Figure 3. High-resolution XPS spectra of Pd 3d of (**a**) fresh Pd/SiO_2 catalysts, (**b**) Pd/SiO_2 catalysts annealed at 250 °C (full CO conversion), and (**c**) Pd/SiO_2 catalysts after the heating/cooling cycle in the CO/O_2 reaction mixture, and the deconvolution of O 1s XPS spectrum: (**d**) Before CO oxidation, (**e**) at 250 °C, and (**f**) after CO oxidation.

To analyze the change in the concentration of Pd^{2+} before and after CO oxidation, we compared the areas under the Pd $3d_{5/2}$ of Pd(0):Pd^{2+} peaks before CO oxidation (Figure 3a) to the ratio of areas of the peaks after CO oxidation (Figure 3c). The ratio of areas under the peaks of Pd(0):Pd^{2+} were 1.8:1 before and 1:1.6 after CO oxidation, respectively, indicating that the PdO concentration increases after CO oxidation, suggesting a lowering of the activation energy and confirming that the active phase for CO oxidation is PdO [29]. The increase in surface oxygen concentration (Pd-O) after the reaction (Figure 3f) could aid the activation of the C-O bond in the CO molecule for other oxidation reactions as reported elsewhere [23]. This increase is also accompanied by a reduction of oxygen concentration (O-Pd-O), which suggests that Pd interacts with the SiO_2 support [30,31]. Oxygen in SiO_2 loses electrons resulting in a shift and bending of the Fermi level. O_2 initially diffuses to the metallic Pd surface, where it is adsorbed to form an active adsorption state. Following this, the oxygen atoms interact with the Pd atoms on the surface to form PdO.

Results obtained from the XPS analysis show that the active sites and the state of Pd can provide valuable information on the metal-support interaction in the Pd/SiO_2 catalyst through monitoring of the electronic modifications of the Pd surface before and after the CO oxidation. This observation is consistent with the XRD [2] and TEM results. The XPS of the freshly prepared Pd/SiO_2 aerogel catalyst confirms that the Pd nanoparticles are attached to the support material through oxygen atoms of either the free silanol or siloxane groups present on the silica-network as oxygen is highly electronegative and can draw more electrons from Pd nanoparticles, resulting in higher BE for the Pd atoms in the Pd/SiO_2 catalyst.

2.1.2. Effect of Annealing Atmosphere on Catalytic Efficiency and Hysteresis of Pd/SiO_2 Aerogel Catalyst

The oxidation state of Pd alone cannot explain the change of the catalytic activity and hysteresis behavior before and after the heat treatment. To examine the thermal stability and catalytic performance of the catalyst aiming at optimizing the best condition for Pd active sites, we conducted several experiments to support our results. It has been reported that the catalyst support modification can contribute to the activity and hysteresis behavior of the catalyst due to its binding to the catalyst metal. This metal–support interaction modifies both the electronic and geometric properties of the catalyst support, which influences the activity of the catalytic sites on the metal surface and enhances active sites [23]. Reportedly, the hysteresis effect depended on the pretreatment of catalysts and was attributed to the changes in the catalyst structure for CO oxidation on partially oxidized Pd nanoparticles, where hysteresis effects were found to depend on the pretreatment of catalyst samples [32]. Pretreatment conditions of the catalyst influence the catalytic activity, and metal-support interaction motivated by the oxidation state of metal and the nature of the reactions. The effect of the catalyst pretreatment on the hysteresis behavior was reported and attributed to the changes in the catalyst structure [17]. Therefore, the pretreatment atmosphere is an essential factor that influences the final state of the catalyst and metal-support interaction. Oxidizing or reducing the atmosphere can yield oxide active or metallic phases depending on the temperature of the treatment.

The effect of the pretreatment atmosphere on catalytic activities, hysteresis, and ignition/extinction curves of Pd/SiO_2 aerogel catalyst was studied, and the results are summarized in Table 4 and Figure 4. The catalytic ignition/extinction curves and hysteresis of Pd/SiO_2 aerogel catalyst freshly prepared and annealed in different atmospheres have been plotted. It is clear that under ignition, the catalytic activities of annealed samples are higher, and the ignition temperatures are shifted to lower temperatures, which are attributed to the efficient removal of adsorbents from the surface, improving the exposure of active sites to fresh adsorbates possibly led by the decomposition of the metal complex to metal or metal oxide. However, the best activity and highest increase in the hysteresis width was observed for the sample treated in air. The results can be explained based on the gas composition of the annealing atmosphere, the reactant gas mixture, and the nature of gas used in the annealing atmosphere (reducing or oxidizing). Although, He and N_2 are inert gasses and do not affect the oxidation state of the freshly prepared Pd/SiO_2 aerogel catalyst, the thermal treatment at high temperature can affect the physio-chemical properties of the catalyst. The Pd/SiO_2-air aerogel sample showed the best catalytic activity compared to the untreated Pd/SiO_2 and Pd/SiO_2-N_2 aerogel samples, a plausible reason being that during the reaction, some of the metallic Pd nanoparticles converts to PdO as observed in XPS results, and the interfaces between Pd and PdO act as active catalytic sites. During extinction, the sample treated in N_2 gas shows an extinction temperature of 123 °C and hysteresis width of 104 °C, while the Pd/SiO_2-air aerogel sample, on the other hand, has an extinction temperature of 79 °C and a wider hysteresis width of 138 °C. This may arise from the formation of PdO in the Pd/SiO_2-air aerogel sample in the whole bulk of the catalyst during heating in the presence of excess O_2 through surface oxidation of metallic Pd preceded by diffusion of oxygen atoms from the bulk of the catalyst [33]. During extinction, the surface PdO is reduced directly to metallic Pd. The appearance of the hysteresis loop can be associated with the slow transition from an oxygen-enriched surface and surface palladium oxide formation, present during extinction, to a CO-covered surface including Pd reduction, resulting in reversible formation of surface Pd oxide. Since Pd in the Pd/SiO_2-air aerogel sample was oxidized entirely, the reduction of Pd oxide will take a longer time than in the Pd/SiO_2 aerogel treated in N_2 samples, resulting in the broadening of the hysteresis curve [34].

Table 4. Catalytic ignition and extinction profile of Pd/SiO_2 aerogel catalysts annealed in different atmospheres.

Gas	T_{ig} (°C)	T_{50} (°C)	T_{100} (°C)	T_{ex} (°C)	Hysteresis Width (°C)
Untreated	214	276	296	138	121
Air	212	223	225	79	138
N_2	212	248	287	123	104

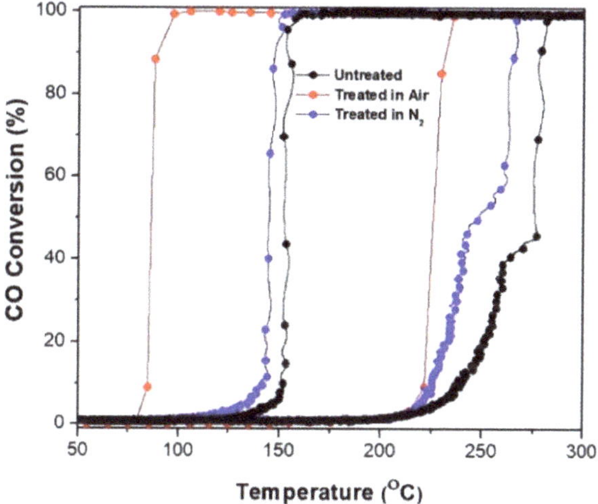

Figure 4. Comparison between the catalytic activity curves of Pd/SiO_2 aerogel catalysts annealed in different atmospheres during ignition and extinction.

To investigate the thermal stability of Pd/SiO_2 aerogel catalyst and the surface oxidation and reduction of Pd/SiO_2, TGA was performed up to 600 °C at a heating rate of 10 °C/min in air and N_2 atmospheres, while the DSC of Pd/SiO_2 aerogel was carried out under heating (red) and cooling (black) at a heating rate of 10 °C/min in air and N_2 atmospheres up to 300 °C, which corresponds to the temperature where full CO conversion is reached. TG-DSC spectra are shown in Figure 5. During heating, TGA and DSC studies for samples treated in an air environment showed two exothermic peaks and one endothermic peak, as shown in Figure 5a,c. The spectra indicate clearly that the catalyst is thermally stable up to 600 °C. A weight loss of ~3 wt% occurs when the samples are heated from 50 to 100 °C due to the loss of physically adsorbed water and ethanol from the porous catalyst.

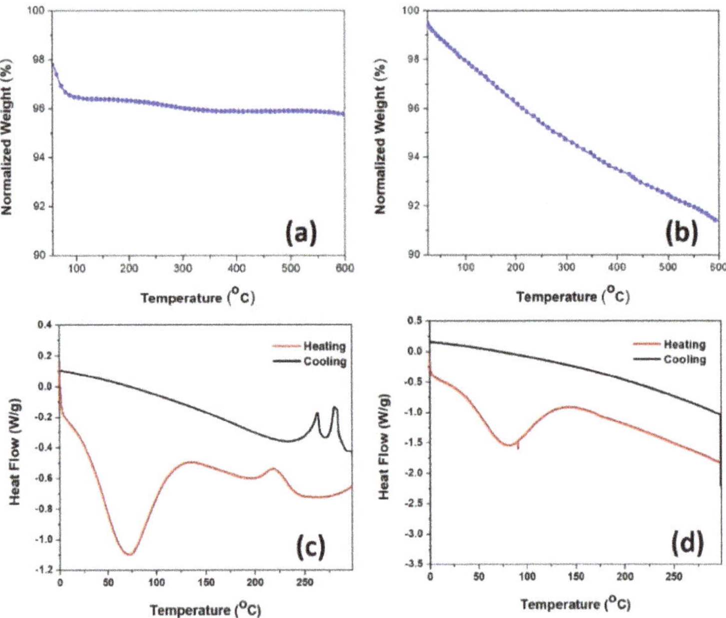

Figure 5. (**a**,**c**) Thermal gravimetric analysis-differential scanning calorimetry (TGA-DSC) of Pd/SiO_2 aerogel catalysts heated in the air atmosphere, (**b**,**d**) TGA-DSC of Pd/SiO_2 aerogel catalysts heated in the N_2 atmosphere.

These results are consistent with the DSC results, which show an endothermic peak in the region from room temperature up to 125 °C. Furthermore, an exothermic peak accompanied by an endothermic peak between 215 and 243 °C can be attributed to the surface oxidation of palladium accompanied with the reduction of silica surface, indicative of the formation of the interface between the palladium and silica support [35]. During the cooling ramp, as shown in Figure 5c, two exothermic peaks at 315 and 225 °C accompanied with an endothermic peak in between, could be attributed to the reduction of the small domain PdO to Pd^0 on the surface of PdO, leading to polycrystalline particles that easily re-oxidize upon cooling due to the lack of Pd nucleation sites on the surface of the metal particles. Moreover, this could be associated with the combustion of unreacted organics such as $Si–CH_3$ groups from the synthesis process. This behavior was observed previously by Datye et al. on the surface of Pd/Al_2O_3 when the catalysts were heated in air [36] and Colussi et al. at higher temperature on Pd/Al_2O_3 and $Pd/CeO_2/Al_2O_3$, where they reported re-dispersion of Pd on the surface of oxide and transformation between Pd and PdO [37]. The temperature of this transformation was reported to strongly depend on the characteristics of the oxide support [38]. The results agree with the XPS analysis and catalytic tests of Pd/SiO_2 treated in air under heating/cooling cycles. However, TGA and DSC studies for samples treated in a nitrogen environment showed only well-resolved steps, Figure 5b,d. The major weight loss (about 5 wt%) was observed between 50 and 125 °C as evident in the DSC, as the broad endothermic peak in the curve in Figure 5d is attributed to the loss of water from the porous catalyst [39]. After 125 °C, the weight loss until 600 °C is attributed to the condensation of silanol groups from the surface of pristine silica aerogel, which was experimentally found to occur between 150 and 500 °C, as reported by Mueller et al. [40]. No peaks were observed during the cooling cycle which suggest that most of Pd in the sample is in the metallic state.

It is well known that the support nature and composition of SiO_2 and pretreatment conditions (oxidation or reduction treatments) have a direct impact on the metal-support interaction and ratio of oxidized and reduced forms of the supported palladium metal. To investigate the thermal effect on the catalytic activities, hysteresis behavior, and surface oxidation and reduction of Pd/SiO_2 heated in air and N_2 atmospheres, we preformed FTIR spectroscopy of Pd/SiO_2 aerogel catalyst annealed in different atmospheres (N_2 and air) and compared it to the untreated catalysts. Figure S2 compares FTIR spectra of freshly prepared Pd/SiO_2 aerogel catalysts and after the heat treatment at 450 °C in air and nitrogen atmospheres. Room-temperature FTIR spectra of Pd/SiO_2 samples measured in spectral range (400–4000 cm^{-1}) were recorded and compared to the untreated samples to determine the changes after annealing in N_2 and air atmospheres. The FTIR spectra of both samples revealed several sharp, well-defined absorption bands within the measured spectral range. The spectra show bands centered at 567, 794.5, and 1049.1 cm^{-1} with a shoulder peak at 1162.9 cm^{-1} corresponding to stretching vibrations of siloxane groups (Si–O–Si bonds), respectively [41], while the peak centered at 954.5 cm^{-1} corresponds to the stretching vibration of silanol groups (Si-OH) in the silica lattice suggesting the presence of a considerable amount of silanol groups on the silica surface or pores in all the samples. A small peak observed at 2987.2 cm^{-1} is assigned to the vibrations of the stretching vibration of -CH_3 and -CH_2 groups indicating the presence of a small amount of Si–OC_2H_5 groups, which can be attributed to an incomplete condensation during gelation [42]. The low-intensity peak at 3367.1 cm^{-1} is assigned to -OH stretching vibrations [43].

The FTIR spectra of the Pd/SiO_2 sample treated in air show that all peaks are shifted to a higher wavenumber compared to the untreated sample indicating the interaction between Pd and SiO_2, which can affect the formation of the Si-O-Si network as observed in Cu/SiO_2 [44] leading to a stronger metal-support interaction. The intensity of the peak at 958.45 cm^{-1} is considerably lower, while that at 794.5 cm^{-1} increases indicating the formation of new Si–O–Si bonds via the reduction of Si-OH bonds as a result of the condensation reaction between Si–O and the Pd metal. This reaction could shrink the SiO_2 network, which might be responsible for decreasing the pore volume and, consequently, surface area, as observed in N_2 adsorption-desorption results suggesting that the SiO_2 framework is formed by the Si–O–Si bonds [19]. Furthermore, the intensity of the peak at 3367.1 cm^{-1}, which is assigned to -OH stretching vibrations decreases, suggesting removal of the adsorbed OH or water. However, the FTIR spectra of Pd/SiO_2 sample treated in a nitrogen atmosphere showed similar spectra of the untreated sample (no shift is observed) except for reducing the intensity of the peak at 1162.9 cm^{-1}, which indicates the reduction of the Si-O-Si bond. The results suggest that the heat treatment in N_2 environment did not affect the interaction between Pd and the SiO_2 network.

2.1.3. Effect of Annealing Temperature on Catalytic Efficiency and Hysteresis of Pd/SiO_2 Aerogel Catalyst

A further impact of annealing temperature on the catalytic activity, hysteresis, and ignition/extinction of Pd/SiO_2 treated in the air aerogel catalyst was studied by annealing some samples at 450 and 750 °C in air as summarized in Table 5 and shown in Figure 6. During the ignition cycle, the air-annealed catalysts demonstrated higher activity compared to the freshly prepared catalysts mainly due to the removal of moisture and hydrocarbons from the surface of the catalyst and the reduction of metal oxide to metallic Pd and improving the active site for the CO oxidation reaction. The catalyst treated at 450 °C in air shows better activity than fresh catalysts due to the removal of silanol groups, which increases the metal dispersion and the catalytic activity of Pd/SiO_2 catalyst [21]. The results indicate that the heat treatment in air does not probably affect the Pd clusters or particles as they are pinned to the silica surface, and the diffusion of ions is difficult, preventing sintering. The Pd clusters showed resistance to sintering upon calcination to 550 °C in air, which was attributed to the confinement of Pd clusters within mesopores [45].

Table 5. Catalytic ignition and extinction profile of Pd/SiO_2 aerogel catalysts annealed in air for 1 h at different temperatures.

Annealing Temperature	T_{ig} (°C)	T_{50} (°C)	T_{100} (°C)	T_{ex} (°C)	Hysteresis Width (°C)
Untreated	214	276	296	138	121
450 °C Air	212	223	225	79	138
750 °C Air	214	251	259	157	70

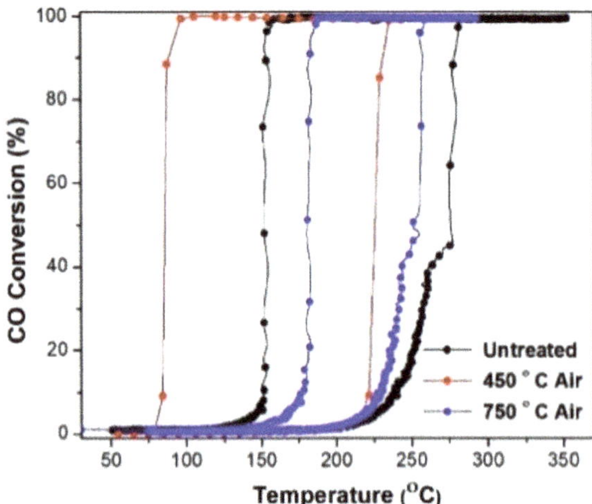

Figure 6. Comparison between the catalytic activity curves of Pd/SiO_2 aerogel catalysts annealed for 1 h in air at different temperatures during ignition and extinction cycles.

The Pd/SiO_2-air aerogel catalyst initially contained a considerable amount of metal oxide; heating in air will ensure that the sample is fully oxidized at 450 °C. In studies on the catalytic activities conducted in the CO/O_2 mixture, PdO will undergo a reduction to Pd, which in-effect would prevent any Pd particle growth (due to sintering) up to 700 °C, except for outermost particles (outside the pores). Upon annealing at 750 °C, all PdO will be reduced to metallic Pd and this will lead to the growth and sintering of metallic Pd particles which would lower catalytic activity [46]. The presence of fully oxidized Pd particles and well-defined active sites in the samples annealed at 450 °C in air ensures higher activity and lower extinction temperature. The as-prepared samples contain a considerable amount of metallic Pd, and most of the active sites are blocked, making it less active even during extinction. For samples annealed at 750 °C, the growth and sintering of the Pd particles lead to a lower activity resulting in narrower hysteresis width and lower ignition temperature compared to the sample treated at 450 °C.

Based on the experimental results, the origin of the high thermal stability and catalytic performance of Pd/SiO_2 aerogel catalysts were attributed to their mesoporous structures as confirmed by the N_2 adsorption-desorption isotherm, XPS, FTIR, and TEM. The results also confirmed that the catalyst structure could protect the Pd nanoparticles from sintering during the thermal treatment and catalytic CO oxidation reaction. The thermal stability of the Pd/SiO_2 catalyst highly depends on the oxidative/reductive nature of the gas environment. Under the CO oxidation reaction in the oxygen atmosphere, a small amount of Pd is converted to PdO. However, under air atmosphere, the porous structure of the silica is still stable, and the formation of small Pd particles inside the SiO_2 pores and on the surface increases as temperature increases.

Moreover, the absence of reducing gases in air leads to the oxidation of Pd particles, which directly impacts thermal stability and catalyst performance. The same behavior was observed under air and H_2 environment [47]. The FTIR results confirmed that the interaction of Pd sites with -OH on the SiO_2 stabilizes the catalyst surface resulting in excellent thermal stability. Consequently, the Pd/SiO_2 catalysts showed a much more stable CO oxidation performance after annealing in air. These results agree well with recent studies, suggesting that the thermal treatment by annealing and catalytic CO oxidation at high temperatures (below 500) enhances the stability of Pd/SiO_2 catalyst under catalytic CO oxidation reactions due to its structure stability [48].

To understand the effect of annealing temperature on the catalytic activities, hysteresis behavior, and surface oxidation and reduction of Pd/SiO_2 catalyst heated in air atmosphere at different temperatures. We performed FTIR spectroscopy of Pd/SiO_2 aerogel catalyst annealed at 450 and 750 °C in air atmospheres than the untreated catalyst. Figure S3 shows FTIR spectra of the freshly prepared Pd/SiO_2 aerogel catalyst and after the heat treatment at 450 and 750 °C in air, respectively. The infrared absorption is similar to the discussion mentioned above in Figure S2 with bands centered at 567, 794.5, and 1049.1 cm^{-1} with a shoulder at 1162.9 cm^{-1} corresponding to stretching vibrations of siloxane groups (Si–O–Si bonds), respectively. A peak centered at 954.5 cm^{-1} corresponds to the stretching vibration of silanol groups (Si-OH) in the silica lattice which suggest the presence of a considerable amount of silanol groups on the silica surface or the pores in all samples. Small peaks observed at 2987.2 and 3367.1 cm^{-1} are assigned to the stretching vibration of -CH_3 and -CH_2 to -OH stretching vibrations or water, as reported earlier. As the annealing temperature increases, the peak at 958.45 cm^{-1} slowly disappears, the intensity of the peak at 2987.2 cm^{-1} decreases, while that of the peak at 794.5 cm^{-1} increases. This suggests the formation of additional Si–O–Si bonds by the condensation reaction. The peaks at 1710 and 3367.1 cm^{-1} that belong to vibrations of water molecules decrease with the increasing annealing temperature due to the removal of water from the SiO_2 network structure [23]. The results suggest that heating to 750 °C could lead to the formation of additional Si–O–Si bonds, which strengthens the SiO_2 network structure [19]. The presence of 958.45 cm^{-1}, which corresponds to the stretching vibration of silanol groups (Si-OH) in the sample heated at 450 °C, enhanced the Pd-silica interaction, facilitating the dispersion of Pd particles.

2.2. Hysteresis Behavior during (Ignition/Extinction) Cycles

The reaction mechanism of CO oxidation during ignition/extinction cycles was reported by many researchers. CO is strongly adsorbed onto Pd, inhibiting the formation of active oxygen needed for low-temperature CO oxidation. Hence, low-temperature CO oxidation over Pd catalysts proceeds via Langmuir–Hinshelwood. Such adsorption has been shown to occur during CO oxidation on Pd metal catalysts under low O_2 pressure, wherein CO and O_2 gas are adsorbed on free adsorption sites on the Pd metal surface, followed by the interaction of the adsorbed CO and O, respectively, on the Pd active sites, resulting in the palladium oxide (PdO) surface. However, the reduction of PdO proceeds via the Mars-van Krevelen mechanism, where CO is adsorbed on the PdO surface [49].

The TEM results of large surface particles in the Pd/SiO_2 catalyst suggest that the as prepared catalyst contains Pd particles and a small amount of PdO particles on the surface, as shown in Figure 7a, while at 250 °C, the surface is almost fully covered with oxide as shown in Figure 7b, suggesting the formation of surface and bulk oxides. The TEM image of the particles after the heating/cooling cycle (Figure 7c) shows that part of the Pd (0) surface is restored, and the PdO surface is destroyed with a small island of 2D PdO observed. Similar observations in the literature suggested the oxidation of Pd and the formation of surface oxide during heating [12]. At higher temperatures, wherein the formation of 3D PdO is preferential on the catalyst surface while during cooling conditions, the 3D PdO is reduced to the Pd surface. The TEM results agree well with the XPS peak fitting, which show a shift of Pd $3d_{5/2}$ peak towards higher BE (Figure 3b) with respect to the metal peak (Figure 3a) due to the interaction with atomic oxygen to form PdO during the CO oxidation reaction. However, during cooling to room temperature (after heating/cooling cycle) and as a result of the reaction medium (Figure 3c), PdO particles are reduced, leading to a decrease in the oxide intensity. In this region, the metallic palladium Pd(0) and the PdO particles co-exist, which serve as intermediates causing the self-sustained oscillation and hysteresis. A similar behavior was reported for Pd/Al_2O_3 systems [12].

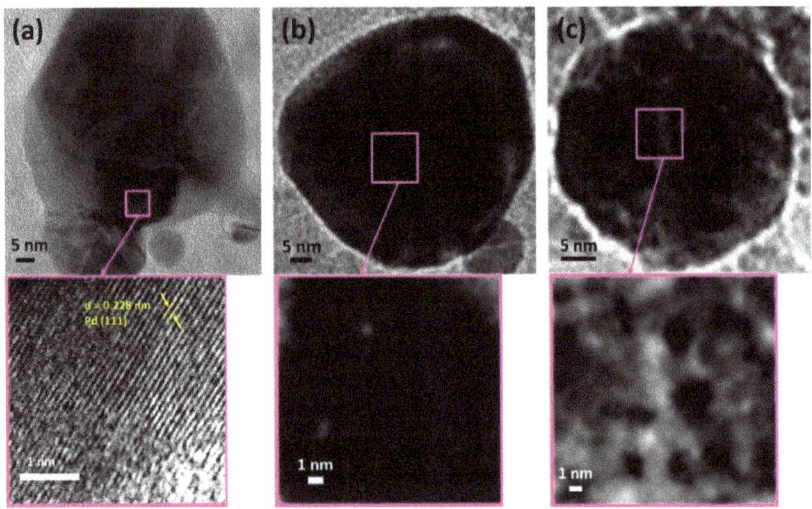

Figure 7. TEM of (**a**) the fresh Pd/SiO_2 catalysts, (**b**) Pd/SiO_2 catalysts annealed at 250 °C in air, and (**c**) after heating/cooling cycle in the CO/O_2 reaction mixture.

In our study, XPS and TEM results suggest that the oxidation state, morphology of the Pd/SiO_2 catalyst, and the large palladium surface particles, changed before, during, and after the CO oxidation reaction. The presence of surface and subsurface Pd (0) and PdO with 2D morphologies, and the interplay between the two phases (reversed oxidation of Pd (0) and reduction or decomposition of PdO) causes the oscillatory behavior and hysteresis during the CO oxidation reaction, which aid the catalytic activity by lowering the activation energy after the first light-off.

3. Experimental Methodology

3.1. Catalyst Synthesis

The catalysts were prepared using the sol-gel method and dried under conditions with supercritical ethanol at 260 °C. This synthesis method described in a previous study [26] ensured the synthesis of well-dispersed Pd nanoparticles on a silica support. The silica gel was impregnated with metal ions before the drying step to replace the pores previously filled with the solvents used in the synthesis, which leads to hierarchical porosity. The advantage of this synthesis method is the possibility of producing highly dispersed and stable Pd catalysts with controlled structures and catalytic performances. The resulting Pd/SiO_2 catalyst is composed of accessible palladium particles located inside the pores or within the network silica particles, which result in sinter-proof Pd particles. The active catalytic sites are easily accessed by the reactants by diffusing through micropores and mesopores with no mass transfer limitations [23].

In a typical synthesis with tetraethyl orthosilicate (TEOS, Sigma-Aldrich, St. Louis, MO, USA, 98%) and ethanol, the resultant solution is aged overnight in a sealed container to obtain a gel [18]. The wet-gel obtained is then transferred to a solution of palladium precursor ($PdCl_2$, Sigma-Aldrich, St. Louis, MO, USA) in ethanol. The silica aerogels are dipped in the solutions of Pd ions in ethanol to ensure the exchange of ethanol in the gel with the Pd ions in the solution, followed by the ethanol supercritical drying at

260 °C. Dried Pd supported silica aerogels were annealed at a heating rate of 10 °C under atmospheric pressure (1 atm) in two different ambient conditions: Air (Pd/SiO_2–air), nitrogen (Pd/SiO_2–N_2), each at a 100 mL/min flow rate. All catalytic experiments and characterization were carried out using powdered aerogels.

3.2. Catalysts Characterization

The Pd/SiO_2 aerogel catalysts were characterized by X-ray photoelectron spectroscopy (XPS; Omicron Nanotechnology, Germany) with a monochromatic Al K_α radiation (energy = 1486.6 eV) working at 15 kV, which was used to study the surface states of the catalyst. The obtained XPS spectra were calibrated with respect to the C 1s feature at 285 eV [49]. The catalyst's surface area and pore sizes were determined using a Rise 1010 surface area and porosity analyzer (Jinan Rise Science and Technology Co., Ltd., Jinan, China). The Brunauer–Emmett–Teller (BET) and Langmuir models based specific surface area were calculated from the nitrogen adsorption-desorption isotherms recorded at 77 K at the relative pressure range of P/P_0 = 0.05–0.2. Pore size distributions were calculated using the desorption branches of the isotherms using the Barrett–Joyner–Halenda (BJH) model [50]. Transmission electron microscopy (TEM) measurements were carried out using a (JEM2100F field emission transmission electron microscope (TEM)) operating at a voltage of 200 kV (JEOL Ltd., Tokyo, Japan). Fourier transform infrared (FTIR) spectroscopy was performed using the FTIR650 spectrometer equipped with a LA-025-1100 universal ATR unit (Labfreez Instruments (Hunan) Co., Ltd., Changsha, China). Thermogravimetric analysis (TGA) and differential scanning calorimetry (DSC) were performed using the TG 2100D (Analytical Technologies Limited, Shanghai, China) analyzer at a ramp rate of 10 °C/min to reach (from 23 to 800 °C) in air (flow rate of 100 cc/min).

3.3. Catalyst Activity Testing

Catalytic activities of the synthesized samples were carried out in a custom-built fixed-bed continuous flow reactor placed inside a programmable tube furnace coupled to an infrared gas analyzer (ACS-CO_2 infrared analyzer) [2]. Twenty milligrams of powdered metal loaded aerogel catalyst was placed in the middle of a quartz tube and sandwiched between two pieces of glass wool to form a cylindrical pellet. The catalyst temperature was recorded using an Omega K-type thermocouple inserted in the middle of the sample. The reactant gas mixture containing 3.5 wt% CO and 20 wt% O_2 in helium was flown through the reactor at 100 mL/min (calculated weight hourly space velocity (WHSV) was approximately 300,000 mL $g^{-1}h^{-1}$. CO conversion was assessed by measuring CO_2 in the outflow with an IR analyzer, and the catalytic activity was expressed by the CO conversion in the effluent gas. The temperature and the concentration data were collected using a National Instruments multifunction USB-6008 and NI-DAQmx (National Instruments, Roscoe, IL, USA) data acquisition system and recorded using a custom-built LabVIEW data acquisition software. The gas flow rate was controlled by a set of digital mass flow controllers. The flow rate of the mixture was maintained at 100 mL/min, while the catalyst was heated to different temperatures (25–600 °C). All the experiments were performed at an atmospheric pressure (1 atm) with a heating rate of 10 °C/min (ignition or activation) until the CO conversion reaches a full conversion (100%), then the furnace was switched off, and the sample was left to cool naturally until the CO conversion became zero (0%) (extinction or relaxation). Catalyst conditioning was achieved by calcination of the catalyst at 450 °C in air atmosphere for 1 h. For the other experiments, the calcination temperature was changed to 750 °C in air or N_2 for 1 h. The process is capable of removing moisture and improving the active sites. The ignition/light-off and extinction/light-out temperatures denoted the temperatures where the CO conversion reaches 3% during ignition and extinction curves. The CO conversion of the catalyst was measured as a function of the catalyst temperature and calculated using the following relation:

$$\text{CO conversion } (\%) = \frac{[\text{CO}]_{\text{in}}\text{vol.\%} - [\text{CO}]_{\text{out}}\text{vol.\%}}{[\text{CO}]_{\text{in}}\text{vol.\%}} \times 100 \tag{1}$$

where $[CO]_{in}$ is the CO concentration in the reaction gas mixture and $[CO]_{out}$ is the concentration in the product gas mixture.

4. Conclusions

In this work, the low temperature CO oxidation and hysteresis behavior as a function of the catalyst temperature under optimum reaction conditions was investigated over a range of pretreatment mixture/conditions and temperatures that directly affect the nature of the active site. Carefully designed control experiments provide strong evidence that the pretreatment temperature and medium, nature of the support (porosity, thermal conductivity), the stability of the active site, and their formation were crucial factors in enhancing the catalytic performance. The best low-temperature CO oxidation performance was achieved by pretreating the catalyst in an air atmosphere at 450 °C. Recent studies suggest that the catalyst activation and the CO conversion hysteresis is attributed to local heating and heat dissipation by the support, respectively. However, the results presented here suggest that pretreatment conditions in N_2 or air, and the pretreatment temperature have a direct impact on the catalyst activation and hysteresis behavior, in addition to the structural and chemical effect of the catalyst. The active site formation is influenced by the presence of oxidative or reductive pretreatment gas, and the surface morphological and chemical changes of the Pd during the reaction due to CO and O_2 adsorption to achieve the lower activation energy. Therefore, the hysteresis arises from the stabilization of these active sites by forming different oxidation states of Pd during the ignition and alteration of PdO to Pd during the extinction. These results were realized by the XPS, which suggest partial oxidation of Pd and the formation of different surface oxides reaching full oxidation at 250 °C. These results were confirmed by TEM, which show the formation and reduction of PdO during heating/cooling cycles. The TGA and DSC results confirm the surface oxidation and reduction behavior of Pd/SiO_2 in air for both the ignition and extinction cycles. Furthermore, the catalyst structure including the particle size distribution, the porosity of the sample, and thermal stability contribute to the catalytic activity and hysteresis phenomenon. This study confirms that the structure-activity relationship is very crucial for the design of a highly active and thermally stable catalyst for CO oxidation as a model reaction, which could aid the design of next-generation catalysts for CO oxidation.

Supplementary Materials: The following are available online at https://www.mdpi.com/2073-4344/11/1/131/s1. Figure S1: Langmuir isotherms of N2 adsorption and desorption (a) and pore diameter distribution (b) of fresh Pd/SiO_2 aerogel catalysts before and after the catalytic CO oxidation (first light-off); Figure S2: Comparison between FTIR spectra of fresh Pd/SiO_2 aerogel catalysts and after the heat treatment at 450 °C in the air and nitrogen atmosphere; Figure S3: Comparison between FTIR spectra of fresh Pd/SiO_2 aerogel catalysts and after the heat treatment at 450 and 750 °C in air.

Author Contributions: R.M.A.S. synthesized materials, performed experiments and characterization, analyzed data, wrote and edited the entire paper; J.D. supervised the work and the quality of the manuscript; K.M.S. participated in characterization and data analysis; M.T.Z.M. and M.A.G. performed and analyzed XPS data. The authors worked together to prepare this manuscript. All authors have read and agreed to the published version of the manuscript.

Funding: We would like to acknowledge partial financial support from Swedish Energy Agency (Energimyndigheten) through a project entitled HESAC (Project No. 45504–1).

Institutional Review Board Statement: Not applicable.

Informed Consent Statement: Not applicable.

Data Availability Statement: Data is contained within the article.

Acknowledgments: We want to show our gratitude to Shaukat Saeed from the Pakistan Institute of Engineering and Applied Sciences (PIEAS), Islamabad, Pakistan, Ayman Samara from the Qatar Environment and Energy Research Institute (QEERI), Hamad Ben Khalifa University, Doha, Qatar, and Karthik Laxman Kunjali from the Royal Institute of Technology (KTH), Stockholm, Sweden who provided insights and expertise that helped our research.

Conflicts of Interest: The authors declare no conflict of interest.

References

1. Freund, H.J.; Meijer, G.; Scheffler, M.; Schlögl, R.; Wolf, M. CO Oxidation as a Prototypical Reaction for Heterogeneous Processes. *Angew. Chem. Int. Ed.* **2011**, *50*, 10064. [CrossRef] [PubMed]
2. Al Soubaihi, R.M.; Saoud, K.M.; Ye, F.; Zar Myint, M.T.; Saeed, S.; Dutta, J. Synthesis of hierarchically porous silica aerogel supported Palladium catalyst for low-temperature CO oxidation under ignition/extinction conditions. *Microporous Mesoporous Mater.* **2020**, *292*, 109758. [CrossRef]
3. Berlowitz, P.J.; Peden, C.H.F.; Goodman, D.W. Kinetics of carbon monoxide oxidation on single-crystal palladium, platinum, and iridium. *J. Phys. Chem.* **1988**, *92*, 5213–5221. [CrossRef]
4. Sawisai, R.; Wanchanthuek, R.; Radchatawedchakoon, W.; Sakee, U. Synthesis, Characterization, and Catalytic Activity of Pd(II) Salen-Functionalized Mesoporous Silica. *J. Chem.* **2017**, *2017*, 1–12. [CrossRef]
5. Kumar, S.; Boro, J.C.; Ray, D.; Mukherjee, A.; Dutta, J. Bionanocomposite films of agar incorporated with ZnO nanoparticles as an active packaging material for shelf life extension of green grape. *Heliyon* **2019**, *5*, e01867. [CrossRef]
6. Forman, A.J.; Park, J.-N.; Tang, W.; Hu, Y.-S.; Stucky, G.D.; McFarland, E.W. Silica-Encapsulated Pd Nanoparticles as a Regenerable and Sintering-Resistant Catalyst. *ChemCatChem* **2010**, *2*, 1318–1324. [CrossRef]
7. Min, B.K.; Santra, A.K.; Goodman, D.W. Understanding Silica-Supported Metal Catalysts: Pd/Silica as a Case Study. *ChemInform* **2004**, *35*. [CrossRef]
8. Gustafsson, T.; Garfunkel, E.; Gusev, E.P.; HÄBerle, P.; Lu, H.C.; Zhou, J.B. Structural studies of oxide surfaces. *Surf. Rev. Lett.* **1996**, *3*, 1561–1565. [CrossRef]
9. Chen, Z.P.; Wang, S.; Ding, Y.; Zhang, L.; Lv, L.R.; Wang, M.Z.; Wang, S.D. Pd catalysts supported on Co3O4 with the specified morphologies in CO and CH4 oxidation. *Appl. Catal. A Gen.* **2017**, *532*, 95–104. [CrossRef]
10. Slavinskaya, E.M.; Stonkus, O.A.; Gulyaev, R.V.; Ivanova, A.S.; Zaikovskii, V.I.; Kuznetsov, P.A.; Boronin, A.I. Structural and chemical states of palladium in Pd/Al_2O_3 catalysts under self-sustained oscillations in reaction of CO oxidation. *Appl. Catal. A Gen.* **2011**, *401*, 83–97. [CrossRef]
11. Imbihl, R.; Ertl, G. Oscillatory Kinetics in Heterogeneous Catalysis. *Chem. Rev.* **1995**, *95*, 697–733. [CrossRef]
12. Lashina, E.A.; Slavinskaya, E.M.; Chumakova, N.A.; Stadnichenko, A.I.; Salanov, A.N.; Chumakov, G.A.; Boronin, A.I. Inverse temperature hysteresis and self-sustained oscillations in CO oxidation over Pd at elevated pressures of reaction mixture: Experiment and mathematical modeling. *Chem. Eng. Sci.* **2020**, *212*, 115312. [CrossRef]
13. Dadi, R.K.; Luss, D.; Balakotaiah, V. Dynamic hysteresis in monolith reactors and hysteresis effects during co-oxidation of CO and C2H6. *Chem. Eng. J.* **2016**, *297*, 325–340. [CrossRef]
14. Assovskii, I.G. Ignition, extinction, and thermal hysteresis of a heterogeneous exothermic reaction. *Combust. Explos. Shock Waves* **1998**, *34*, 163–169. [CrossRef]
15. Abedi, A.; Hayes, R.; Votsmeier, M.; Epling, W.S. Inverse Hysteresis Phenomena During CO and C_3H_6 Oxidation over a Pt/Al_2O_3 Catalyst. *Catal. Lett.* **2012**, *142*, 930–935. [CrossRef]
16. An, W.; Liu, P. The complex behavior of the Pd-7 cluster supported on TiO_2(110) during CO oxidation: Adsorbate-driven promoting effect. *Phys. Chem. Chem. Phys.* **2016**, *18*, 30899–30902. [CrossRef]
17. Koutoufaris, I.; Koltsakis, G. Heat- and mass-transfer induced hysteresis effects during catalyst light-off testing. *Can. J. Chem. Eng.* **2014**, *92*, 1561–1569. [CrossRef]
18. Al Soubaihi, R.; Saoud, K.; Dutta, J. Critical Review of Low-Temperature CO Oxidation and Hysteresis Phenomenon on Heterogeneous Catalysts. *Catalysts* **2018**, *8*, 660. [CrossRef]
19. Casapu, M.; Fischer, A.; Gänzler, A.M.; Popescu, R.; Crone, M.; Gerthsen, D.; Türk, M.; Grunwaldt, J.-D. Origin of the Normal and Inverse Hysteresis Behavior during CO Oxidation over Pt/Al_2O_3. *ACS Catal.* **2016**, *7*, 343–355. [CrossRef]
20. Daniell, W.; Landes, H.; Fouad, N.E.; Knözinger, H. Influence of pretreatment atmosphere on the nature of silica-supported Pd generated via decomposition of Pd(acac)$_2$: An FTIR spectroscopic study of adsorbed CO. *J. Mol. Catal. A Chem.* **2002**, *178*, 211–218. [CrossRef]
21. Kwon, J.S.; Kim, J.S.; Lee, H.S.; Lee, M.S. Surface Modification of SiO_2 for Highly Dispersed Pd/SiO_2 Catalyst. *J. Nanosci. Nanotechnol.* **2019**, *19*, 882–887. [CrossRef] [PubMed]
22. Panpranot, J.; Pattamakomsan, K.; Goodwin, J.G.; Praserthdam, P. A comparative study of Pd/SiO_2 and Pd/MCM-41 catalysts in liquid-phase hydrogenation. *Catal. Commun.* **2004**, *5*, 583–590. [CrossRef]
23. Raj, R.; Harold, M.P.; Balakotaiah, V. Steady-state and dynamic hysteresis effects during lean co-oxidation of CO and C_3H_6 over Pt/Al_2O_3 monolithic catalyst. *Chem. Eng. J.* **2015**, *281*, 322–333. [CrossRef]
24. Inoue, S.; Hanzawa, Y.; Kaneko, K. Prediction of Hysteresis Disappearance in the Adsorption Isotherm of N_2 on Regular Mesoporous Silica. *Langmuir* **1998**, *14*, 3079–3081. [CrossRef]
25. Gage, S.H.; Engelhardt, J.; Menart, M.J.; Ngo, C.; Leong, G.J.; Ji, Y.; Trewyn, B.G.; Pylypenko, S.; Richards, R.M. Palladium Intercalated into the Walls of Mesoporous Silica as Robust and Regenerable Catalysts for Hydrodeoxygenation of Phenolic Compounds. *ACS Omega* **2018**, *3*, 7681–7691. [CrossRef]

26. Heinrichs, B.; Lambert, S.; Alié, C.; Pirard, J.P.; Beketov, G.; Nehasil, V.; Kruse, N. Cogelation: An effective sol-gel method to produce sinter-proof finely dispersed metal catalysts supported on highly porous oxides. In *Studies in Surface Science and Catalysis*; Gaigneaux, E., De Vos, D.E., Grange, P., Jacobs, P.A., Martens, J.A., Ruiz, P., Poncelet, G., Eds.; Elsevier: Amsterdam, The Netherlands, 2000; Volume 143, pp. 25–33.
27. Qi, T.; Sun, J.; Yang, X.; Yan, F.; Zuo, J. Effects of Chemical State of the Pd Species on H2 Sensing Characteristics of PdOx/SnO_2 Based Chemiresistive Sensors. *Sensors* **2019**, *19*, 3131. [CrossRef]
28. Zemlyanov, D.; Aszalos-Kiss, B.; Kleimenov, E.; Teschner, D.; Zafeiratos, S.; Hävecker, M.; Knop-Gericke, A.; Schlögl, R.; Gabasch, H.; Unterberger, W.; et al. In situ XPS study of Pd(111) oxidation. Part 1: 2D oxide formation in 10−3mbar O_2. *Surf. Sci.* **2006**, *600*, 983–994. [CrossRef]
29. Wang, Z.; Li, B.; Chen, M.; Weng, W.; Wan, H. Size and support effects for CO oxidation on supported Pd catalysts. *Sci. China Chem.* **2010**, *53*, 2047–2056. [CrossRef]
30. Juszczyk, W.; Karpiński, Z.; Łomot, D.; Pielaszek, J. Transformation of Pd/SiO_2 into palladium silicide during reduction at 450 and 500 °C. *J. Catal.* **2003**, *220*, 299–308. [CrossRef]
31. Riley, J.D.; Ley, L.; Azoulay, J.; Terakura, K. Partial densities of states in amorphousPd0.81Si0.19. *Phys. Rev. B* **1979**, *20*, 776–783. [CrossRef]
32. Dubbe, H.; Bühner, F.; Eigenberger, G.; Nieken, U. Hysteresis Phenomena on Platinum and Palladium-based Diesel Oxidation Catalysts (DOCs). *Emiss. Control Sci. Technol.* **2016**, *2*, 137–144. [CrossRef]
33. Shipilin, M.; Gustafson, J.; Zhang, C.; Merte, L.R.; Stierle, A.; Hejral, U.; Ruett, U.; Gutowski, O.; Skoglundh, M.; Carlsson, P.-A.; et al. Transient Structures of PdO during CO Oxidation over Pd(100). *J. Phys. Chem. C* **2015**, *119*, 15469–15476. [CrossRef]
34. Bychkov, V.Y.; Tulenin, Y.P.; Slinko, M.M.; Khudorozhkov, A.K.; Bukhtiyarov, V.I.; Sokolov, S.; Korchak, V.N. Self-oscillations during methane oxidation over Pd/Al_2O_3: Variations of Pd oxidation state and their effect on Pd catalytic activity. *Appl. Catal. A Gen.* **2016**, *522*, 40–44. [CrossRef]
35. Newton, M.A.; Belver-Coldeira, C.; Martínez-Arias, A.; Fernández-García, M. "Oxidationless" Promotion of Rapid Palladium Redispersion by Oxygen during Redox CO/(NO+O_2) Cycling. *Angew. Chem.* **2007**, *119*, 8783–8785. [CrossRef]
36. Datye, A.K.; Bravo, J.; Nelson, T.R.; Atanasova, P.; Lyubovsky, M.; Pfefferle, L. Catalyst microstructure and methane oxidation reactivity during the Pd↔PdO transformation on alumina supports. *Appl. Catal. A Gen.* **2000**, *198*, 179–196. [CrossRef]
37. Colussi, S.; Trovarelli, A.; Vesselli, E.; Baraldi, A.; Comelli, G.; Groppi, G.; Llorca, J. Structure and morphology of Pd/Al_2O_3 and Pd/CeO_2/Al_2O_3 combustion catalysts in Pd–PdO transformation hysteresis. *Appl. Catal. A Gen.* **2010**, *390*, 1–10. [CrossRef]
38. Yue, B.; Zhou, R.; Wang, Y.; Zheng, X. Study of the methane combustion and TPR/TPO properties of Pd/Ce–Zr–M/Al_2O_3 catalysts with M=Mg, Ca, Sr, Ba. *J. Mol. Catal. A Chem.* **2005**, *238*, 241–249. [CrossRef]
39. Zhuravlev, L.T. The surface chemistry of amorphous silica. Zhuravlev model. *Colloids Surf. A Physicochem. Eng. Asp.* **2000**, *173*, 1–38. [CrossRef]
40. Mueller, R.; Kammler, H.K.; Wegner, K.; Pratsinis, S.E. OH Surface Density of SiO_2 and TiO_2 by Thermogravimetric Analysis. *Langmuir* **2003**, *19*, 160–165. [CrossRef]
41. Bertoluzza, A.; Fagnano, C.; Antonietta Morelli, M.; Gottardi, V.; Guglielmi, M. Raman and infrared spectra on silica gel evolving toward glass. *J. Non-Cryst. Solids* **1982**, *48*, 117–128. [CrossRef]
42. Lamber, R.; Jaeger, N.; Schulz-Ekloff, G. Metal-support interaction in the Pd/SiO_2 system: Influence of the support pretreatment. *J. Catal.* **1990**, *123*, 285–297. [CrossRef]
43. Shewale, P.M.; Venkateswara Rao, A.; Parvathy Rao, A.; Bhagat, S.D. Synthesis of transparent silica aerogels with low density and better hydrophobicity by controlled sol–gel route and subsequent atmospheric pressure drying. *J. Sol-Gel Sci. Technol.* **2009**, *49*, 285–292. [CrossRef]
44. Li, G.; Zhu, T.; Deng, Z.; Zhang, Y.; Jiao, F.; Zheng, H. Preparation of Cu-SiO_2 composite aerogel by ambient drying and the influence of synthesizing conditions on the structure of the aerogel. *Chin. Sci. Bull.* **2011**, *56*, 685–690. [CrossRef]
45. Yuranov, I.; Moeckli, P.; Suvorova, E.; Buffat, P.; Kiwi-Minsker, L.; Renken, A. Pd/SiO_2 catalysts: Synthesis of Pd nanoparticles with the controlled size in mesoporous silicas. *J. Mol. Catal. A Chem.* **2003**, *192*, 239–251. [CrossRef]
46. Grunwaldt, J.-D.; Vegten, N.v.; Baiker, A. Insight into the structure of supported palladium catalysts during the total oxidation of methane. *Chem. Commun.* **2007**, 4635. [CrossRef]
47. Baaziz, W.; Bahri, M.; Gay, A.S.; Chaumonnot, A.; Uzio, D.; Valette, S.; Hirlimann, C.; Ersen, O. Thermal behavior of Pd@SiO_2 nanostructures in various gas environments: A combined 3D and in situ TEM approach. *Nanoscale* **2018**, *10*, 20178–20188. [CrossRef]
48. Hu, Y.; Tao, K.; Wu, C.; Zhou, C.; Yin, H.; Zhou, S. Size-Controlled Synthesis of Highly Stable and Active Pd@SiO_2 Core–Shell Nanocatalysts for Hydrogenation of Nitrobenzene. *J. Phys. Chem. C* **2013**, *117*, 8974–8982. [CrossRef]
49. Taylor, A. Practical surface analysis, 2nd edn., vol I, auger and X-ray photoelectron spectroscopy. Edited by D. Briggs & M. P. Seah, John Wiley, New York, 1990, 657 pp., price: £86.50. ISBN 0471 92081 9. *J. Chem. Technol. Biotechnol.* **2007**, *53*, 215. [CrossRef]
50. Barrett, E.P.; Joyner, L.G. Determination of Nitrogen Adsorption-Desorption Isotherms. *Anal. Chem.* **1951**, *23*, 791–792. [CrossRef]

 catalysts

MDPI

Article

Biojet Fuel Production from Waste of Palm Oil Mill Effluent through Enzymatic Hydrolysis and Decarboxylation

Papasanee Muanruksa [1,2,3], James Winterburn [2] and Pakawadee Kaewkannetra [3,4,*]

[1] Department of Biotechnology, Graduate School of Khon Kaen University, Khon Kaen 40002, Thailand; m.papasanee@kkumail.com
[2] Department of Chemical Engineering and Analytical Science (CEAS), The University of Manchester, Manchester M13 9PL, UK; James.Winterburn@manchester.ac.uk
[3] Centre for Alternative Energy Research and Development (AERD), Faculty of Engineering, Khon Kaen University, Khon Kaen 40002, Thailand
[4] Research Centre for Environmental and Hazardous Substance Management (EHSM), Faculty of Engineering, Khon Kaen University, Khon Kaen 40002, Thailand
* Correspondence: paknar@kku.ac.th

Abstract: Palm oil mill effluent (POME), wastewater discharged from the palm oil refinery industry, is classified as an environmental pollutant. In this work, a heterogeneous catalytic process for biojet fuel or green kerosene production was investigated. The enzymatic hydrolysis of POME was firstly performed in order to obtain hydrolysed POME (HPOME) rich in free fatty acid (FFA) content. The variations of the water content (30 to 50), temperature (30 to 60 °C) and agitation speed (150 to 250 rpm) were evaluated. The optimal condition for the POME hydrolysis reaction was obtained at a 50% v/v water content, 40 °C and 200 rpm. The highest FFA yield (Y_{FA}) of 90% was obtained. Subsequently, FFA in HPOME was converted into hydrocarbon fuels via a hydrocracking reaction catalysed by Pd/Al_2O_3 at 400 °C, 10 bars H_2 for 1 h under a high pressure autoclave reactor (HPAR). The refined-biofuel yield (94%) and the biojet selectivity (57.44%) were achieved. In this study, we are the first group to successfully demonstrate the POME waste valorisation towards renewable biojet fuel production based on biochemical and thermochemical routes. The process can be applied for the sustainable management of POME waste. It promises to be a high value-added product parallel to the alleviation of wastewater environmental issues.

Keywords: palm oil mill effluent; enzymatic hydrolysis; decarboxylation; green gasoline; green kerosene and green diesel

Citation: Muanruksa, P.; Winterburn, J.; Kaewkannetra, P. Biojet Fuel Production from Waste of Palm Oil Mill Effluent through Enzymatic Hydrolysis and Decarboxylation. *Catalysts* **2021**, *11*, 78. https://doi.org/10.3390/catal11010078

Received: 10 December 2020
Accepted: 5 January 2021
Published: 8 January 2021

Publisher's Note: MDPI stays neutral with regard to jurisdictional claims in published maps and institutional affiliations.

1. Introduction

Biojet fuel, classified as green kerosene derived from biomass, is generally blended with petroleum-based kerosene. The range of the jet fuel carbon number is strictly controlled to be between 8 and 16 in order to obtain desirable fuel properties and meet the strict standard specifications of the American Society for Testing and Materials (ASTM) D1655. Since the primary function of biojet fuel is to power an aircraft, the energy content and combustion quality are significant key factors. Other important properties are stability, lubricity, fluidity, volatility, noncorrosivity and cleanliness [1]. Typically, there are four routes to produce green kerosene from bio-feedstocks: (i) Hydro-processing of fatty acids and natural triglycerides; (ii) Hydrothermal liquefaction of algal biomass and pyrolysis of lignocellulosic materials; (iii) Gas-to-jet fuel platform: the gasification reaction of biomass in order to obtain syngas, which is subsequently converted into hydrocarbon fuel via the Fischer–Tropsch reaction; and (iv) Alcohol-to-jet platform: bio-based alcohols are sequentially dehydrated, oligomerized and hydrogenated in order to generate biojet fuel. Currently, only the hydroprocessing of oil/fat is selected for renewable jet fuel production at the industrial scale [2]. The key factors to decide on suitable raw materials for this process are availability, potential yield and especially the price. It is quite cheap, classified as waste, and it also plays an important role in the total production cost. Thus, palm oil can be considered as feedstock for bio-aviation fuel production because it is the least expensive vegetable oil [3]. However, the major challenge of first-generation feedstock is food competition.

In palm oil milling plants, large quantities of palm oil mill effluent (POME) are produced during the milling process. According to its high pH values, biochemical oxygen demand (BOD) and chemical oxygen demand (COD), it was classified as an industrial waste. It was discharged as wastewater containing high BOD (>25,000 mg·L^{-1}), which accounted for 100 times more pollution than municipal wastes [4,5]. Nowadays, the biological process named anaerobic digestion is the most common method for POME treatment when creating value-added products, including fertilizer and biogas. In addition, applications of biogas capture and membrane separation were reported in an integrated reactor for the production of biogas from POME, leading to a higher energy yield and lower greenhouse gas (GHG) emissions [6,7].

Considering new applications of POME, the production of hydrocarbon fuel from waste triglyceride is an interesting possible route for delivering clean fuel and as a response to environmental issues. The use of free fatty acid (FFA) as a feedstock for biojet fuel production has been reported as a cost-effective process due to a low hydrogen pressure, mild cracking reaction, short retention time and high yield, as compared to oil/fats [8–12]. In addition, the noble metals palladium (Pd) and platinum (Pt) were also proven to be high potential catalysts in decarboxylation reactions. They showed a high catalytic efficiency for biofuel production from triglycerides (TG) and free fatty acids (FFA) due to their high hydrogenation ability [13–15]. Accordingly, consecutive processes for hydrocarbon fuel have been developed, and triglycerides were hydrolysed to FFA under a high pressure and temperature, followed by the decarboxylation of FFA [16,17].

In this study, a new strategy to decrease an environmental pollutant of POME was presented. A combination of biochemical and thermochemical processes was firstly applied to generate renewable biojet fuel from POME. The enzymatic hydrolysis of POME was carried out in order to obtain FFA. After that, it was converted to hydrocarbon fuels via a hydrocracking reaction under mild conditions.

2. Results and Discussion

2.1. Influence of Temperature on Hydrolysis Reaction

The results showed that the hydrolysis degrees of POME increased considerably from 30 °C to 40 °C and presented a downward trend when the reaction temperature was higher than 40 °C (see Figure 1A). This can be explained by the fact that the viscosity of the substrate can be reduced by a higher temperature, resulting in a high catalytic rate of lipase. As a result, a higher FFA yield was observed in the mixture reaction. However, an excessively high temperature caused the enzyme structure to be denatured and to eventually deactivate. Thus, by increasing the reaction temperature to more than 40 °C, the hydrolysis degree of the three raw materials decreased slightly, as shown in Figure 1. Meanwhile, the highest hydrolysis degree of POME (89.51 ± 0.04%) was obtained at 40 °C. Therefore, the reaction temperature of 40 °C was selected to investigate the effect of the water content on the hydrolysis reaction in the following experiment. Furthermore, the results obtained were in agreement with the previous work [18], which reported an enzymatic POME hydrolysis. They found that the maximum hydrolytic rate of lipase was achieved at 40 °C.

2.2. Influence of Water Content on Hydrolysis Reaction

The results illustrated that the hydrolysis degree of POME kept continually rising as the water content increased from 30 to 50% v/v (see Figure 1B). This can be explained by the fact that the enzymatic hydrolysis reaction is a reversible reaction. It is controlled by the amount of water in the mixture solution. Normally, excess water affects the hydrolysis reaction in a positive way; however, a high water content (60% v/v) resulted in the hydrolysis degree decreasing dramatically [19]. When the water in the mixture solution was 50% v/v, the maximum FFA yield of 89.48 ± 0.05% was obtained. Therefore, the water content was fixed at 50% v/v for the following experiments. The results obtained were in agreement with the previous work [20], which reported that as the molar ratio of oil to water increased from 1:1 to 1:4, the fatty acid yield also increased continuously. However, the fatty acid yield decreased as the ratio of oil to water increased up to 1:5.

2.3. Influence of Agitation Rate on Hydrolysis Reaction

The results observed showed that the agitation rate affected the hydrolysis reaction in a positive way, as it was clearly seen that the hydrolysis degree of POME significantly increased between the agitation rates of 100 rpm and 200 rpm (see Figure 1C). This might due to a higher agitation rate creating more oil-water interface, where the catalysis of the hydrolysis reaction via lipase occurred; this led to a high FFA yield. However, increasing the agitation rate up to 250 rpm caused a lower hydrolysis degree of POME. The agitation can reduce the droplet size and increase the interface area between oil and water, leading to a higher activity of lipase. However, the enzyme activity can be gradually decreased under a high agitation rate due to the shear stress [21]. The highest FFA yield (90 ± 0.04%) was observed under an agitation rate of 200 rpm. Therefore, this rate was chosen as the optimal rate for the enzymatic hydrolysis of POME.

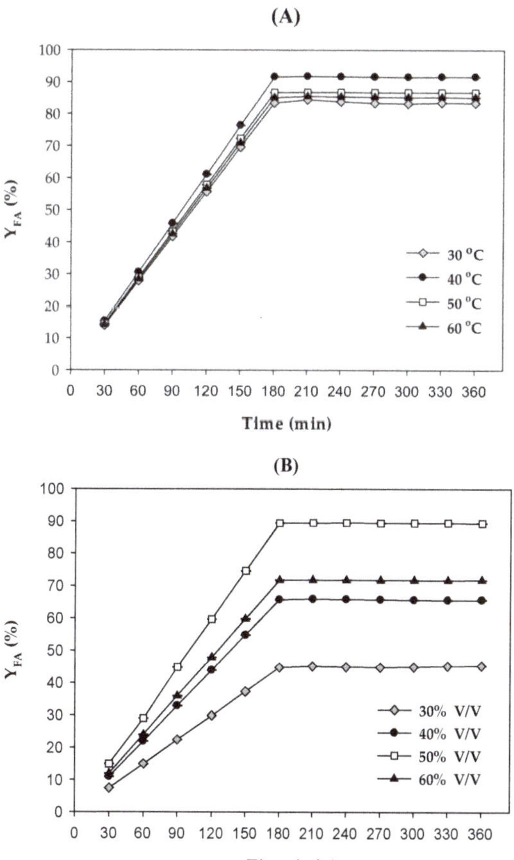

Figure 1. *Cont.*

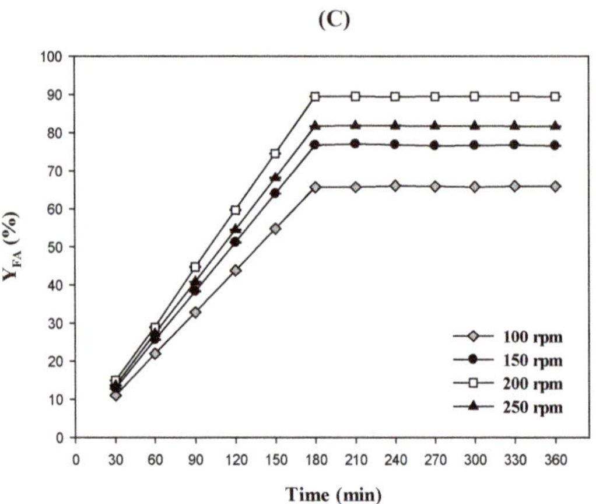

Figure 1. Enzymatic hydrolysis of POME, (**A**) the effect of the temperature, (**B**) the effect of the water content, and (**C**) the effect of the agitation rate (standard deviation (SD) $\leq$ 0.05).

2.4. Fatty Acid Composition of Hydolysed Palm Oil Mill Effluent (HPOME)

In Table 1, the major fatty acid compositions of HPOME were palmitic acid (50%), oleic acid (34%), steric acid (13%) and others. An equal proportion between saturated fatty acids, palmitic acid (C16:0) and stearic acid (C18:0), and unsaturated fatty acid of oleic acid (C18:1) was also found.

Table 1. Fatty acid composition of hydrolysed palm oil mill effluent (HPOME).

Fatty Acid	Formula	Molecular Weight	Structure	% wt.
Capric acid	$C_{10}H_{20}O_2$	172.26	C10:0	0.07 ± 0.06
Lauric acid	$C_{12}H_{24}O_2$	200.32	C12:0	0.09 ± 0.03
Myristic acid	$C_{14}H_{28}O_2$	228.37	C14:0	1.78 ± 0.06
Palmitic acid	$C_{16}H_{32}O_2$	256.42	C16:0	50.03 ± 0.04
Stearic acid	$C_{18}H_{36}O_2$	284.48	C18:0	13.04 ± 0.03
Oleic acid	$C_{18}H_{34}O_2$	282.47	C18:1	34.02 ± 0.05

2.5. Production of Hydrocarbon Fuels via Hydrocracking Reaction

Following the hydrocracking reaction of fatty acid in HPOME, a crude bio-fuel yield of 96 $\pm$ 0.04% was obtained. Based on the volume fraction, crude biofuel could be classified into three kinds: green kerosene (54 $\pm$ 0.03%), green diesel (30 $\pm$ 0.04%) and green gasoline (10 $\pm$ 0.04%), respectively. In addition, their selectivity was also reached at 57.44 $\pm$ 0.06%, 31.91 $\pm$ 0.03% and 10.64 $\pm$ 0.05%, as illustrated in Figure 2.

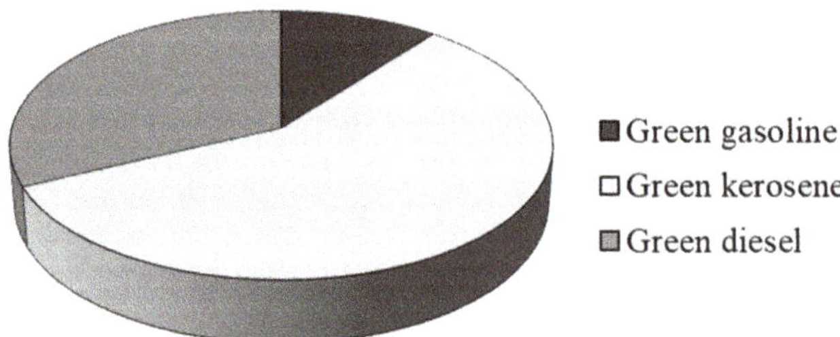

Figure 2. Volume fraction of green gasoline/kerosene/diesel obtained from the hydrocracking reaction of HPOME (SD less than 1.00).

2.6. Characterisation of Refined Biofuels

The characteristic of refined biofuels in terms of the carbon number was analysed by Gas chromatography- Mass spectroscopy (GC–MS). According to the result in Figure 3, the carbon atom distribution of refined biofuels was mainly in the ranges of kerosene (C9–C15) and diesel (C15–C18). This was due to three major fatty acid compositions of HPOME: palmitic acid (C16:0), stearic acid (C18:0) and oleic acid (C18:1). One carbon atom of the carboxyl group (–COOH) was eliminated during decarboxylation [22]. It should be noted that the refined biofuels were produced and measured in triplicate for reproducibility.

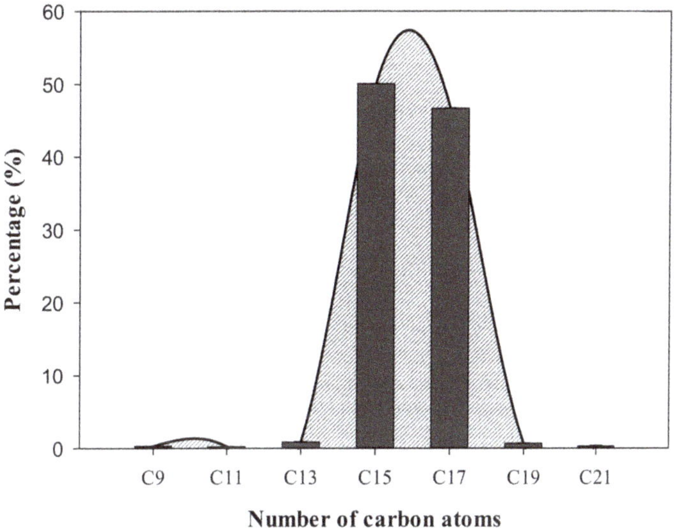

Figure 3. Carbon distribution of refined biofuel obtained from the hydrocracking reaction of HPOME (SD less than 1.00).

2.7. Comparative Green Kerosene Yields Obtained from Hydrocracking Reaction of Different Waste Oils

Based on the results in Table 2, enzymatic hydrolysis coupled with the hydrocracking reaction of POME could serve the highest green kerosene yield when compared to the single hydrocracking reaction of other waste oils. The production of green kerosene from waste cooking oil [23] was operated at the lowest temperature (380 °C), and a high green kerosene yield of 40.50% was also observed. However, this took a longer reaction time than in other works. The results clearly indicated that HPOME could be efficiently transformed into hydrocarbon fuel by using only a 1% catalyst loading, low H_2 pressure (10 bar), moderate temperature (400 °C) and short reaction time (1 h). This was due to the decarboxylation of fatty acids having a higher efficiency than triglycerides [17]. Therefore, the heterogeneous catalytic process could be considered as a high potential process for renewable jet fuel production in the future.

Table 2. Comparative green kerosene production from waste oils through the hydrocracking reaction in batch mode.

Feedstocks	Catalysts	Operating Conditions	Green Kerosene Yield (%)	References
HPOME	0.3% Pd/Al_2O_3	1% catalyst loading, 10 bar H_2, 400 °C and 1 h	54.00	This work
Waste cooking oil	10% Ni/Meso-Y	5% catalyst loading, 30 bar H_2, 380 °C and 8 h	40.50	[23]
Waste lubricant oil	5% Fe/Al_2O_3	4% catalyst loading, 6.8 bar H_2, 450 °C and 1.25 h	24.16	[24]
Waste lubricant oil	0.5% Fe/SiO_2-Al_2O_3	4% catalyst loading, 6.8 bar H_2, 430 °C and 1 h	11.41	[24]

3. Materials and Methods

3.1. Materials

Samples of low-cost raw material of palm oil mill effluent (POME) were collected from the waste water lagoon of E-san palm oil industry, Sakonakhorn, Thailand. A chemical catalyst, Pd/Al_2O_3, was purchased from Sigma (Missouri, MO, USA). An immobilised lipase of *Rhizopus oryzae* on sodium alginate (SAL beads) was prepared by a simple entrapment technique in a $CaCl_2$ solution. First, sodium alginate (2 g) was added into 100 mL of phosphate buffer (pH 7.0) solution. Then, it was heated at 40 °C until a homogeneous solution was obtained. After that, it was cooled down to room temperature. Subsequently, lipase (2 g) was mixed with alginate solution (100 mL). The mixture solution was dropped into 0.1 M $CaCl_2$ by a sterile syringe and stored at 4 °C for 12 h to form an alginate bead (Diameter size of 3 mm.). Finally, the immobilised bead was washed three times with distilled water and was kept in distilled water at 4 °C before being used as a biocatalyst in the hydrolysis reaction. The hydrolytic activity of the SAL beads was 360 U/mg. This was determined by the spectrophotometric method using p-nitrophenyl laurate (p-NPL) as the substrate [25].

3.2. Pretreatment of POME via Enzymatic Hydrolysis

The contaminants contained in POME, such as trunks barks and leaves, were separated by gravity method using a separating funnel [20]. Consequently, SAL beads (2 g), POME (100 mL) and distilled water (50% v/v) were loaded into a 250-mL Erlenmeyer flask. The hydrolysis reaction was conducted in an incubator shaker at 40 °C, 200 rpm for 4 h. The samples were further analysed to determine the FFA content, as described by a previous study [26]. The fatty acid yield (Y_{FA}) was calculated as shown in Equation (1). The fatty profile of hydrolysed POME (HPOME) was analysed by gas chromatography–spectrometry (GC–MS) following the standard method EN 1403. The HPOME sample (50 mg) and a 0.5 M methanolic sodium hydroxide solution (5 mL) were added into a reflux flask connected to the condenser. The reaction mixture in the reflux flask was heated to 140–160 °C for 5 min for a saponification reaction. Subsequently, a boron trifluoride-methanol solution (5 mL) was added and heated to 140–160 °C for 5 min for a methylation reaction. The mixture solution was cooled down to room temperature, followed by adding hexane (5 mL) and saturated sodium chloride solution (10 mL) in order to obtain fatty acid methyl ester (FAME). The bi-layer of the mixture reaction was observed. The FAME in the hexanic phase (upper layer) was dried by adding anhydrous sodium sulfate (1 g) and was filtrated through Whatman paper No. 1. Last, the samples were dissolved in heptane and filtrated through a microfilter (0.45 μm) before being injected into GC–MS equipped with a FID detector in order to identify the fatty acid composition, as described in previous studies [27]. The operating condition for the fatty acids' profile analysis was explained by Muanruksa et al. [28]. First, a 250-mg sample was filled into a 10-mL vial, followed by the addition of 5 mL of internal standard (methyl haptadecanoate solution, 10 mg/mL). Subsequently, the sample was analysed by gas chromatography–mass spectroscopy (GC–MS) (GC-2010, Shimadzu, Tokyo Japan) equipped with a 30-m long and 0.25-mm diameter capillary column lined with a 0.25-μm(Rtx-5 ms, Rextex). Samples were injected in a split/column flow ratio of 24:1. Helium was used as the carrier gas at a flow rate of 1 mL/min. The injection temperature was 250 °C, and the column oven's temperature was 250 °C (programmed to start at 120 °C, held at this temperature for 5 min and heated at a rate of 3 °C/min to 250 °C).

$$YFA\ (\%) = \frac{FA2\ - FA1}{FA2} \times 100\% \tag{1}$$

when FA1 is the FFA content of the oil sample before the hydrolysis reaction, and FA2 is the FFA content of the oil sample after the hydrolysis reaction.

3.3. Decarboxylation of Fatty Acid in HPOME

The hydrocarbon fuel production from hydrolysed POME (HPOME) was carried out in a high pressure batch reactor (HPAR) (See Figure 4). Raw materials of HPOME (100 mL) and Pd/Al_2O_3 (1 g) were loaded in a chamber of the reactor. Then, hydrogen was flushed three times to remove the oxygen in the reaction. Subsequently, it was fed to the reactor until a hydrogen pressure of 10 bars was reached at the initial stage, and the reaction temperature was set at 400 °C and kept constant for 1 h. Finally, the crude biofuel was refined by using a fractional distillation technique (ASTM D86).

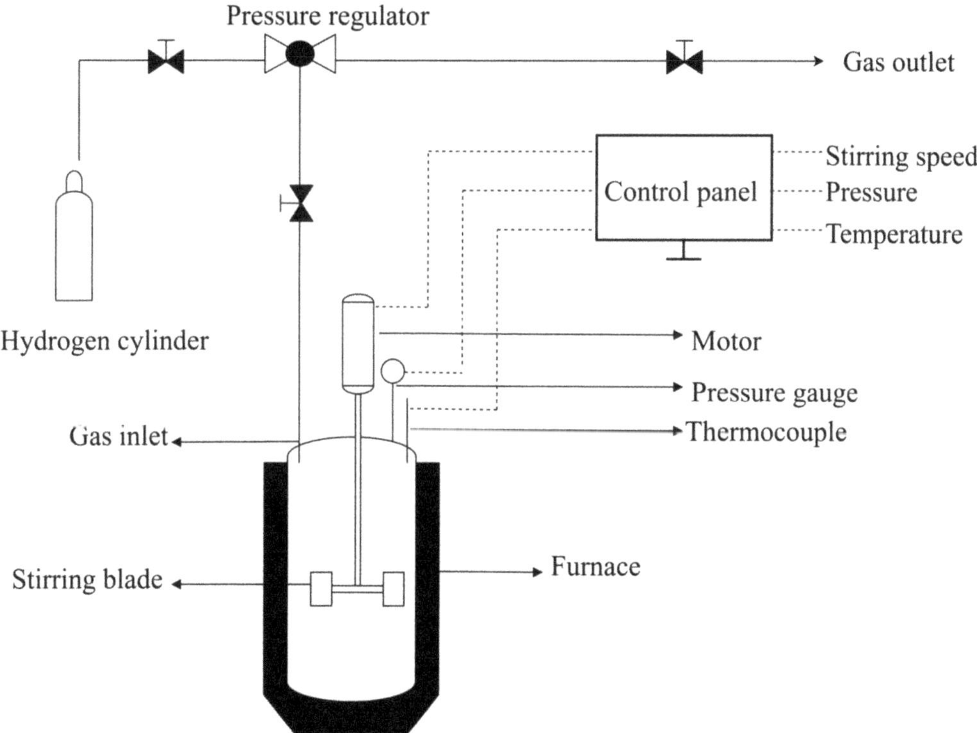

Figure 4. Schematic diagram of high pressure autoclave reactor (HPAR).

3.4. Fractional Distillation

A commercially manual distillation apparatus (Koehler model K45200, New York, NY, USA) was used to conduct fractional distillation in this research work. The apparatus components met the standard of the ASTM D86 specifications, including a 125-mL sidearm distillation flask (borosilicate glass), a 100-mL graduated cylinder with 1.0-mL graduation intervals and a mercury thermometer. The graduate cylinder was centered under the condenser tube exit to allow condensate drops to fall to the bottom of the cylinder. The HPOME (100 mL) was taken to be used as a liquid sample in the fractional distillation at 1 atm. The refined-biofuel products could be classified into three types based on their boiling points, including gasoline (50–150 °C), kerosene (150–280 °C) and diesel (280–360 °C), respectively. The percentages of crude-biofuel yield, refined-biofuel yield, green diesel selectivity, green kerosene selectivity and green gasoline selectivity were determined as described in Equations (2)–(6).

The liquid products were analysed by gas chromatography–mass spectroscopy (GC–MS) using a flame ionization detector (GC-FID-QP2010 Shimadzu, Tokyo, Japan) equipped with HT5 capillary columns (length of 15 m, diameter of 0.25 mm, film of 0.1 µm), as described by a previous study [29]. The liquid samples (40 mg) were diluted via 4 mL hexane (high performance liquid chromatography (HPLC) grade). Then, 0.5 µL of diluted sample was injected into the column. Helium was used as the carrier gas, and the flow rate was set at 1.5 mL/min. The oven temperature was first operated at 50 °C, held for 1 min, and then increased to 220 °C at the rate of 9 °C/min, followed by an increase of 20 °C/min to 350 °C, before being increased to 380 °C at the rate of 10 °C/min and held for 5 min. The n-alkanes and n-alkenes (C8–C20) were used as the chemical standard for the calibration curves to identify the composition of the liquid product. The internal standard was 1-bromohexane [27].

$$\%\ Crude - biofuel\ yield = \frac{Total\ liquid\ product\ (mL)}{HPOME\ (mL)} \times 100\% \tag{2}$$

$$\%Refined - biofuel\ yield = \frac{Total\ distilled\ biofuels\ (mL)}{Crude\ biofuel\ (mL)} \times 100\% \tag{3}$$

$$\%Green\ diesel\ selectivity = \frac{Distilled\ green\ diesel\ (mL)}{Total\ refined - biofuel\ (mL)} \times 100\% \tag{4}$$

$$\%\ Green\ kerosene\ selectivity = \frac{Distilled\ green\ kerosene\ (mL)}{Total\ refined - biofuel\ (mL)} \times 100\% \tag{5}$$

$$\%Green\ gasoline\ selectivity = \frac{Distilled\ green\ gasoline\ (mL)}{Total\ refined - biofuel\ (mL)} \times 100\% \tag{6}$$

4. Conclusions

A combination of biochemical and thermochemical processes for the production of renewable biojet fuel from POME has successfully been performed in this study. The process provided a high efficiency in terms of a high green kerosene yield and a high selectivity of green kerosene. Additionally, the hydrocracking reaction could be carried out entirely using a low amount of catalyst, low H_2 pressure, moderate temperature and short reaction time. It was apparently indicated that the pretreatment of POME via the hydrolysis reaction was catalysed by immobilised lipase, resulting in positive effects on the hydrocracking reaction. Therefore, the POME showed a high potential use as a second-generation feedstock for producing biojet fuel. The heterogeneous catalytic process can be classified as a possible pathway for figuring out environmental issues related to POME, as well as for the development of a sustainable renewable energy production.

Author Contributions: This article was performed as collaborative research between P.M., J.W. and P.K.; the work included: conceptualization, methodology, resources, formal analysis, data curation, visualisation and writing—original draft preparation, P.M. and J.W. considered the software, validation and investigation, while P.K. performed the supervision, project administration, funding acquisition, review and corrections. All authors have read and agreed to the published version of the manuscript.

Funding: This research was co-funded by Royal Golden Jubilee (RGJ) PhD Programme (Contract no. PHD/0018/2557) for P. Muanruksa and Newton Fund Institutional Links 2019/2020 as well as EHSM for P. Kaewkannetra.

Institutional Review Board Statement: Not applicable.

Informed Consent Statement: Not applicable.

Data Availability Statement: Data sharing is not applicable to this article.

Acknowledgments: All authors sincerely thank all sponsors: The Royal Golden Jubilee (RGJ) PhD Programme (Contract no. PHD/0018/2557), Bangkok, Thailand, for the main research support; Centre for Alternative Energy Research and Development (AERD) and Research Centre for Environmental and Hazardous Substance Management (EHSM), Faculty of Engineering, Khon Kaen University, Khon Kaen, Thailand; and Newton Fund Institutional Links 2019/2020, British Council, London, UK, for the collaborative research, travel bursary and PhD researcher exchange.

Conflicts of Interest: The authors declare no conflict of interest.

References

1. Yang, J.; Xin, Z.; He, Q. (Sophia); Corscadden, K.; Niu, H. An overview on performance characteristics of bio-jet fuels. *Fuel* **2019**, *237*, 916–936. [CrossRef]
2. Roth, A.; Riegel, F.; Batteiger, V. Potentials of Biomass and Renewable Energy: The Question of Sustainable Availability. *Biokerosene Status Prospect* **2017**, *1*, 95–122. [CrossRef]
3. Dujjanutat, P.; Kaewkannetra, P. Production of bio-hydrogenated kerosene by catalytic hydrocracking from refined bleached deodorised palm/ palm kernel oils. *Renew. Energy* **2020**, *147*, 464–472. [CrossRef]
4. Madaki, Y.S.; Seng, L. Palm Oil Mill Effluent (Pome) From Malaysia Palm Oil Mills: Waste or Resource. *Int. J. Sci. Environ. Technol.* **2013**, *2*, 1138–1155.
5. Kamyab, H.; Chelliapan, S.; Din, M.F.M.; Rezania, S.; Khademi, T.; Kumar, A. Palm Oil Mill Effluent as an Environmental Pollutant. *Palm Oil* **2018**, *13*, 13–28. [CrossRef]
6. Abdurahman, N.H.; Azhari, N.H. An integrated UMAS for POME treatment. *J. Water Reuse Desalin.* **2018**, *8*, 68–75. [CrossRef]
7. Harsono, S.S.; Grundmann, P.; Soebronto, S. Anaerobic treatment of palm oil mill effluents: Potential contribution to net energy yield and reduction of greenhouse gas emissions from biodiesel production. *J. Clean. Prod.* **2014**, *64*, 619–627. [CrossRef]
8. Fu, J.; Lu, X.; Savage, P.E. Catalytic hydrothermal deoxygenation of palmitic acid. *Energy Environ. Sci.* **2010**, *3*, 311–317. [CrossRef]
9. Fu, J.; Lu, X.; Savage, P.E. Hydrothermal Decarboxylation and Hydrogenation of Fatty Acids over Pt/C. *ChemSusChem* **2011**, *4*, 481–486. [CrossRef]
10. Fu, J.; Shi, F.; Thompson, L.T.; Lu, X.; Savage, P.E. Activated Carbons for Hydrothermal Decarboxylation of Fatty Acids. *ACS Catal.* **2011**, *1*, 227–231. [CrossRef]
11. Hossain, M.Z.; Chowdhury, M.B.I.; Jhawar, A.K.; Xu, W.Z.; Charpentier, P.A. Continuous low pressure decarboxylation of fatty acids to fuel-range hydrocarbons with in situ hydrogen production. *Fuel* **2018**, *212*, 470–478. [CrossRef]
12. Yang, C.; Nie, R.; Fu, J.; Hou, Z.; Lu, X. Production of aviation fuel via catalytic hydrothermal decarboxylation of fatty acids in microalgae oil. *Bioresour. Technol.* **2013**, *146*, 569–573. [CrossRef] [PubMed]
13. Lestari, S.; Mäki-Arvela, P.; Simakova, I.; Beltramini, J.; Lu, G.Q.M.; Murzin, D. Catalytic Deoxygenation of Stearic Acid and Palmitic Acid in Semibatch Mode. *Catal. Lett.* **2009**, *130*, 48–51. [CrossRef]
14. Rozmysłowicz, B.; Mäki-Arvela, P.; Lestari, S.; Simakova, O.A.; Eränen, K.; Simakov, A.; Murzin, D.; Salmi, T. Catalytic Deoxygenation of Tall Oil Fatty Acids Over a Palladium-Mesoporous Carbon Catalyst: A New Source of Biofuels. *Top. Catal.* **2010**, *53*, 1274–1277. [CrossRef]
15. Simakova, I.; Simakova, O.; Mäki-Arvela, P.; Murzin, D. Decarboxylation of fatty acids over Pd supported on mesoporous carbon. *Catal. Today* **2010**, *150*, 28–31. [CrossRef]
16. Sugami, Y.; Minami, E.; Saka, S. Renewable diesel production from rapeseed oil with hydrothermal hydrogenation and subsequent decarboxylation. *Fuel* **2016**, *166*, 376–381. [CrossRef]
17. Sugami, Y.; Minami, E.; Saka, S. Hydrocarbon production from coconut oil by hydrolysis coupled with hydrogenation and subsequent decarboxylation. *Fuel* **2017**, *197*, 272–276. [CrossRef]
18. Syaima, M.T.S.; Ong, K.H.; Mohd Noor, I.; Zamratul, M.I.M.; Brahim, S.A.; Hafizul, M.M. The synthesis of bio-lubricant based oil by hydrolysis and non-catalytic of palm oil mill effluent (POME) using lipase. *Renew. Sustain. Energy Rev.* **2015**, *44*, 669–675. [CrossRef]
19. You, Q.; Yin, X.; Zhao, Y.; Zhang, Y. Biodiesel production from jatropha oil catalyzed by immobilized Burkholderia cepacia lipase on modified attapulgite. *Bioresour. Technol.* **2013**, *148*, 202–207. [CrossRef]
20. Zenevicz, M.C.P.; Jacques, A.; Furigo, A.F.; Oliveira, J.V.; De Oliveira, D. Enzymatic hydrolysis of soybean and waste cooking oils under ultrasound system. *Ind. Crops Prod.* **2016**, *80*, 235–241. [CrossRef]

21. Salimon, J.; Salih, N.; Yousif, E. Biolubricants: Raw materials, chemical modifications and environmental benefits. *Eur. J. Lipid Sci. Technol.* **2010**, *112*, 519–530. [CrossRef]
22. Srihanun, N.; Dujjanutat, P.; Muanruksa, P.; Kaewkannetra, P. Biofuels of Green Diesel–Kerosene–Gasoline Production from Palm Oil: Effect of Palladium Cooperated with Second Metal on Hydrocracking Reaction. *Catalysts* **2020**, *10*, 241. [CrossRef]
23. Li, T.; Cheng, J.; Huang, R.; Zhou, J.; Cen, K. Conversion of waste cooking oil to jet biofuel with nickel-based mesoporous zeolite Y catalyst. *Bioresour. Technol.* **2015**, *197*, 289–294. [CrossRef] [PubMed]
24. Makvisai, W.; Promdee, K.; Tanatavikorn, H.; Vitidsant, T. Catalytic cracking of used lubricating oil over Fe/Al2O3 and Fe/SiO2-Al2O3. *Pet. Coal* **2016**, *58*, 83–94.
25. Kabbashi, N.A.; Mohammed, N.I.; Alam, Z.; Mirghani, M.E.S. Hydrolysis of Jatropha curcas oil for biodiesel synthesis using immobilized Candida cylindracea lipase. *J. Mol. Catal. B Enzym.* **2015**, *116*, 95–100. [CrossRef]
26. Muanruksa, P.; Kaewkannetra, P. Combination of fatty acids extraction and enzymatic esterification for biodiesel production using sludge palm oil as a low-cost substrate. *Renew. Energy* **2020**, *146*, 901–906. [CrossRef]
27. Dujjanutat, P.; Muanruksa, P.; Kaewkannetra, P. Techniques for analysing and monitoring during continuous bio-hydrogenation of kerosene from palm oils. *MethodsX* **2020**, *7*, 101128. [CrossRef]
28. Muanruksa, P.; Winterburn, J.; Kaewkannetra, P. A novel process for biodiesel production from sludge palm oil. *MethodsX* **2019**, *6*, 2838–2844. [CrossRef]
29. Chen, R.-X.; Wang, W.-C. The production of renewable aviation fuel from waste cooking oil. Part I: Bio-alkane conversion through hydro-processing of oil. *Renew. Energy* **2019**, *135*, 819–835. [CrossRef]

MDPI

Article

Morphology-Controlled Synthesis of ZnO Nanostructures for Caffeine Degradation and *Escherichia coli* Inactivation in Water

Shaila Thakur [1], Sudarsan Neogi [1] and Ajay K. Ray [2,*]

[1] Department of Chemical Engineering, Indian Institute of Technology, Kharagpur 721302, India; shaila.thakur@hotmail.com (S.T.); sneogi@che.iitkgp.ernet.in (S.N.)
[2] Department of Chemical and Biochemical Engineering, Western University, London, ON N6A 3K7, Canada
* Correspondence: aray@eng.uwo.ca

Abstract: Photocatalytic and antibacterial activity of nanoparticles are strongly governed by their morphology. By varying the type of solvent used, one can obtain different shapes of ZnO nanoparticles and tune the amount of reactive oxygen species (ROS) and metal ion (Zn^{2+}) generation, which in turn dictates their activity. ZnO nanostructures were fabricated via facile wet chemical method by varying the type of solvents. Solar light assisted photocatalytic degradation of caffeine and antibacterial activity against *E. coli* were examined in presence ZnO nanostructures. In addition to an elaborate nanoparticle characterization, adsorption and kinetic experiments were performed to determine the ability of nanostructures to degrade caffeine. Zone of inhibition, time kill assay and electron microscopy imaging were carried out to assess the antibacterial activity. Experimental findings indicate that ZnO nanospheres generated maximum ROS and Zn^{2+} ions followed by ZnO nanopetals and ZnO nanorods. As a result, ZnO nanospheres exhibited highest degradation of caffeine as well as killing of *E. coli*. While ROS is mainly responsible for the photocatalytic activity of nanostructures, their antibacterial activity is mostly due to the combination of ROS, metal ion, physical attrition and cell internalization.

Keywords: zinc oxide; nanosphere; nanorod; nanopetal; photocatalytic; antibacterial; caffeine; reactive oxygen species (ROS); degradation; pathogen

Citation: Thakur, S.; Neogi, S.; Ray, A.K. Morphology-Controlled Synthesis of ZnO Nanostructures for Caffeine Degradation and *Escherichia coli* Inactivation in Water. *Catalysts* **2021**, *11*, 63. https://doi.org/10.3390/catal11010063

Received: 9 December 2020
Accepted: 31 December 2020
Published: 5 January 2021

1. Introduction

Caffeine is one of the most abundant xenobiotics that causes water pollution due to its high daily consumption across the globe [1]. Apart from food industry, caffeine is also extensively used in the pharmaceutical industry [2,3]. Another pollutant found in wastewater effluents that causes serious health concerns are pathogenic microorganisms. *Escherichia coli* (*E. coli*) is a common pathogen found in aquatic environment and is known to cause several enteric diseases even at low concentrations [4]. Hence, it is imperative to treat these detrimental effluents before they can be discharged to water bodies. The conventional techniques employed by researchers for wastewater treatment are easy to handle and in most cases reusable [5]. However, most of the processes such as activated sludge-based technique, ion-exchange and coagulation have low treatment efficiencies and they are not cost effective in the long run. Even the advanced techniques such as membrane filtration faces a serious drawback of fouling that increases energy consumption and lowers the separation efficiency. To overcome these obstacles, nanotechnology offers a versatile and promising solution for the degradation of organic matter and elimination of microbes from wastewater in a cost-effective way.

A variety of nanomaterials have been investigated as photocatalysts for treating emerging contaminants from wastewater [6,7]. ZnO was selected for this study because of its low cost, simple synthesis technique and it can safely be used as antibacterial agent in food products [8–11]. It has been reported that ZnO nanoparticles can degrade a variety of contaminants and are effective against a wide range of bacteria but ZnO morphology largely governs its photocatalytic and antibacterial properties [12–16]. A comparison of how different shape of ZnO nanoparticles affect its photocatalytic and antibacterial activity (against *E. coli*) is shown in Table 1. Despite the fact that morphology governs the photocatalytic activity of ZnO, there has been limited research to study the shape dependent photocatalytic and antibacterial activity of nanomaterials. A fundamental understanding of difference in activities arising from different nanoparticle morphologies and their mechanism of action is still unknown. Moreover, most of the previous studies have utilized UV irradiated nanoparticles for the photocatalytic degradation of caffeine [17–19].

Table 1. Comparison of photocatalytic and antibacterial activity of different shapes of ZnO nanoparticles.

	Particle Size (nm)	Model Pollutant	Photocatalytic Activity (% Degradation)	Maximum %Reduction in *E. coli* Growth	Ref
ZnO sphere	9.6–25.5	Methylene blue (MB)	82.1% at 180 min UV exposure	69.2% at 100 μg/mL	[20]
	133.7–260.2	MB	18% and 29% after 1 h UV exposure (1 g/L)	30 and 35% after 1 h exposure (1 g/L)	[21]
	53.99	MB	95.45% after 180 min	Not reported	[22]
	4.35	4-nitrophenol	78% in 100 min	85% in 5 h at 100 μg/mL	[13]
	65.00	Acid Orange 74	80% after 80 min	99.93% at 20 min at 20 ppm NP	[23]
ZnO petals	214.38 × 178.22	MB	96.52 after 180 min	Not reported	[22]
	45.0	Acid Orange 74	90% after 80 min	99.97% after 20 min at 20 ppm NP	[23]
	(1.41–1.8) × (0.33–0.4)	methylene blue and Congo red	81% for CR and 67% for MB after 80 min	90% for 150 mg/mL after 6 h	[24]
ZnO rod	155.0	MB	87.12 after 180 min	Not reported	[22]
	20.0	Orange II	100% after 150 minsolar irradiation with 1 mg/mL	100% in >3 h (1 mg/mL)	[25]
	76.0	Acid Orange 74	70% after 80 min	99.8% after 20 min t 20 ppm NP	[23]

This study aims at determining the photocatalytic dependence of different nanostructures using solar light which is relatively inexpensive compared to the UV light. There are studies that individually report different nanostructures for their photocatalytic and antibacterial activity, but to the best of our knowledge, this is the first report to study the comparative solvent induced morphology-dependent photocatalytic and antibacterial mechanism of action of ZnO nanostructures. The goals of this study are: (i) To investigate the adsorption of caffeine on different morphology of nanoparticles (in dark) (ii) To evaluate the effect of initial caffeine concentration, amount of nanostructures and solar light intensity on the rate of caffeine degradation and estimate their kinetic rate constants, and (iii) study the antibacterial activity of ZnO nanostructures on *E. coli*.

2. Results and Discussion

2.1. Nanoparticle Characterization

2.1.1. Electron Microscopy

From the SEM images (Figure 1a–c), ZnO appears as well-defined nanospheres, nanorods and nanopetals under the influence of different solvents. PEG400 gives rise to spherical morphology; for water, we obtain petals attached to each other and for toluene, we obtain nanorod like structure. The morphological parameters are presented in Table 2. From the TEM images (Figure 1d–f), it can be seen that nanospherical ZnO are the smallest of all structures with an average diameter of 10.18 nm. Nanorods exhibited a mean width of 157 nm and a length of 1.43 μm while nanopetals have an average thickness of 31.85 nm. The EDX analysis of nanoparticles is shown in the Supplementary Information (Table S1). It can be seen that the elemental composition does not change significantly with the change in ZnO morphology.

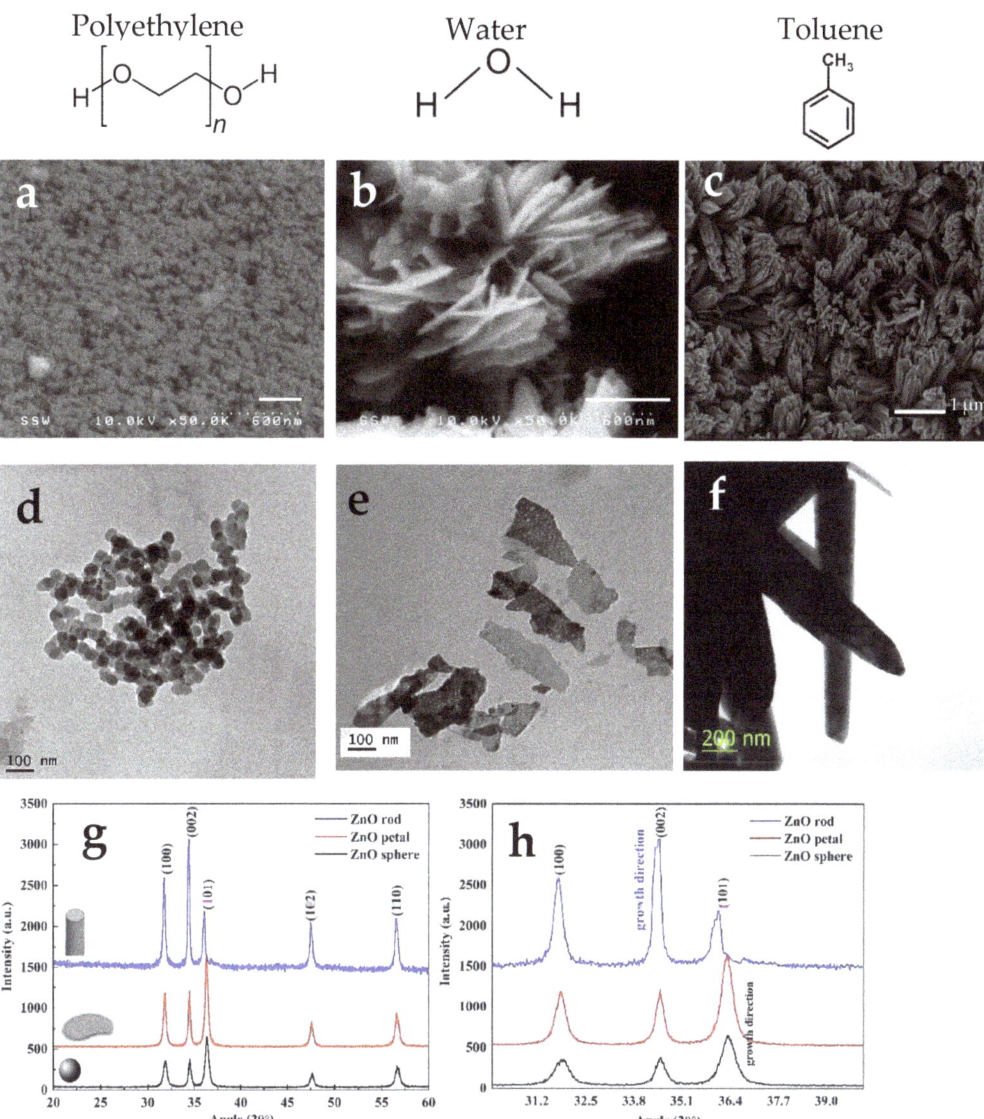

Figure 1. SEM (**a**–**c**) and TEM (**d**–**f**) images of ZnO nanomaterials synthesized in different solvents (**a**,**d**) PEG400 (**b**,**e**) water (**c**,**f**) toluene. (**g**) XRD spectra of different ZnO shapes (**h**) XRD spectra in the 2θ range 30–40° showing the crystal growth direction.

Table 2. Characterization of ZnO nanostructures.

Sample	Average Particle Diameter (nm) (TEM)	BET Specific Surface Area (m^2/g)	Total Pore Volume (cc/g)	Average Pore Diameter (nm)	Point of Zero Charge
ZnO sphere	10.18	92.22	0.15	6.64	4.90
ZnO petal	31.85 (petal thickness)	12.02	0.03	10.70	6.00
ZnO rod	157.00 (diameter)	6.60	0.017	10.26	6.80

2.1.2. BET

The BET parameters for different morphologies of ZnO nanoparticles is given in Table 2. According to these results, the synthesized nanostructures have a mesoporous nature [26]. ZnO sphere had the largest specific surface area, highest pore volume and lowest average pore diameter. ZnO petals and rods has different specific surface areas and total pore volume but the average pore diameters were similar. These parameters are used to determine the effectiveness of nanoparticles.

2.1.3. X-ray Diffraction (XRD)

Figure 1g,h shows the XRD spectra of ZnO nanoparticles prepared in different solvents. The peaks at 31.85°, 34.52°, 36.23°, 47.50°, 56.50°, 62.9°, 66.29°, 67.92° and 68.82° indicate wurtzite hexagonal ZnO structure (JCPDS 89-1397). The preferred growth direction for ZnO were (101) for spheres and petals and (002) for ZnO rods. It can be seen that the (101) peak for ZnO nanorods is shifted towards lower diffraction angle. This may be attributed to the change in growth direction from (101) to (002) thereby reducing the strain along (101) direction.

2.1.4. Zeta Potential and Zn^{2+} Generation

The point of zero charge is indicative of the net charge of particles in the solution. A plot of zeta potential for different morphologies of ZnO is presented in Figure 2a. It was observed that the point of zero charge (pH_{ZC}) varied for different nanostructures. The pH_{ZC} for ZnO spheres was ~4.9 (4.86) while it shifted to 6.0 for petals and 6.8 for rods. The lower pH_{ZC} for spheres indicate better stability of spheres compared to other nanostructures at neutral pH. The lower pH_{ZC} of ZnO spheres is due to the presence of PEG molecules on ZnO nanoparticles. It has been shown in literature that the presence of carbonates shifts the peak in the direction of lower pH [27]. Different pH_{ZC} may be a result of differences in surface energy arising from the shape controlling process. Different surface energy of nanoparticle shapes leads to dissimilar adsorption of protons and hydroxyl ions on nanoparticles [28].

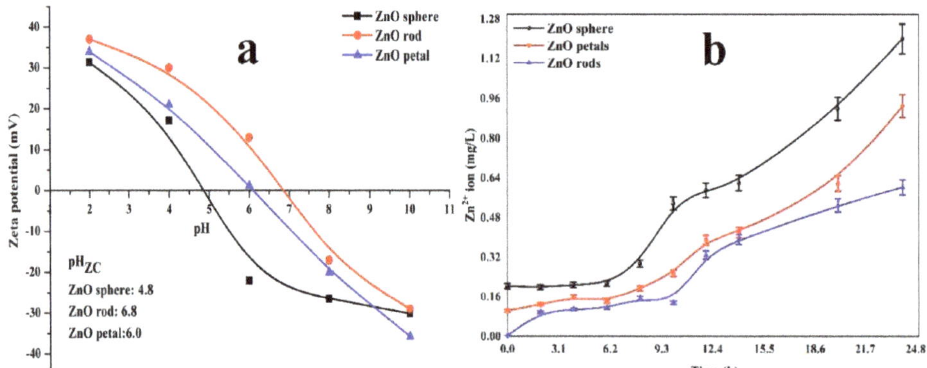

Figure 2. *Cont.*

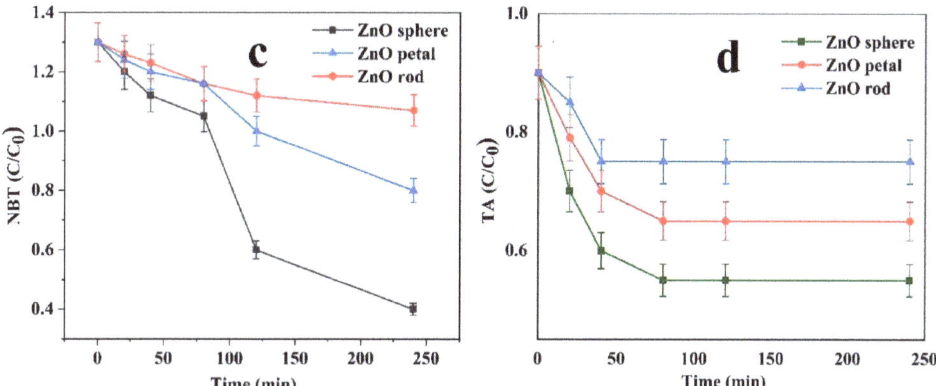

Figure 2. (**a**) Zeta potential and (**b**) Zn^{2+} release curves of ZnO nanostructures synthesized using different solvents. Degradation kinetics of (**c**) NBT and (**d**) TA in presence of ZnO nanostructures for the quantification of superoxide ($\bullet O_2^-$) and hydroxyl radicals (•OH), respectively.

The Zn^{2+} release profile is presented in Figure 2b. The ion release rates followed an order ZnO spheres > ZnO petals > ZnO rods.

2.1.5. Generation of Hydroxyl (OH•) and Superoxide Radicals ($\bullet O_2^-$)

The formation of superoxide radicals by ZnO nanostructures was analysed by nitroblue tetrazolium (NBT) assay and the plots are shown in Figure 2c. The reduction in absorbance of NBT with time was recorded and correlated with the concentration using standard calibaration curve. The extent of degradation can be directly correlated to the extent of ROS generatio which follows a simlar trend like metal ion release: ZnO spheres > ZnO petals> ZnO rods.

Terephthalic acid degradation is a commonly used technique for the detection of hydroxyl radicals [29,30]. Terephthalic acid does not show any fluorescence peak but the product of terephthalic acid and hydroxyl radicals (2-hydroxyterephthalic acid) shows a characteristic fluorescent peak at 425 nm. The degradation rate of terephthalic acid by hydroxyl radicals is plotted in Figure 2d. It can be noted that the rate of generation of OH• is maximum in the first 40 min after which it slows down and stops. The fluorescence spectra of terephthalic acid in absence and presence of nanoparticles after 40 min is presented in the Supplementary Information (Figure S1). The intensity of fluorescent peak is indicative of the amount of OH• generated by the nanostructures and thus it is evident that ZnO sphere produces maximum number of OH• followed by ZnO petals and ZnO rods. This is similar to the trend showed by $\bullet O_2^-$.

2.2. Mechanism of Morphology Change with the Choice of Solvent

The main factors that lead to change in nanoparticle morphology are: seeding and growth direction, hydrolysis and solvent templating [31–33]. Spherical nanoparticles were formed in case of PEG because glycol chains surround Zn^{2+} ions and inhibit particle agglomeration. Moreover, PEG consists of two polar hydroxyl groups that bind to (002) plane (*c* direction) and promote growth along the (101) plane thus leading to spherical morphology [31]. On the contrary when water is used as a solvent, the rate of hydrolysis increases and the amount of ($[Zn(OH)_4]^{2-}$) seeds increases. To reduce the overall surface energy of the system, petal like morphology is formed that attach to each other in a "template-like" fashion so that the system moves towards a more energetically favored low energy state [31]. Rod like ZnO nanostructures are formed when the crystals selectively grow along normal of the (002) planes. For non-polar aromatic molecules such as toluene, solvent attachment to the polar (002) plane is ineffective which directs the formation of ZnO along the perpendicular (002) plane giving rise to nanorods [31].

2.3. Caffeine Degradation

2.3.1. Adsorption Study

A calibration curve for caffeine has been shown in the Supplementary Information (Figure S2). Figure 3 shows that the equilibrium adsorption capacity (q_e) increased with the increase in initial concentration of caffeine. At high caffeine concentrations, the adsorption sites get saturated and q_e either remains constant or decreases due to desorption. The increase in adsorption with the increase in caffeine concentration can be attributed to the high concertation gradient which promoted the mass transfer of aqueous caffeine towards solid ZnO nanostructures. We also studied the time dependence of adsorption process and found that the adsorption of caffeine on ZnO nanostructures was a two-stage process and it reached equilibrium in almost 45 min (unaffected by the concentrations of ZnO and caffeine and type of nanoparticles). The first 20 min witnessed a rapid adsorption followed by a slower phase of 25 min of gradual uptake. A large number of unoccupied adsorption sites on ZnO nanostructures promoted an initial rapid adsorption and the rate slowed down when most of these sites were occupied.

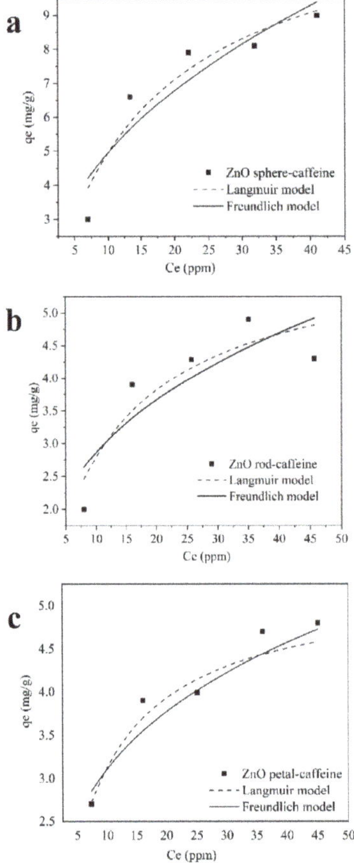

Figure 3. Adsorption isotherms of caffeine on ZnO nanostructures: (**a**) ZnO sphere-caffeine; (**b**) ZnO rod-caffeine; (**c**) ZnO petal-caffeine (—Freundlich model, - - -Langmuir model).

Langmuir and Freundlich isotherms (Equations (2) and (3)) were used to fit the adsorption data and the plots are shown in Figure 3. The equilibrium adsorption capacity of ZnO nanostructures was calculated at different initial concentration of caffeine and the isotherm parameters are presented in Table 3. The coefficient of regression (R^2) and chi-square values (χ^2) were also tabulated for the assessment of goodness-of-fit for these models. We note that although both these models describe the experimental data well, yet Langmuir model provides a better fit since the correlation coefficient (R^2) values for Langmuir adsorption isotherm is higher than that for Freundlich isotherm. Langmuir isotherm is based on the assumption that the molecules to be adsorbed form a monolayer on the adsorbent surface and there is ideally no chemical reaction involved between the two (physisorption). In this study, the experimentally determined adsorption capacity is similar to the theoretical one, thus supporting the Langmuir isotherm.

Table 3. Isotherm parameters for Langmuir and Freundlich adsorption models.

Sample	Langmuir Model				Freundlich Model			
	q_m (mg/g)	b (L/mg)	R^2	χ^2	$1/n$	K_f ($mg^{1-1/n}$ (L) $^{1/n}$($g)^{-1}$)	R^2	χ^2
ZnO-sphere	12.57	0.06	0.89	0.61	0.45	1.74	0.80	0.85
ZnO-petal	5.28	0.14	0.78	0.14	0.28	1.64	0.77	0.14
ZnO-rod	6.04	0.09	0.78	0.26	0.36	1.26	0.64	0.44

Freundlich isotherm is based on the assumption that there are chemical interactions involved between the molecules to be adsorbed and a heterogeneous adsorbing surface (chemisorption) leading to the formation of a multiple layers of adsorbing molecule. In our case, Freundlich isotherm fits the data as well, suggesting that caffeine molecules occupy heterogeneous adsorption sites on ZnO nanostructure surfaces and chemical interaction between caffeine molecules and ZnO may also contribute to the adsorption process. From Table 3, it can be seen that the values for b in Langmuir isotherm fall in between 0 and 1 which implies that the adsorption of caffeine on ZnO nanostructures is favorable [32]. Similarly, the parameter $1/n < 1$ in Freundlich isotherm also implies a favorable adsorption.

It was found that the maximum adsorption capacity (Q_m) for ZnO spheres (12.57 mg/g) was double compared to ZnO rods and petals. This difference in maximum adsorption capacities can be attributed to the different shapes, sizes and porosity of ZnO nanostructures.

2.3.2. Photocatalytic Experiments

We first tried to rule out the effect of photolysis from photocatalysis and for this purpose, we studied the degradation of caffeine in presence of 250 mW/m^2 visible light and absence of nanoparticles. The results shown in Supplementary Information (Figure S3) indicate that caffeine is only slightly affected by photolysis and thus we can exclude the effect of photolysis and perform photocatalytic study to describe the degradation. Next, we estimated the effect of different initial caffeine concentrations, ZnO concentration and morphologies and visible light intensities on the photocatalytic degradation of caffeine.

Nanoparticle Dosage

The effect of nanoparticle dosage on caffeine degradation was analyzed to find the lowest dosage at which the interaction of nanoparticles with caffeine can be maximized. We tested the degradation of caffeine at four different concentrations of ZnO (0.5, 1.0, 1.5 and 2.0 g/L) and it was seen that caffeine degraded faster when nanoparticle concentration was increased from 0.5 to 1.0 g/L for all morphologies of nanoparticles but it decreased at higher concentrations. This phenomenon can be explained by the tendency of nanoparticles to agglomerate at higher concentrations leading to a decrease in overall available surface area and an increase in diffusion time. Hence a nanoparticle dosage of 1.0 g/L was chosen for all subsequent experiments.

Initial Concentration of Caffeine and Nanoparticle Morphology

After the adsorption equilibrium was reached, we subjected the samples to 250 mW/cm^2 of simulated solar light and 1.0 g/L nanoparticle dosage for the photocatalytic experiments. The photocatalytic data was fitted to pseudo-first order and pseudo-second order kinetic model (Equations (4) and (5)) as shown in Figure 4 and the value of rate constants were calculated from the above-mentioned equations using non-linear regression (Levenberg–Marquardt algorithm, 10^{-5} tolerance). From the R^2 values (shown in Table 4), it can be concluded that the experimental data can be suitably represented by both the kinetic models with high correlation coefficients. The rate constants for ZnO sphere falls with the rise in amount of caffeine. ZnO spheres exhibited the highest value of degradation rate constant at the lowest caffeine concertation of 10 ppm (k_1-1.323 min^{-1}, k_2-1.74 g/mg min). This is about 1.1 times (k_1 and k_2) higher than that achieved by petals whereas 1.6-(k_1) and 2.2-(k_2) times higher than the degradation achieved by ZnO rods. At highest concentration, ZnO spheres are 1.4-(first order) and 2.8-fold (second order) higher than ZnO petals whereas 2.6-(first order) and 2.5-fold (second order) higher compared to ZnO rods. Thus, the differences in the rate of caffeine degradation is more pronounced at higher concentrations irrespective of the morphology.

Table 4. Kinetic rate parameters for caffeine degradation by ZnO nanostructures (Experimental conditions: [ZnO] = 1.0 g L^{-1}, light intensity = 250 mW/cm^2).

Sample	Parameter	Caffeine Concentration (ppm)					Light Intensity (mW/cm^2)		
		10	20	30	40	50	50	150	250
ZnO sphere	0.1 k_1 * (min^{-1})	1.32	1.33	0.69	0.48	0.43	0.21	0.23	0.48
	R^2	0.97	0.97	0.96	0.96	0.98	0.97	0.99	0.99
	0.01 k_2 * (g/mg min)	1.74	0.97	0.26	0.16	0.10	0.07	0.06	0.09
	R^2	0.94	0.99	0.99	0.98	0.99	0.97	0.99	0.99
ZnO petal	0.1 k_1 (min^{-1})	1.24	0.61	0.59	0.24	0.32	0.11	0.26	0.43
	R^2	0.99	0.95	0.96	0.98	0.97	0.99	0.95	0.97
	0.01 k_2 (g/mg min)	1.65	0.34	0.22	0.05	0.07	0.03	0.05	0.07
	R^2	0.98	0.97	0.99	0.99	0.98	0.99	0.97	0.99
ZnO rod	0.1 k_1 (min^{-1})	0.81	0.35	0.15	0.12	0.16	0.10	0.03	0.12
	R^2	0.96	0.98	0.98	0.98	0.99	0.96	0.99	0.99
	0.01 k_2 (g/mg min)	0.77	0.15	0.03	0.02	0.04	0.02	0.01	0.02
	R^2	0.96	0.99	0.98	0.99	0.99	0.96	0.99	0.99

* k_1 is pseudo-first order rate constant, and k_2 is pseudo-second order rate constant.

In terms of percentage degradation, it can be seen that ZnO spheres could degrade 72.5% of 30 ppm caffeine in 30 min and complete degradation was achieved in 120 min while the degradation rates for ZnO petals was 69% in 30 min and complete degradation in 135 min and rods was only 37% after 30 min and complete degradation in and complete degradation in 180 min. The corresponding degradation rates for different morphologies of ZnO can be listed as follows: ZnO spheres > ZnO petals > ZnO rods. Based on these findings, it could be perceived that the enhanced generation of ROS (Figure 2) in case of ZnO spheres might contribute to the increased photocatalytic activity.

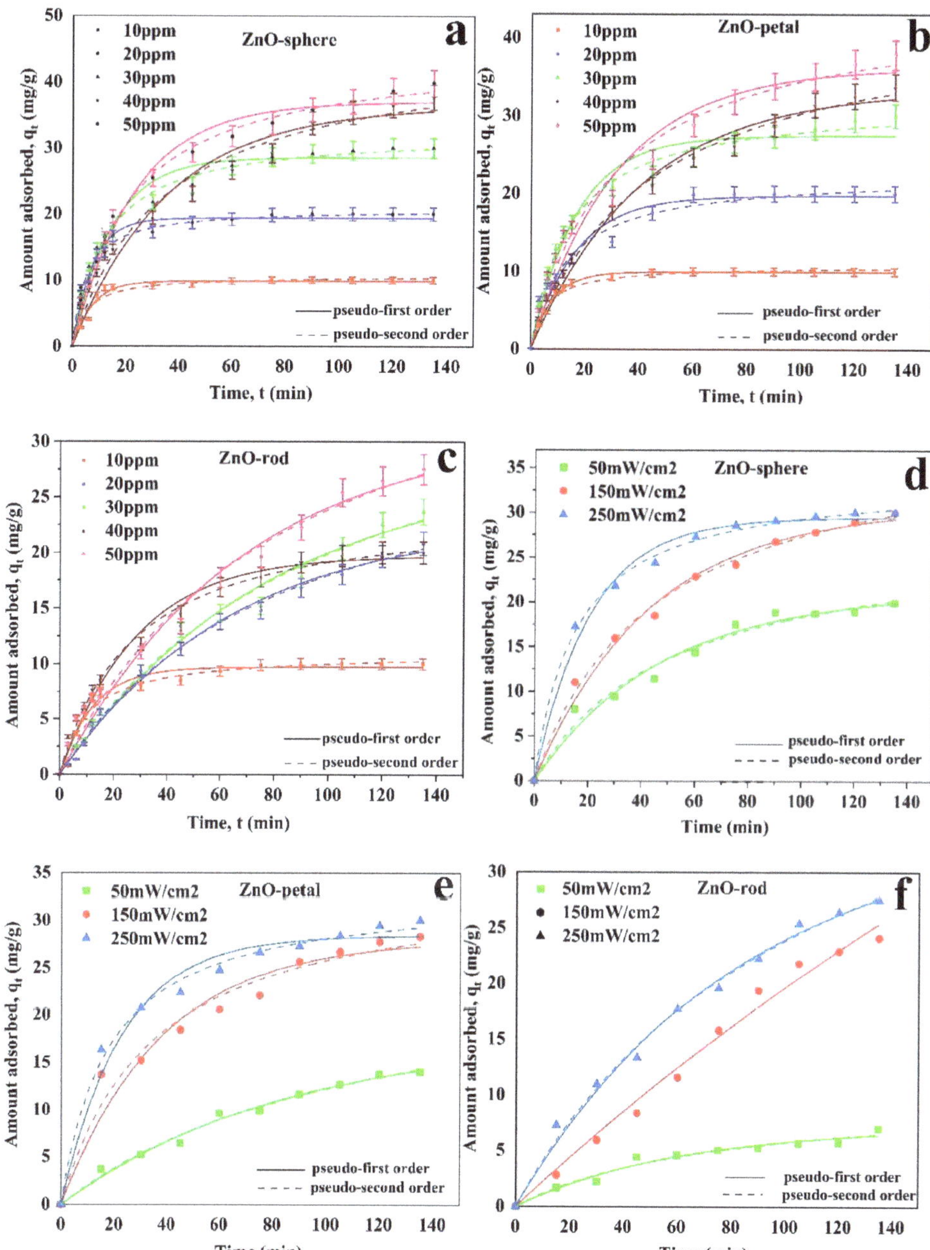

Figure 4. Kinetic model for the adsorption of caffeine on ZnO nanostructures at (**a–c**) various caffeine concentrations and (**d–f**) different light intensity. (—) and (- - -) represent pseudo-first-order and pseudo-second-order model, respectively.

Light Intensity

The degradation constant for caffeine increased with the increase in light intensity as a result of ROS generation and subsequent oxidation of caffeine (Figure 4d–f). The rate constants ae presented in Table 4 and it was evident that a light intensity of 250 mW/cm^2 resulted in the highest rate of degradation.

When the solar light is incident on these nanostructures, photons are adsorbed on the surface of nanostructures causing electron-hole pair to form. These pairs reach the surface of nanostructures, the electrons (in conduction band) are scavenged by the O_2 molecules present in the medium leading to the generation of ROS ($\bullet O_2^-$) and the holes remain in the valance band. The holes also contribute to the generation of other ROS species (such as $\bullet$OH) by reacting with water molecules. These radicals react with the caffeine molecules, subsequently degrading it into harmless products such as CO_2 and H_2O.

2.4. Antibacterial Activity

2.4.1. Minimum Inhibitory Concentration (MIC)

The minimum inhibitory concentration (MIC) of nanostructures for *E. coli* is listed in Table 5. MIC values for different nanostructures (for both *S.aureus* and *E. coli*) under consideration is as follows: ZnO sphere < ZnO rod < ZnO petal. Maximum value of MIC among all the nanostructures (90 μg/mL) was chosen for all subsequent experiments.

Table 5. MIC values for different morphologies of ZnO against *E. coli.*

Sample	MIC Value (μg/mL)	Reference
ZnO-sphere	60	[33,34]
	78	[14]
	65	This study
ZnO-petals	25	[34,35]
	5	[35,36]
	72	This study
ZnO-rods	2	[36,37]
	512	[37,38]
	90	This study

2.4.2. Zone of Inhibition (Disc Diffusion Assay)

The antibacterial property of ZnO nanostructures was assessed against *E. coli* by standard disc diffusion assay. The diameter of zone of inhibitions obtained after overnight incubation with nanostructures (different concentrations) was measured and shown in Figure 5a,b. Among the different shapes of ZnO, ZnO spheres showed the best activity whereas ZnO petals and rods produced almost similar zone of inhibitions.

2.4.3. Cell Viability Assay (CFU/mL)

To quantitatively assess the interaction of nanostructures with bacteria, the antibacterial tests were carried out in by suspending nanostructures (90 μg/mL) in liquid media contaning bacteria under constant shaking in dark at 37 °C. The number of viable bacteria were noted by counting the number of colonies (CFU/mL) and shown in Figure 5c. It was found that ZnO sphere and petals reduced the growth of bacteria to nearly 2.0 log CFU and 2.8 log CFU within 12 h, respectively. However, ZnO rods showed a much lower antibacterial activity of 5.0 log CFU reduction after 12 h treatment time. Thus among the different shapes of ZnO, ZnO spheres showed the highest reduction in cell viability followed by ZnO petals and ZnO rods.

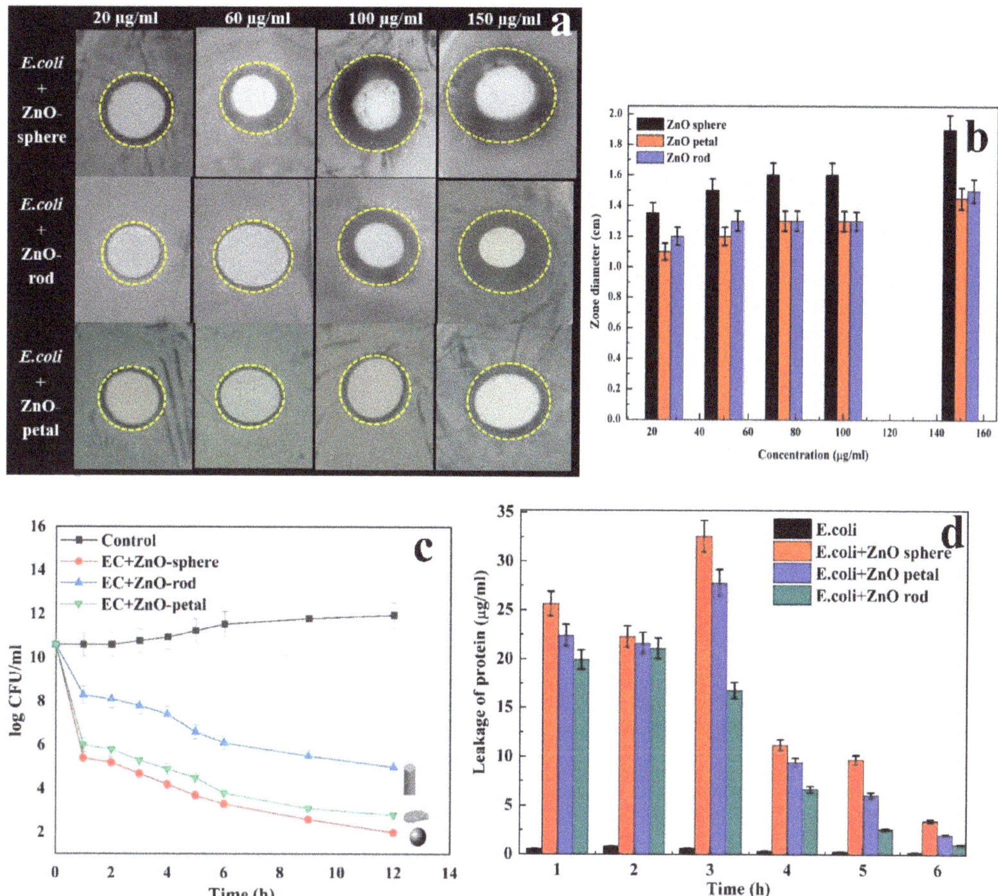

Figure 5. (**a**,**b**) Zone of inhibition (cm) formed in disc diffusion assay against *E. coli* for different concentration of nanostructures (20, 60, 100, 150 µg/mL) (**c**) Cell viability assay, and (**d**) Protein leakage analysis for *E. coli* treated with different morphologies of ZnO nanostructures.

2.4.4. Protein Leakage Analysis

In order to further confirm the rupture of membrane and leakage of cellular contents, the concentration of protein leaked out of the cell was monitored with time. Higher protein leakge is associated with a greated degree of membrane disruption. Leakage of proteins was detected in the medium containing nanostructure-exposed cells whereas almost negligible protein leakage was observed for untreated *E. coli* (Figure 5d). Maximum protein leakage was observed after 3 h for ZnO spheres and petals (32.59 and 22.41 µg/mL, respectively) and 2 h for rods (21.08 µg/mL). The permeability of cell membranes increase after proteins are leaked out in the soution which leads to oxidative stress in the cells [39].

2.4.5. Imaging

The FESEM images and EDS spectra of control and nanoparticle treated *E. coli* are shown in Figure 6a,c,e,g,i–k. Before the antibacterial treatment, *E. coli* had a well-defined rod shape and an integrated cell wall structure. On exposure to nanostructures, the original shape of bacteria was damaged with holes and ruptured cell membrane and the nanostructures were seen to be in close proximity with the cells. Drainage of cell contents from *E. coli* was also evident. *E. coli* following nanoparticle exposure revealed the presence of ZnO petals near the cells (Figure 6g). EDS spectra of bacteria-nanoparticle system indicates ZnO peaks which indicates ZnO around the damaged bacteria specimen.

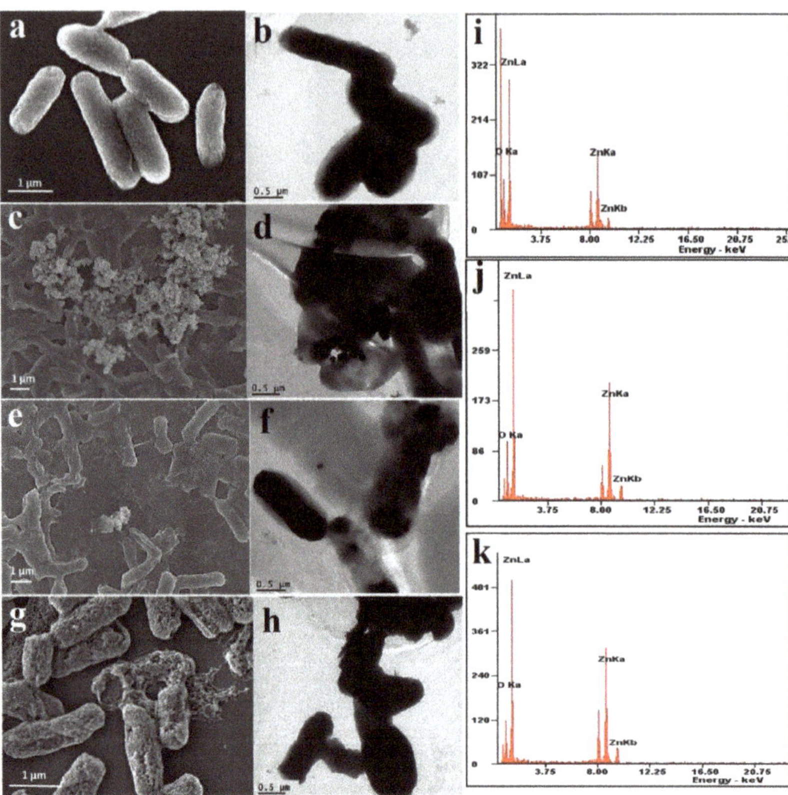

Figure 6. Scanning electron microscope images (**a**,**c**,**e**,**g**), Transmission electron microscope images (**b**,**d**,**f**,**h**) and EDS spectra for (**i**–**k**) for (**a**,**b**) Control (**c**,**d**,**i**) *E. coli* treated with ZnO sphere (**e**,**f**,**j**) *E. coli* treated with ZnO petals (**g**,**h**,**k**) *E. coli* treated with ZnO rod.

To further visualize the direct interaction between nanostructures and bacteria, TEM imaging experiments were performed. ZnO spheres penetrated the cell walls and internalize the cells leading to cell death (Figure 6e). ZnO petals did not internalize the cells (Figure 6f) and the antibacterial activity can be attributed to physical attrition. No internalization was seen in case of ZnO rods as well (Figure 6h), which implies that the sharp edges of ZnO rods pierced the cells and physical attrition led to the cell damage. Apart from physical attrition and cell internalization, ROS generation and Zn^{2+} toxicity are also responsible for the antibacterial activity of nanostructures. All the structures were found to adhere to the bacteria cell walls but except for ZnO spheres, no internalization was observed for other particles. Thus, cell wall damage and cytoplasm leakage for rods and petals was caused by other mechanisms such as physical attrition, ROS generation and cation release. The small size of ZnO spheres enable them to penetrate the bacterial cell resulting in cell death. The ROS and cations react with the intracellular contents of the bacteria and inactivate the cell. Studies have shown that ROS released from bigger nanostructures interacts with the cell membrane of these bacteria and oxidize the phospholipid contents of the membrane [13,40]. However, for smaller structures, the internalized contents are more damaged compared to the cell membrane. Thus, the rigidity of cell membrane, generation of holes and pits on the surface and leakage of internal contents are mostly determined by the morphology of nanostructures. Another theory that may find its applicability in this case is the abrasiveness of different nanostructures. ZnO rods have rough edges and sharp ends that causes mechanical damage to the membranes (physical attrition and piercing action) creating holes and pits causing localized leakage of cell contents.

All the antibacterial experiments with ZnO nanostructures were conducted in dark, suggesting that the cell membrane distortion by physical interaction with nanostructures and reaction of cell contents with cations might be a bigger contributor to cell damage compared to ROS [41,42].

Being a chemical process, there is no directly visible effect of nanoparticle shape on caffeine degradation and it is mostly governed by the ROS and Zn^{2+} released by nanostructures. On the contrary, shape of nanoparticles were visually evident to play a major role in *E. coli* disinfection since physical interaction of nanoparticles with bacteria is one of the ways that lead to cell death. For example, nanoparticles with sharp edges have a different mechanism of killing bacteria as compared to smooth surfaces. However, in addition to shape, we cannot also rule out the role of surface area in the photocatalytic and antibacterial activity of ZnO nanoparticles. For example, nanoparticles with larger surface area (ZnO sphere) were found to generate more ROS and Zn^{2+} as compared to the ones with lower surface area (ZnO petals and rods), thus affecting caffeine degradation in different ways. Additionally, smaller nanoparticles could internalize the bacterial cell and distort the cellular contents as compared to larger nanostructures that were limited to outer cell wall damage. Thus, the effect of surface area must also be considered in addition to shape to gain a clear understanding of the photocatalytic and antibacterial activity of these nanostructures.

3. Experimental Details

3.1. Materials

All the materials used in this work were of analytical grade and used without further purification. Zinc nitrate hexahydrate [99%] was purchased from JT Baker Chemicals, Phillipsburg, NJ, USA and sodium hydroxide [NaOH] was obtained from Caledon Laboratories Ltd., Georgetown, ON, Canada. Caffeine and polyethylene glycol (Kollisolv PEG E 400) were procured from Sigma Aldrich, Oakville, ON, Canada. Pure agar and Luria Bertani (LB) media were provided by HI MEDIA Laboratories, Bengaluru, Karnataka, India. Bacteria *Escherichia coli* (NCIM 2137) was purchased from NCIM, NCL Pune, Pune, Maharashtra, India.

3.2. Method

3.2.1. Synthesis of Nanoparticles

Different morphologies of ZnO nanoparticles were obtained by the hydrolysis reaction of zinc nitrate hexahydrate $Zn(NO_3)_2{\cdot}6H_2O$ in different solvents. The details of the synthesis procedure can be found elsewhere [14]. Briefly, 0.1 M $Zn(NO_3)_2{\cdot}6H_2O$ in 100 mL solvent (either PEG400 or water or toluene) was kept under magnetic stirring followed by the addition of 2 M NaOH. After 45 min, the nanoparticles were collected by ultracentrifugation, washing, overnight drying for 12 h and calcining at 300 °C.

3.2.2. Characterization of Nanomaterials

The synthesized nanoparticles were characterized using scanning electron microscope (SEM, Hitachi, Tokyo, Japan) for surface morphology. Nanoparticles were coated with platinum before imaging to avoid the charge accumulation near the surface of the particles. The shape and size of the nanomaterials were assessed using a JEOL High-Resolution Transmission Electron Microscope (HRTEM, Tokyo, Japan). BET analysis was performed for the estimation of porosity and specific surface area of nanomaterials using a Quantachrome ASiQwin-Automated Gas Sorption Data Acquisition and Reduction instrument (outgassing time: 1.5 h, temperature: 270 °C, Boynton Beach, FL, USA) with nitrogen as the adsorption gas. The phase composition of the samples was determined using a Panalytical High Resolution XRD-I, PW 3040/60 (Chennai, Tamil Nadu, India). The 2-theta angle was from 30° to 80° using a CuKα radiation (λ = 1.5418 Å). Zeta potential of particles was measured by Zetasizer Nano-ZS-90 (Malvern Instruments, Malvern, UK) with the scattering angle in the range of 90° to 25°. Nanoparticles (10 µg/mL) were ultrasonically dispersed in distilled water and measured under automatic mode. Dissolution of metal ions in water was studied for all the nanoparticles at a concentration of 90µg/mL. Aliquots from the supernatant of the suspension were collected and nanoparticles were separated by centrifugation. The supernatant was analyzed by PinAAcle 900 H atomic absorption spectrometer (AAS), Waltham, MA, USA. The standard deviation of triplicate measurements has been reported for all the experiments. The superoxide radicals ($\bullet O_2^-$) from ZnO nanoparticles were measured using nitroblue tetrazolium (NBT) assay [14]. One mM NBT was added to 100 mL of methanol (4%) and nanoparticles (90 µg/mL) under magnetic stirring, regular samples were withdrawn and UV-vis spectra was recorded from 200–400 nm. The hydroxyl radicals (OH•) were estimated by measuring the reaction product of terephthalic acid in solution. Briefly, a solution of 0.5 mM of terephthalic acid was prepared in NaOH and ZnO nanoparticles (90 µg/mL) were ultrasonically dispersed in the solution. Samples were withdrawn, centrifuged and the excitation fluorescence spectra was taken at 315 nm using.

3.2.3. Photocatalytic Degradation of Caffeine

Adsorption Study

The adsorption of caffeine on different morphology of ZnO nanoparticles was studied using UV–vis spectrophotometer (UV-3600, Shimadzu Scientific Instruments, Columbia, MD, USA) and the spectra were recorded in a range of 250–350 nm. Batch adsorption experiments were conducted with 1 g/L of ZnO nanoparticles with different caffeine concentrations (~10, 20, 30, 40, 50 ppm) under shaking in dark at 150 rpm for 2 h. The adsorption studies were performed to determine the time required to reach equilibrium and to calculate the adsorption coefficients. The equilibrium adsorption capacity of caffeine on nanoparticles (q_e, mg/g) was calculated by Equation (1):

$$q_e = (C_0 - C_e) \times \frac{V}{W} \tag{1}$$

where C_0 and C_e are the initial and equilibrium concentrations of caffeine (ppm), respectively, V is the volume of caffeine solution (L) and W is the weight of nanoparticles used (g).

Adsorption isotherms are fundamental for describing the adsorption capability of adsorbents. The equilibrium adsorption isotherms were developed by plotting the equilibrium concentration of caffeine and its corresponding uptake. The Langmuir isotherm is based on the assumption of monolayer adsorption of adsorbate on the adsorbent surface, and is given by the following expression:

$$q_e = \frac{bq_mC_e}{1 + bC_e} \tag{2}$$

where, C_e (mg/L) and q_e (mg/g) represent the concentration and amount of adsorbed material, respectively and q_m (mg/g) and b (L/mg) are the maximum adsorption capacity and Langmuir constant, respectively. The Freundlich isotherm is valid for adsorption on a heterogeneous surface and is given by

$$q_e = K_F(C_e)^{\frac{1}{n}} \tag{3}$$

where, K_F (mg/g) and n are capacity and adsorption favorability, respectively.

Kinetic Study

In a typical kinetic study, the important variables that affect the photocatalytic process are photocatalyst concentration, initial concentration of pollutant and light intensity. Batch experiments were conducted using nanoparticle/caffeine suspension of required concentration at neutral pH. The solution was kept under shaking at 150 rpm and the samples were analyzed in UV–vis spectrophotometer at regular intervals. The aliquots were filtered through a 0.45 μm filter prior to the absorbance measurements. A solar simulator (SS1 KW, ScienceTech, London, ON, Canada) equipped with a 1000 W Xe arc lamp and an Air-Mass (AM) 1.5 G filter was used to irradiate the sample with visible light.

The optimum amount of ZnO nanoparticles required for caffeine degradation was determined first. For this, four different concentrations of nanoparticles (0.5, 1, 1.5, 2 g/L). To investigate the effect of light intensity on caffeine degradation four different light intensities (50, 150 and 250 mW/cm^2) were used at optimum dosage of nanoparticles (1 g/L) and 30 ppm caffeine concentration. The initial concentration of caffeine (10, 20, 30, 40, 50 ppm) was subsequently varied at 250 mW/cm^2 light intensity (which provides highest degradation). All the experiments were carried out under constant magnetic stirring at room temperature. From the adsorption studies, it was seen that the adsorption equilibrium was reached after 1.5 h and accordingly, the photocatalytic experiments were carried out after 1.5 h of dark reaction.

The kinetics of caffeine adsorption by nanoparticles was investigated using a pseudo-first-order and pseudo-second-order kinetics, given by the following equations:

$$q_t = q_e(1 - \exp(k_1 t)) \tag{4}$$

$$q_t = \frac{k_2 {q_e}^2 t}{1 + k_2 q_e t} \tag{5}$$

where, q_t and q_e are the adsorption capacities of nanoparticles (mg/g) at time t and equilibrium respectively and k_1 (min^{-1}) and k_2 (g/mg min) are the pseudo-first-order and pseudo-second order rate constants, respectively.

3.2.4. Bacterial Toxicity Assessment

E. coli was incubated overnight in Luria Bertani (LB) broth at 37 °C under constant shaking (120 rpm). The CFU/mL of bacteria was adjusted to 10^8 for all the experiments unless otherwise stated.

Minimum Inhibitory Concentration (MIC)

The MIC value is the minimum concentration of nanomaterials that prevents any visible development of pathogen colonies. To obtain the MIC values, overnight grown cultures of *E. coli* and nanostructures (10–150 μg/mL) were mixed and incubated for 24 h. The optical density (OD) was recorded at 590 nm and the MIC value was reported as the one that inhibited the growth of 99% bacteria. The MIC value obtained from experiments (90 μg/mL) will be used for all subsequent experiments unless otherwise stated.

Zone of Inhibition

For the disc diffusion assay, different concentrations of nanoparticles (25, 50, 75, 100 and 150 μg/mL) were ultrasonically coated on paper discs (8 mm). Ten μL bacteria (10^8 CFU/mL) was spread on solidified agar plates and the paper disk was positioned above the bacteria coated agar and kept at 37 °C for 12 h. The zone diameter (mm) after incubation was measured and a mean value of triplicate experiments was noted.

Cell Viability Assay (CFU/mL)

Change in bacterial CFU with time in absence and presence of nanoparticles provides an estimate of the antibacterial strength of nanoparticles. Bacteria was incubated at 37 °C with nanoparticles and the CFU was recorded at intermediate times for a duration of

12 h. The CFU value obtained from the average of three experiments falling within 95% confidence interval were recorded.

Protein Leakage Analysis (Bradford Assay)

The protein released by nanoparticle treated deformed cells was quantified by Bradford assay [43]. Briefly, cells were washed and re-dispersed in PBS along with nanoparticles and incubated under shaking at 120 rpm and 30 °C. Aliquots of sample (1 mL) were withdrawn every hour, centrifuged and a solution of 200 µL supernatant and 800 µL Bradford's reagent was prepared and incubated in dark for 10 min. The absorbance of the samples was measured at 595 nm with the help of a spectrophotometer.

Imaging of Bacteria-Nanoparticle Interaction

Damage to the bacteria cell morphology before and after exposure to nanoparticles was observed in SEM. Samples were platinum coated prior to the SEM analysis. Bacteria treated with nanoparticles for 10 h were dropped on a coverslip, kept in 2.5% glutaraldehyde solution for 12 h at 4 °C and then dehydrated using ethanol. And the samples were vacuum dried prior to imaging. For the TEM imaging, a drop of untreated and treated bacteria was dropped on a carbon coated copper grid, dried and imaged.

4. Conclusions

Morphology-controlled ZnO were successfully synthesized by wet chemical process. The nature of solvent used for synthesis had a significant impact on the size and direction of crystal growth. Electron microscopy images showed that sphere, petal and rod-like ZnO were formed in PEG400, water and toluene, respectively. These nanostructures were studied for caffeine degradation and *E. coli* disinfection. The ROS and Zn^{2+} ions generated by these nanostructures were mainly responsible for the nanoparticle activity. The ROS consisted of O_2^- and OH• while H_2O_2 was found to be insufficient to show any photocatalytic or antibacterial activity. These nanostructures were very effective for the degradation of caffeine using solar light which has a relatively low operating cost compared to UV photocatalytic processes. The nanostructures were also effective in killing 99% of bacteria at low concentrations. However, some questions still remain unanswered and further studies are needed to determine the intermediate products of caffeine degradation by different nanostructures. Future studies should also focus on how photocatalytic and antibacterial activity are affected by other parameters such as water chemistry, pH, temperature, etc. and the toxicity of different nanostructures in environment.

Supplementary Materials: The supplementary materials are available online at https://www.mdpi.com/2073-4344/11/1/63/s1.

Author Contributions: Conceptualization: S.T., S.N., A.K.R.; methodology: S.T.; resources: S.N. and A.K.R.; writing—original draft preparation: S.T.; writing—review and editing: S.T. and A.K.R.; supervision: S.N. and A.K.R. All authors have read and agreed to the published version of the manuscript.

Funding: This research did not receive any specific grant from funding agencies in the public, commercial, or not-for-profit sectors.

Institutional Review Board Statement: Not applicable.

Informed Consent Statement: Not applicable.

Data Availability Statement: The data presented in this study are available on request from the corresponding author. The data are not publicly available due to privacy restrictions.

Acknowledgments: We thank the research facilities of IIT Kharagpur, India and Western University, Canada for providing the required necessary equipment. The first author would like to express her hearty gratitude to Malini Ghosh for her help in data analysis.

Conflicts of Interest: The authors declare no conflict of interest.

References

1. Kinnear, C.; Moore, T.L.; Rodriguez-Lorenzo, L.; Rothen-Rutishauser, B.; Petri-Fink, A. Form Follows Function: Nanoparticle Shape and Its Implications for Nanomedicine. *Chem. Rev.* **2017**, *117*, 11476. [CrossRef]
2. Korekar, G.; Kumar, A.; Ugale, C. Occurrence, fate, persistence and remediation of caffeine: A review. *Envrion. Sci. Pollut. Res.* **2020**, *27*, 34715. [CrossRef]
3. Sui, Q.; Cao, X.; Lu, S.; Zhao, W.; Qiu, Z.; Yu, G. Occurrence, sources and fate of pharmaceuticals and personal care products in the groundwater: A review. *Emerg. Contam.* **2015**, *1*, 14. [CrossRef]
4. Griffin, P.M.; Tauxe, R.V. The Epidemiology of Infections Caused by Escherichia coli O157:H7, Other Enterohemorrhagic *E. coli*, and the Associated Hemolytic Uremic Syndrome. *Epidemiol. Rev.* **1991**, *13*, 60. [CrossRef]
5. Rajasulochana, P.; Preethy, V. Comparison on efficiency of various techniques in treatment of waste and sewage water—A comprehensive review. *Resour. Effic. Technol.* **2016**, *2*, 175. [CrossRef]
6. Rodriguez-Narvaez, O.M.; Peralta-Hernandez, J.M.; Goonetilleke, A.; Bandal, E.R. Treatment technologies for emerging contaminants in water: A review. *Chem. Eng. J.* **2017**, *323*, 361. [CrossRef]
7. Zhao, L.; Denga, J.; Sun, P.; Liu, J.; Ji, Y.; Nakada, N.; Qiao, Z.; Tanak, H.; Yang, Y. Nanomaterials for treating emerging contaminants in water by adsorption and photocatalysis: Systematic review and bibliometric analysis. *Sci. Total Environ.* **2018**, *627*, 1253. [CrossRef]
8. Espitia, P.J.P.; Otoni, C.G.; Soares, N.F.F. Zinc Oxide Nanoparticles for Food Packaging Applications. In *Antimicrobial Food Packaging*; Academic Press: Cambridge, MA, USA, 2016; p. 425. [CrossRef]
9. Hoffmann, M.R.; Martin, S.T.; Choi, W.; Bahnemann, D.W. Environmental applications of semiconductor photocatalysis. *Chem. Rev.* **1995**, *95*, 69. [CrossRef]
10. Muggli, D.S.; Ding, L. Photocatalytic performance of sulfated TiO_2 and Degussa P-25 TiO_2 during oxidation of organics. *Appl. Catal. B* **2001**, *32*, 181. [CrossRef]
11. Adams, L.K.; Lyon, D.Y.; Alvarez, P.J.J. Comparative ecotoxicity of nanoscale TiO_2, SiO_2, and ZnO water suspensions. *Water Res.* **2006**, *40*, 3527. [CrossRef]
12. Li, G.R.; Hu, T.; Pan, G.L.; Yan, T.Y.; Gao, X.P.; Zhu, H.Y. Morphology−Function Relationship of ZnO: Polar Planes, Oxygen Vacancies, and Activity. *J. Phys. Chem. C* **2008**, *112*, 11859. [CrossRef]
13. Talebian, N.; Amininezhad, S.M.; Doudi, M. Controllable synthesis of ZnO nanoparticles and their morphology-dependent antibacterial and optical properties. *J. Photochem. Photobiol. B Biol.* **2013**, *120*, 66. [CrossRef]
14. Thakur, S.; Neogi, S. Effect of doped ZnO nanoparticles on bacterial cell morphology and biochemical composition. *Appl. Nanosci.* **2020**. [CrossRef]
15. Jones, N.; Ray, B.; Ranjit, K.T.; Manna, A.C. Antibacterial activity of ZnO nanoparticle suspensions on a broad spectrum of microorganisms. *FEMS Microbiol. Lett.* **2008**, *279*, 71. [CrossRef] [PubMed]
16. Sundrarajan, M.; Ambik, S.; Bharathi, K. Plant-extract mediated synthesis of ZnO nanoparticles using Pongamia pinnata and their activity against pathogenic bacteria. *Adv. Powder Technol.* **2015**, *26*, 1294. [CrossRef]
17. Arfanis, M.K.; Adamou, P.; Moustakas, N.G.; Triantis, T.M.; Kontos, A.G.; Falaras, P. Photocatalytic degradation of salicylic acid and caffeine emerging contaminants using titania nanotubes. *Chem. Eng. J.* **2017**, *310*, 525. [CrossRef]
18. Elhalil, A.; Elmoubarki, R.; Farnane, M.; Machrouhi, A.; Sadiq, M.; Mahjoubi, F.Z.; Qourzal, S.; Barka, N. Photocatalytic degradation of caffeine as a model pharmaceutical pollutant on Mg doped ZnO-Al_2O_3 heterostructure. Environmental Nanotechnology. *Monit. Manag.* **2018**, *10*, 63. [CrossRef]
19. Vaiano, V.; Matarangolo, M.; Sacco, O. UV-LEDs floating-bed photoreactor for the removal of caffeine and paracetamol using ZnO supported on polystyrene pellets. *Chem. Eng. J.* **2018**, *350*, 703. [CrossRef]
20. Bhuyan, T.; Mishra, K.; Khanuja, M.; Prasad, R.; Varma, A. Biosynthesis of zinc oxide nanoparticles from Azadirachta indica for antibacterial and photocatalytic applications. *Mater. Sci. Semicond. Process.* **2015**, *32*, 55. [CrossRef]
21. Barnes, R.J.; Molina, R.; Xu, J.; Dobson, P.J.; Thompson, I.P. Comparison of TiO_2 and ZnO nanoparticles for photocatalytic degradation of methylene blue and the correlated inactivation of gram-positive and gram-negative bacteria. *J. Nanopart. Res.* **2013**, *15*, 1432. [CrossRef]
22. Singh, R.; Verma, K.; Patyal, A.; Sharma, I.; Barman, P.B.; Sharma, D. Nanosheet and nanosphere morphology dominated photocatalytic & antibacterial properties of ZnO nanostructures. *Solid State Sci.* **2019**, *89*, 1–14. [CrossRef]
23. Kumar, S.; Nann, T. Shape Control of II–VI Semiconductor Nanomaterials. *Small* **2006**, *2*, 316. [CrossRef] [PubMed]
24. Achouri, F.; Merlin, C.; Corbel, S.; Alem, H.; Mathieu, L.; Balan, L.; Medjahdi, G.; Said, M.B.; Ghrabi, A.; Schneider, R. ZnO Nanorods with High Photocatalytic and Antibacterial Activity under Solar Light Irradiation. *Materials* **2018**, *11*, 2158. [CrossRef] [PubMed]
25. Thakur, S.; Mandal, S.K. Morphology engineering of ZnO nanorod arrays to hierarchical nanoflowers for enhanced photocatalytic activity and antibacterial action against *Escherichia coli*. *New J. Chem.* **2020**, *44*, 11796. [CrossRef]
26. Sing, K.S.W.; Everett, D.H.; Haul, R.H.W.; Moscou, L.; Pierotti, R.A.; Rouquerol, J.; Siemieniewska, T. Reporting physisorption data for gas/solid systems—With special reference to the determination of surface area and porosity. *Pure Appl. Chem.* **1985**, *57*, 603. [CrossRef]
27. Kosmulski, M.; Prochniak, P.; Rosenholm, J. Letter: The role of carbonate-free Neodymium (III) oxide. *J. Dispers. Sci. Technol.* **2009**, *30*, 589. [CrossRef]

28. Liao, D.; Wu, G.; Liao, B.-Q. Zeta potential of shape-controlled TiO_2 nanoparticles with surfactants. *Colloids Surf. A Phys. Eng. Asp.* **2009**, *348*, 270–275. [CrossRef]
29. Saran, M.; Summer, K.H. Assaying for hydroxyl radicals: Hydroxylated terephthalate is a superior fluorescence marker than hydroxylated benzoate. *Free Radic. Res.* **1999**, *31*, 429–436. [CrossRef]
30. Sultana, K.A.; Islam, T.; Silva, J.A.; Turley, R.S.; Hernandez-Viezcas, J.A.; Gardea-Torresdey, J.L.; Noveron, J.C. Sustainable synthesis of zinc oxide nanoparticles for photocatalytic degradation of organic pollutant and generation of hydroxyl radical. *J. Mol. Liq.* **2020**, *307*, 112931. [CrossRef]
31. Ayudhya, S.K.N.; Tonto, P.; Mekasuwandumrong, O.; Pavarajarn, V.; Praserthdam, P. Solvothermal Synthesis of ZnO with Various Aspect Ratios Using Organic Solvents. *Cryst. Growth Des.* **2006**, *6*, 2446. [CrossRef]
32. Salehi, R.; Arami, M.; Mahmoodi, N.M.; Bahrami, H.; Khorramfar, S. Novel biocompatible composite (Chitosan-zinc oxide nanoparticle): Preparation, characterization and dye adsorption properties. *Colloids Surf. B Biointerfaces* **2010**, *80*, 86. [CrossRef] [PubMed]
33. Thongam, D.D.; Gupta, J.; Sahu, N.K.; Bahadur, D. Investigating the role of different reducing agents, molar ratios, and synthesis medium over the formation of ZnO nanostructures and their photo-catalytic activity. *J. Mater. Sci.* **2018**, *53*, 1110–1122. [CrossRef]
34. Dutta, R.K.; Nenavathu, B.P.; Gangishetty, M.K.; Reddy, A. Antibacterial effect of chronic exposure of low concentration ZnO nanoparticles on *E. coli*. *J. Envrion. Sci. Health Part A* **2013**, *48*, 871–878. [CrossRef] [PubMed]
35. Kumar, K.M.; Mandal, B.K.; Naidu, E.A.; Sinha, M.; Kumar, K.S.; Reddy, P.S. Synthesis and characterisation of flower shaped Zinc Oxide nanostructures and its antimicrobial activity. *Spectrochim. Acta Part A Mol. Biomol. Spectrosc.* **2013**, *104*, 171–174. [CrossRef]
36. Amna, T. Shape-controlled synthesis of three-dimensional zinc oxide nanoflowers for disinfection of food pathogens. *Z. Für Nat. C* **2018**, *73*, 297–301. [CrossRef]
37. Salah, N.; Al-Shawafi, W.M.; Alshahrie, A.; Baghdadi, N.; Soliman, Y.M.; Memic, A. Size controlled, antimicrobial ZnO nanostructures produced by the microwave assisted route. *Mater. Sci. Eng. C* **2019**, *99*, 1164–1173. [CrossRef]
38. Amornpitoksuk, P.; Suwanboon, S.; Sangkanu, S.; Sukhoom, A.; Muensit, N. Morphology, photocatalytic and antibacterial activities of radial spherical ZnO nanorods controlled with a diblock copolymer. *Superlattices Microstruct.* **2012**, *51*, 103–113. [CrossRef]
39. Smith, A.; McCann, M.; Kavanagh, K. Proteomic analysis of the proteins released from Staphylococcus aureus following exposure to Ag(I). *Toxicol. In Vitro* **2013**, *27*, 1644–1648. [CrossRef]
40. Maness, P.-C.; Smolinski, S.; Blake, D.M.; Huang, Z.; Wolfrum, E.J.; Jacoby, W.A. Bactericidal Activity of Photocatalytic TiO_2 Reaction: Toward an Understanding of Its Killing Mechanism. *Appl. Envrion. Microbiol.* **1999**, *65*, 4094–4098. [CrossRef]
41. Padmavathy, N.; Vijayaraghavan, R. Enhanced bioactivity of ZnO nanoparticles—An antimicrobial study. *Sci. Technol. Adv. Mater.* **2008**, *9*, 035004. [CrossRef]
42. Brayner, R.; Ferrari-Iliou, R.; Brivois, N.; Djediat, S.; Benedetti, A.M.F.; Fiévet, F. Toxicological Impact Studies Based onEscherichiacoliBacteria in Ultrafine ZnO Nanoparticles Colloidal Medium. *Nano Lett.* **2006**, *6*, 866–870. [CrossRef] [PubMed]
43. Das, B.; Dadhich, P.; Pal, P.; Thakur, S.; Neogi, S.; Dhara, S. Carbon nano dot decorated copper nanowires for SERS-Fluorescence dual-mode imaging/anti-microbial activity and enhanced angiogenic activity. *Spectrochim. Acta Part A Mol. Biomol. Spectrosc.* **2020**, *227*, 117669. [CrossRef] [PubMed]

Article

N-Donor Ligand Supported "ReO_2^+": A Pre-Catalyst for the Deoxydehydration of Diols and Polyols

Jing Li [1], Martin Lutz [2] and Robertus J. M. Klein Gebbink [1,*]

[1] Organic Chemistry and Catalysis, Debye Institute for Nanomaterials Science, Utrecht University, Universiteitsweg 99, 3584CG Utrecht, The Netherlands; j.li3@uu.nl

[2] Crystal and Structural Chemistry, Bijvoet Centre for Biomolecular Research, Faculty of Science, Utrecht University, Padualaan 8, 3584 CH Utrecht, The Netherlands; m.lutz@uu.nl

* Correspondence: r.j.m.kleingebbink@uu.nl

Received: 2 June 2020; Accepted: 26 June 2020; Published: 7 July 2020

Abstract: A selected number of tetradentate N_2Py_2 ligand-supported ReO_2^+ complexes and a monodentate pyridine-supported ReO_2^+ complex have been investigated as catalysts for the deoxydehydration (DODH) of diols and polyols. In situ ^{1}H NMR experiments showed that these N-donor ligand-supported ReO_2^+ complexes are only the pre-catalyst of the DODH reaction. Treatment of (N_2Py_2) ReO_2^+ with an excess amount of water generates an active species for DODH catalysis; use of the Re-product of this reaction shows a much shorter induction period compared to the pristine complex. No ligand is coordinated to the "water-treated" complex indicating that the real catalyst is formed after ligand dissociation. IR analysis suggested this catalyst to be a rhenium-oxide/hydroxide oligomer. The monodentate pyridine ligand is much easier to dissociate from the metal center than a tetradentate N_2Py_2 ligand, which makes the $Py_4ReO_2^+$-initiated DODH reaction more efficient. For the $Py_4ReO_2^+$-initiated DODH of diols and biomass-based polyols, both PPh_3 and 3-pentanol could be used as a reductant. Excellent olefin yields are achieved.

Keywords: deoxydehydration reactions; rhenium-based pre-catalyst; biomass conversion

---

1. Introduction

Cellulosic biomass is currently being considered as a potential renewable feedstock for the chemical industry [1]. Unlike petroleum-based resources, the typical platform molecules that can be obtained after depolymerization of cellulosic biomass are overfunctionalized with oxygen-based functional groups. Accordingly, one of the challenges of using cellulosic biomass for chemical production is its defunctionalization to building blocks of lower oxygen content. Several pathways, for example, dehydration [2], deoxygenation [3,4], and deoxydehydration [5,6] have been reported for this purpose during the past two decades. Among these, deoxydehydration (DODH) reactions are known as an efficient way to remove vicinal hydroxyl groups to form the corresponding olefins. Metal complexes, such as vanadium complexes [7–10], molybdenum complexes [11–18], and rhenium complexes [5,19] have been reported to be able to catalyze the DODH of diols and polyols. In terms of activity and selectivity, rhenium complexes have shown unique properties amongst these complexes. Trioxo-rhenium complexes are known as catalysts for DODH reactions since the first Cp^*ReO_3-catalyzed DODH reaction was described by Cook and Andrews [20]. Later on, new catalysts based on Cp-ligands were reported by Klein Gebbink et al. [21,22]. Next to the development of new Re-based DODH catalysts, investigations in this field have also focused on the mechanism by which the rhenium catalysts operate in DODH reactions [23]. It is generally agreed that the catalytic cycle of these high-valent rhenium catalysts consists of three steps: a. reduction of the Re(VII) trioxo complex to a Re(V) dioxo species, b. condensation of the diol substrate to form a Re-diolate intermediate, and c. olefin extrusion from the reduced diolate intermediate to form the organic reaction product and regenerate the Re(VII) trioxo complex (Scheme 1). The reduction step can proceed either before or after the condensation step, and the order of these steps is highly dependent on the combination of metal, reductant, and substrate. From these mechanistic considerations it becomes clear that only two oxo ligands are involved in actual bond making and breaking in the catalytic cycle, and that one oxo ligand (shown in red in Scheme 1) can in principle be considered as a spectator ligand. The question thus arises whether or not Re(V) dioxo complexes are also capable of catalyzing DODH reactions of diols and polyols.

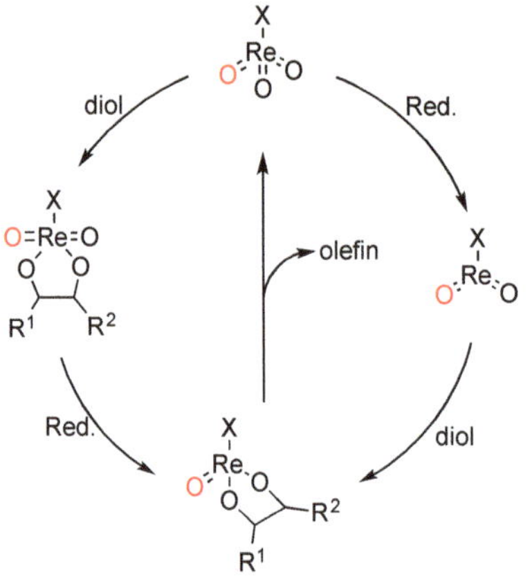

Scheme 1. Proposed catalytic cycle of the trioxo-rhenium catalyzed deoxydehydration (DODH) of diols.

Using a Re(V) dioxo complex as the catalyst, catalysis could proceed through a Re(V)↔Re(III) cycle instead of a Re(VII)↔Re(V) one. In 2013, a Re(V)↔Re(III) catalytic cycle was proposed by Abu-Omar et al. for a DODH reaction catalyzed by methyltrioxorhenium (MTO) using a secondary alcohol as reductant (Scheme 2) [24]. In this mechanism, MTO is reduced to methyldioxorhenium (MDO) via reduction by the secondary alcohol or via oxidative cleavage of an aromatic diolate ligand of a Re(VII)-diolate formed through condensation of the diol with MTO. MDO then condenses with a next diol to form a Re(V)-diolate intermediate, which would be further reduced by a secondary alcohol to form a Re(III)-diolate species, followed by olefin extrusion to regenerate MDO. In this reaction sequence, the Re(V) dioxo species MDO is considered as the true catalyst of the DODH reaction.

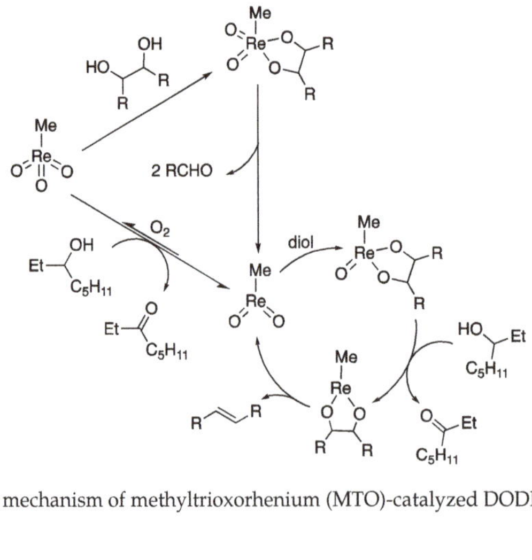

Scheme 2. Proposed mechanism of methyltrioxorhenium (MTO)-catalyzed DODH via a Re(V)↔Re(III) cycle. [24].

Besides the mechanistic proposal by Abu-Omar, Nicholas et al. have investigated *trans*-$[(Py)_4ReO_2]^+$ dioxo-rhenium complexes, i.e., *trans*-$[(Py_4)ReO_2]PF_6$ and *trans*-$[(Py_4)ReO_2]Cl$, as a catalyst for the deoxydehydration of 1,2-decanediol using zinc as reductant [25]. Even though this reaction starts from a Re(V) dioxo species, a Re(V)↔Re(VII) catalytic cycle was proposed. In a reaction of 1,2-decanediol (1.6 equiv.) and *trans*-$[(Py_4)ReO_2]Cl$ (1.0 equiv.), 98% of 1-decene (with respect to Re) formed and no oxidation products of the diol (aldehydes, ketones) were detected. This result indicated that the starting Re(V) complex is not reduced by the diol, i.e., a Re(V)↔Re(VII) redox change is involved in the reaction and a $[(Py)_nRe(VII)O_3]^+$ intermediate is generated during the reaction (Scheme 3). In other words, catalytic cycles involving either a Re(V)↔Re(III) or a Re(V)↔Re(VII) interconversion have been proposed for DODH catalysis starting from a Re(V) dioxo species.

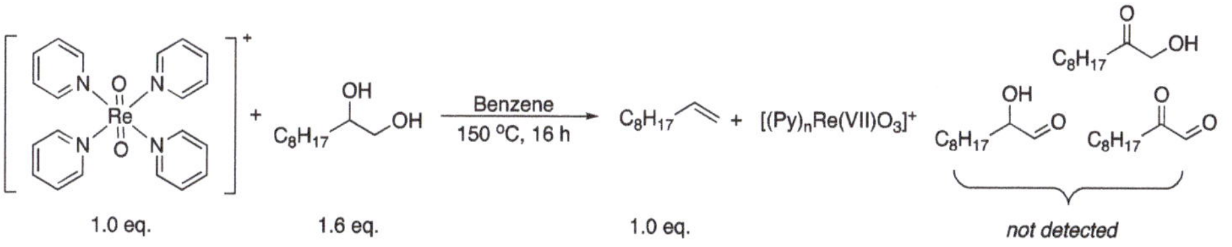

Scheme 3. Stochiometric reaction of 1,2-decanediol and *trans*-$[(Py_4)ReO_2]Cl$ [25].

Recently, a series of *cis*-dioxo-rhenium(V/VI) complexes containing tetradentate N_2Py_2 ligands were reported by Che and coworkers [26]. The authors showed, amongst other, that the $[(BPMCN)ReO_2]^{2+}$ ion (BPMCN = N^1,N^2-dimethyl-N^1,N^2-bis(pyridin-2-ylmethyl)cyclohexane-1,2-diamine), which was prepared by constant potential oxidation of $[(BPMCN)ReO_2]^+$, is able to oxidize hydrocarbons with weak C–H bonds (75.5–76.3 kcal mol^{-1}) via hydrogen atom abstraction. In view of the discussion above, Re(V) dioxo complexes like the ones reported by Che could be interesting catalyst candidates for deoxydehydration reactions. Accordingly, we set out to synthesize *cis*-dioxo-rhenium(V) complexes supported by tetradentate N_2Py_2 ligands and to investigate these as catalysts for the DODH of vicinal diols.

2. Results and Discussion

2.1. Synthesis of Rhenium Complexes

For our initial study, *cis*-[(*S*,*S*-BPBP)ReO_2]PF_6 (**1**) (*S*,*S*-BPBP = (2*S*,2′*S*)-1,1′-bis(pyridin-2-ylmethyl)-2,2′-bipyrrolidine), *cis*-[(BPMEN)ReO_2]PF_6 (**2**) (BPMEN = N^1,N^2-dimethyl-N^1,N^2-bis(pyridin-2-ylmethyl)ethane-1,2-diamine), and *cis*-[(B^{mdm}PMEN)ReO_2]PF_6 (**3**) (B^{mdm}PMEN = N^1,N^2-bis((4-methoxy-3,5-dimethylpyridin-2-yl)methyl)-N^1,N^2-dimethyl ethane-1,2-diamine) (Scheme 4) were synthesized according to the protocol developed by Che et al. for the synthesis of **1** [25]. To this end, reaction of $IReO_2(PPh_3)_2$ with the corresponding N_2Py_2 ligand in dichloromethane afforded crude *cis*-[(N_2Py_2)ReO_2]I, which was then reacted with NH_4PF_6 to afford **2** and **3** in 40% and 31% yields, respectively. The structures of **2** and **3** were established by X-ray crystal structure determination (Figure 1). Diffraction-quality crystals of **2** were obtained by vapor diffusion of diethyl ether into an acetonitrile solution of the complex, while diffraction-quality crystals of **3** were obtained by vapor diffusion of diethyl ether into a methanol solution of the complex.

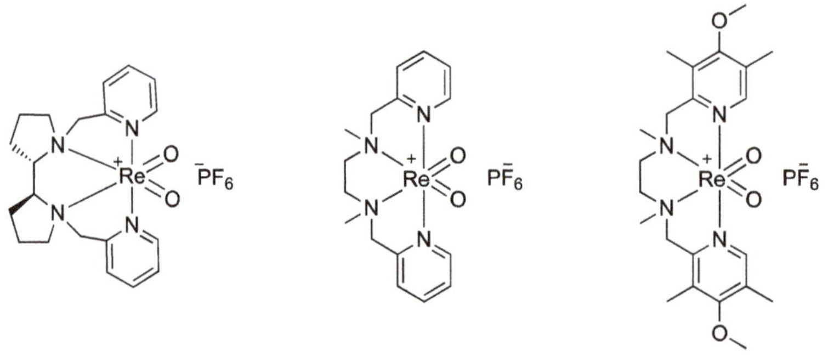

cis-[(S,S-BPBP)ReO_2]PF_6 (**1**) *cis*-[(BPMEN)ReO_2]PF_6 (**2**) *cis*-[(B^{mdm}PMEN)ReO_2]PF_6 (**3**)

Scheme 4. Dioxo-rhenium complexes **1**, **2**, and **3**.

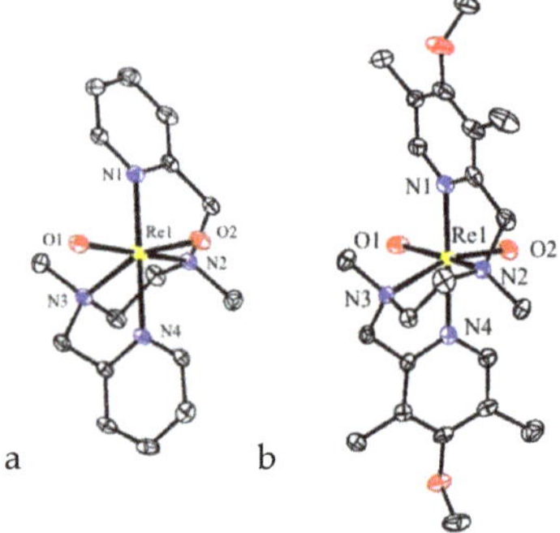

Figure 1. Molecular structures of (**a**). *cis*-[(BPMEN)ReO_2]$^+$ (**2**) and (**b**). *cis*-[(B^{mdm}PMEN)ReO_2]$^+$ (**3**) in the crystal, drawn at the 50% probability level. Hydrogen atoms and the PF_6 counterions are omitted for clarity. For compound **3**, the cocrystallized methanol solvent molecule is omitted as well.

These two new *cis*-dioxo-rhenium(V) complexes adopt a pseudo-octahedral geometry with a *cis*-α configuration of the ligand, similar to the geometry of **1**. Also the bond distances and angles are very similar to that of *cis*-[(*S*,*S*-BPBP)ReO_2]$^+$ reported by Che [26]. Selected bond distances and angles are given in Table 1. For all of these three structures, the N(pyridine)–Re distances are shorter than the N(amine)–Re distances and the planes of the pyridine rings are tilted under a small angle with respect to the N(pyridine)–Re bonds (Table 2). Complexes **1**, **2**, and **3** are chiral at the amine nitrogens. Complexes **1** and **2** are enantiopure in the crystal, while complex **3** is racemic in the crystal. The tetradentate N_2Py_2 ligands of these complexes are well-known ligands for non-heme iron and manganese oxidation catalysts. Interestingly, for both N_2Py_2 supported high-valent rhenium complexes and N_2Py_2 supported low-valent iron/manganese complexes, the same pseudo-octahedral geometry is observed. Further characterization of complexes **2** and **3** included ^{1}H NMR, ^{13}C NMR, ESI-MS, and elemental analysis (see Supplementary Material).

Table 1. Selected bond distances (Å) and angles (°) of *cis*-dioxo-rhenium complexes, *cis*-[(S,S-BPBP)ReO_2]PF_6 (**1**) [27], *cis*-[(BPMEN)ReO_2]PF_6 (**2**), and *cis*-[(BP^{mdm}MEN)ReO_2]PF_6 (**3**).

	1	2	3
Re-O1	1.765(4)	1.750(3)	1.745(3)
Re-O2	1.749(4)	1.747(4)	1.749(3)
Re-N1	2.122(5)	2.126(4)	2.122(4)
Re-N4	2.132(5)	2.124(4)	2.116(4)
Re-N2	2.258(5)	2.286(4)	2.266(4)
Re-N3	2.252(4)	2.284(4)	2.279(4)
O1-Re-O2	122.5(2)	123.1(2)	122.60(16)
N1-Re-N4	176.3(2)	174.53(16)	174.85(13)
N2-Re-N3	75.47(19)	75.89(15)	76.69(13)

Table 2. Angle between pyridine ring and N-Re bond [°].

	1	2	3
N1-Re	2.0(3)	6.8(2)	2.67(19)
N4-Re	1.8(3)	6.7(2)	0.93(18)

2.2. Initial Catalytic Activity Investigation of Complexes 1–3

Next, we investigated the use of **1**, **2**, and **3** as catalysts in the DODH of vicinal diols by using 1,2-octanediol as a benchmark substrate and triphenylphosphine (PPh_3) as reductant. Only trace amounts of 1-octene formed when the reaction was performed at 135 °C using 2 mol% **1** as catalyst and 1.1 equiv. of PPh_3 as reductant. The reaction temperature was then gradually increased from 135 °C to 180 °C. The yield of 1-octene jumped from 11% to quantitative when the temperature was increased from 165 °C to 180 °C (Table 3, entries 1–4). Similarly, when **2** was used as catalyst, an elevated reaction temperature was also necessary; the yield of 1-octene dramatically increased when the reaction temperature was increased from 150 °C to 180 °C (Table 3, entries 5, 6). For all three complexes, full conversion and quantitative 1-octene yield were achieved when the reaction was performed at 180 °C for 15 h under N_2 (Table 3, entries 4, 5, and 7).

Table 3. Deoxydehydration of 1,2-octanediol catalyzed by dioxo-rhenium complexes. [a].

Entry	Catalyst	Time	Temperature	Yield [b]	Conversion
1	**1**	15 h	135 °C	2%	2%
2	**1**	15 h	150 °C	6%	7%
3	**1**	15 h	165 °C	11%	12%
4	**1**	15 h	180 °C	>99%	>99%
5	**2**	15 h	150 °C	16%	16%
6	**2**	15 h	180 °C	>99%	>99%
7	**3**	15 h	180 °C	>99%	>99%
8	**1**	3 h	180 °C	27%	27%
9	**2**	3 h	180 °C	34%	34%
10	**3**	3 h	180 °C	41%	41%
11	$IReO_2(PPh_3)_2$	3 h	180 °C	14%	16%
12	*trans*-[(Py_4)ReO_2]PF_6	3 h	180 °C	>99%	>99%
13 [c]	**1**	7 h	180 °C	75%	75%
14 [c]	**2**	7 h	180 °C	90%	90%
15 [c]	**3**	7 h	180 °C	84%	84%

[a] Reaction conditions: 1,2-octanediol (0.5 mmol), PPh_3 (0.55 mmol, 1.1 equiv.), catalyst (0.01 mmol, 2 mol%), PhCl (5 mL), N_2. [b] 1-Octene yield was determined by ^1H NMR using mesitylene (0.5 mmol) as an internal standard. [c] For the initial 3 h, the reaction was performed under N_2; for the next 4 h, the reaction was performed under air.

In order to compare the reactivity of **1**, **2**, and **3**, a shorter reaction time was chosen. After 3 h of reaction at 180 °C, significantly lower olefin yields (27–41%) were found than for a reaction time of 15 h at this reaction temperature, albeit at 100% 1-octene selectivity. These findings show that the nature of N_2Py_2 ligand has an effect on the proficiency of the dioxo-rhenium complexes in catalysis and that these complexes do not show any isomerization of the olefin product under the current conditions. Remarkably, in DODH reactions catalyzed by Cp-based trioxo-rhenium complexes, olefin isomers are formed when the reactions are performed at higher temperatures (180 °C) [27].

The precursor of these dioxo complexes, i.e., ododioxobis(triphenylphosphine)rhenium ($IReO_2(PPh_3)_2$), was also investigated for its DODH activity and gave a 1-octene yield of only 14% with 2% of octene isomers being formed. *trans*-[$(Py_4)ReO_2$]PF_6 was also tested for this DODH reaction, aiming at providing information on the effect of the configuration of the oxo ligands (*cis* or *trans*) on catalysis. Surprisingly, this complex showed the best activity among all dioxo-rhenium complexes tested in our study (Table 3, compare entries 8–11). After the reaction mixture was heated at 180 °C for 3 h, a full substrate conversion and quantitative 1-octene product yield was achieved with *trans*-[$(Py_4)ReO_2$]PF_6 (entry 11), while the reactions using the three N_2Py_2-ligated *cis*-dioxo-rhenium complexes reached much lower conversions (entries 8–10). As mentioned in the introduction section, Nicholas et al. have carried out a stochiometric reaction between 1,2-decanediol (1.6 equiv.) and *trans*-[$(Py_4)ReO_2$]Cl (1.0 equiv.) to form 1.0 equiv. of 1-decene and no diol oxidation products. From this observation, they concluded that the starting Re(V) complex is converted to a $[(Py)_nRe(VII)O_3]^+$ species, which would also involve pyridine ligand(s) dissociation, and that DODH catalysis using this Re(V) dioxo complex would proceed through a Re(V)↔Re(VII) cycle [25]. On the basis of these considerations, we assume that $[(N_2Py_2)ReO_2]^+$-catalyzed DODH reactions can also proceed through a Re(V)↔Re(VII) cycle. In this case though, the formal oxidation to a Re(VII)-trioxo species (resulting from a single DODH reaction) would be more difficult compared to *trans*-[$(Py_4)ReO_2$]PF_6, since the tetradentate N_2Py_2 ligand is more strongly coordinated to the Re center than the monodentate pyridine donors in *trans*-[$(Py_4)ReO_2$]PF_6. This could then explain why *trans*-[$(Py_4)ReO_2$]PF_6 has a higher catalytic activity compared to complexes **1**–**3**.

As a last aspect of these initial studies, we investigated the sensitivity of DODH catalysis by complexes **1**–**3**. Accordingly, the reaction mixtures of entries 7–9 were then heated for another 4 h under air in the closed reaction vessel. Interestingly, the presence of O_2 does not seem to hamper the reactivity and might even slightly promote the reactions (Table 3, entries 13–15).

2.3. Initial Mechanistic Studies

After the initial study of the catalytic DODH activity of complexes **1**–**3**, the time-course profiles of the DODH of 1,2-octanediol by these N_2Py_2-ligated dioxo-rhenium complexes were investigated. In situ ^1H NMR experiments were carried out using 1,2-octanediol (0.05 mmol), PPh_3 (0.055 mmol, 1.1 equiv.), Re-catalyst (0.005 mmol, 10 mol%), and mesitylene (0.05 mmol, 1.0 equiv., internal standard) in toluene-D_8 (0.5 mL) at 180 °C under an inert N_2 atmosphere.

Except *cis*-[$(BPMEN)ReO_2$]PF_6 (**2**), *cis*-[$(BP^{mdm}MEN)ReO_2$]PF_6 (**3**), and *trans*-[$(Py_4)ReO_2$]PF_6, the dioxo-rhenium complex *cis*-[$(BPMEN)ReO_2$]BPh_4 (**6**) was also included in these studies in order to investigate a possible effect of the counter anion on catalyst activity. For complexes **2**, **3**, and **6**, an induction period was observed. For the reaction catalyzed by **2**, a typical sigmoidal trend for 1-octene formation was found. No 1-octene formation was observed during the initial 5 min., followed by a gradual increase in 1-octene formation until 120 min. After this induction period, formation of 1-octene proceeded at a much higher rate, with the highest yield (86%) reached after 300 min (Scheme 5a). Modification of the N_2Py_2 ligand framework lead to changes in this reaction profile. When *cis*-[$(BP^{mdm}MEN)ReO_2$]PF_6 (**3**) was used as catalyst, the sigmoidal reaction curve showed a much shorter induction period. In this case the highest product yield (90%) was reached after 210 min (Scheme 5b). Furthermore, changing the counter anion also resulted in a different reaction profile, i.e., the induction period was significantly shorter for the BPh_4^- salt **6** than for PF_6^- salt **2** (Scheme 5a,c). However, the three complexes **2**, **3**, and **6** give a very similar final 1-octene yield and reached this yield within approximately the same time.

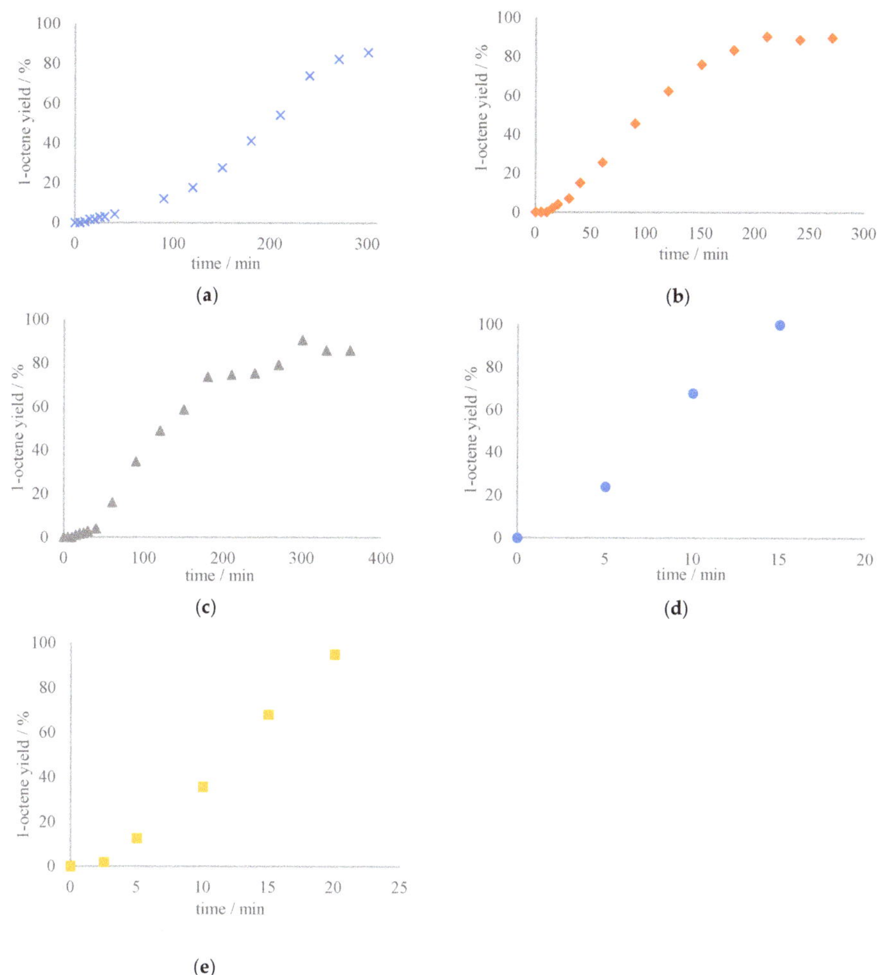

Scheme 5. Time-course profiles of the formation of 1-octene in Re-catalyzed DODH of 1,2-octanediol with PPh_3 as reductant using different Re-catalysts: (**a**) *cis*-[(BPMEN)ReO_2]PF_6 (**2**); (**b**) *cis*-[(BP^{mdm}MEN)ReO_2]PF_6 (**3**); (**c**) *cis*-[(BPMEN)ReO_2]BPh_4 (**6**), (**d**) *trans*-[(Py_4)ReO_2]PF_6 (**4**); (**e**) pyridinium perrhenate. *Reaction conditions*: 1,2-octanediol (0.05 mmol), PPh_3 (0.055 mmol, 1.1 equiv.), Re-catalyst (0.005 mmol, 10 mol%), mesitylene (0.05 mmol, 1.0 equiv., internal standard), toluene-D_8 (0.5 mL), 180 °C, N_2.

In the reaction profile for the DODH reaction catalyzed by *trans*-[(Py_4)ReO_2]PF_6 no induction period was observed and 24% of 1-octene had formed after 5 min, while full conversion was achieved after 15 min. The reaction rate in this case was much higher than that for the other three complexes. In all of these four reactions, no formation of olefin products resulting from 1-octene isomerization was observed.

After the investigation of the time-course profiles of these "ReO_2"-catalyzed deoxydehydrations, the reaction mixtures were analyzed by means of ESI-MS measurements. For the DODH reaction of 1,2-octanediol catalyzed by **1**, both protonated ligand (ES+ *m/z* 323.2225; calc. 323.2230) and ReO_4^- (ES– *m/z* 250.8320; calc. 250.9360) signals were detected by ESI-MS measurements on the crude reaction mixture after the reaction (Figure 2). The signal for the protonated ligand signal was also detected when subjecting pristine complex **1** to ESI-MS analysis, in addition to the signal of the

[(*S*,*S*-BPBP)ReO_2] cation (*m/z* 541.1534; calcd. 541.1608). The signal for the intact [(*S*,*S*-BPBP)ReO_2] cation was, however not observed in the analysis of the crude DODH reaction mixture. The absence of **1**, as well as the appearance of protonated ligand and ReO_4^-, indicates that the starting dioxo-rhenium complex decomposes under the conditions of the DODH reaction and that perrhenate is formed during the reaction. In case of the DODH of 1,2-octanediol catalyzed by **1**, besides the signal for the protonated ligand, signals at *m/z* 480.2455, 576.2290, 669.2780, 857.2693, were detected in positive ion mode. While the signals at *m/z* 480.2455, 576.2290, 857.2693 did not match the isotope distribution of either mono- or multi-rhenium species, the signal at *m/z* 669.2780 can be assigned to the [(*S*,*S*-BPBP)ReO(1,2-octanediolate)]$^+$ ion (calcd. 669.2809), which represents the product formed upon the condensation of [(*S*,*S*-BPBP)ReO_2]$^+$ and 1,2-octanediol. This diolate could either form 1-octene by olefin extrusion and generate [(*S*,*S*-BPBP)ReO_3]$^+$, or be reduced by PPh_3 to form [(*S*,*S*-BPBP)Re(1,2-octanediolate)]$^+$, followed by olefin extrusion to regenerate [(*S*,*S*-BPBP)ReO_2]$^+$. In the former case, a 20e$^-$ species [(*S*,*S*-BPBP)ReO_3]$^+$ would be formed, which could set the stage for N_2Py_2 ligand dissociation and formal decomposition of the starting dioxo-complex.

These combined MS and reaction profile observations indicate that complexes **2**, **3**, and **6** act as pre-catalysts in DODH reactions. Our initial assumption was that the N_2Py_2-supported complexes show slower kinetics due the more difficult formation of a putative Re(VII) trioxo species as a result of stronger ligand chelation (*vide supra*). The MS data now show the presence of protonated ligand and the perrhenate anion ReO_4^- in the crude DODH reaction mixtures of these complexes, suggesting that full ligand dissociation and rhenium oxidation may take place as part of precatalyst activation.

Based on these observations, we decided to investigate the DODH activity of the pyridinium perrhenate salt [PyH][ReO_4]. Pyridinium perrhenate was earlier reported as catalyst for the DODH of diols by Love and coworkers [28]. In their work, aromatic vicinal diols were converted to the corresponding olefins with moderate to excellent yields (22% to quantitative) at a relatively low temperature (90 °C) in chloroform using 5 mol% lutidinium perrhenate as catalyst and 1.09 equiv. PPh_3 as reductant. For aliphatic vicinal diol substrates, a higher reaction temperature of 140 °C was needed for the DODH reaction to proceed, and only moderate olefin yields (21–51%) were obtained after 16 h. When carrying out the DODH reaction of 1,2-octanediol with [PyH][ReO_4] as catalyst under our reaction conditions, we found quantitative 1-octene formation within approx. 20 min without the formation of octane isomers. The time-course profile of this reaction showed a very short induction period and indicated that the reaction rate was not as high as for *trans*-[(Py_4)ReO_2]PF_6 (Scheme 5d,e). This comparison between time-course profiles seems to suggest that perrhenate is the active species that is formed when *trans*-[(Py_4)ReO_2]PF_6 is used in DODH catalysis. On the other hand, the difference in reaction rates does not rule out the possible involvement of yet another active species.

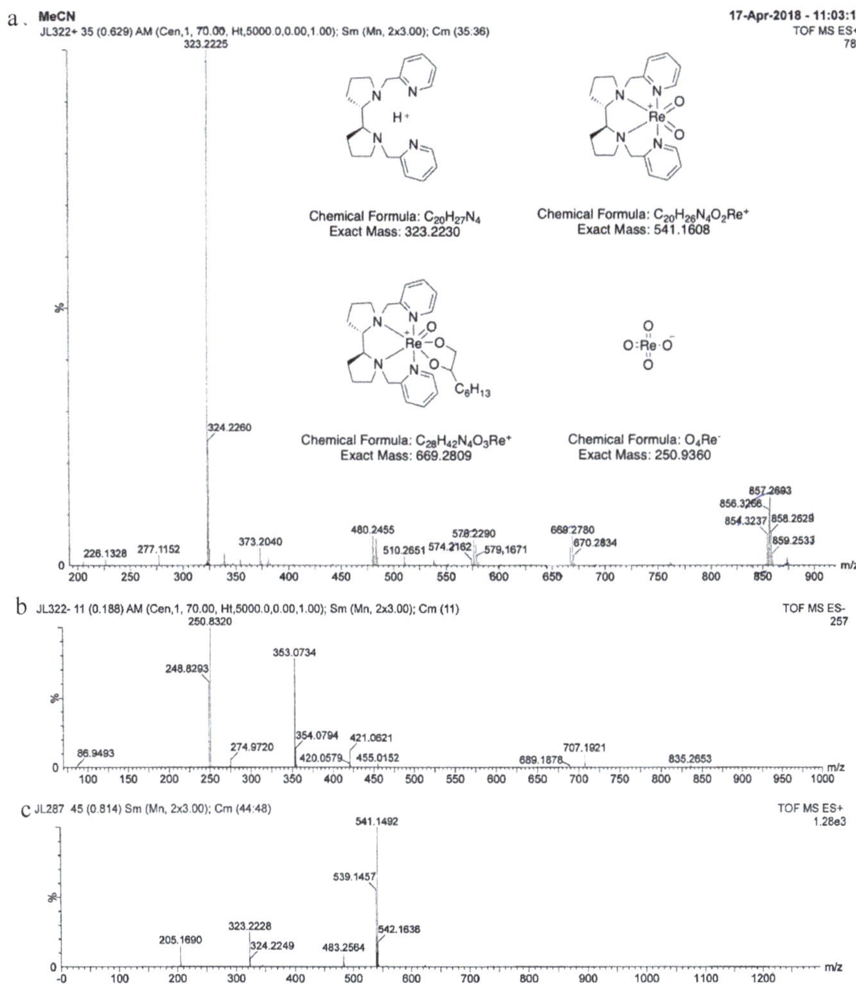

Figure 2. ESI-MS analysis: (**a**) MS ES+ trace of the crude reaction mixture of the DODH reaction of 1,2-octanediol by **1**; (**b**) MS ES- trace of the crude reaction mixture of the DODH reaction of 1,2-octanediol by **1**; (**c**) MS ES+ trace of **1**.

2.4. Investigation of the Active Species

In order to further probe the nature of the active species in DODH reactions catalyzed by dioxo-rhenium complexes, we decided to explore the DODH reaction of erythritol using these complexes. Erythritol is an interesting substrate in DODH chemistry, since it can lead to the formation of 1,3-butadiene as the reaction product starting from a bio-based source. In addition, this tetraol substrate can engage in a number of side-reactions, including dehydrative cyclisation and internal deoxydehydration, and its product distribution profile has been used to investigate the involvement of different active species during DODH catalysis. For the $Cp^{*}ReO_3$-catalyzed DODH of erythritol using PPh_3 as reductant and PhCl as solvent, instead of 2,5-dihydrofuran, *cis*-2-butene-1,4-diol and 3-butene-1,2-diol were detected as byproduct. Also, for the $Cp^{ttt}ReO_3$-catalyzed DODH of erythritol using PPh_3 as reductant and PhCl as solvent, both *cis*-2-butene-1,4-diol and *trans*-2-butene-1,4-diol were detected as byproduct while 2,5-dihydrofuran was not detected [29]. However, the formation of

2,5-dihydrofuran is highly dependent on the solvent and reductant; when 3-octanol was used as solvent and reductant, 2,5-dihydrofuran was the only byproduct detected with either $Cp^{ttt}ReO_3$, $Cp^{tt}ReO_3$, or MTO as catalyst. [21,22,30]

cis-[(BPMEN)ReO_2]PF_6 (**2**), *trans*-[(Py_4)ReO_2]PF_6 (**4**), and pyridinium perrhenate were then tested for the DODH of erythritol using PPh_3 as reductant and PhCl as solvent. The reactions were performed at 180 °C for 1.5 h under N_2. The total yield of 1,3-butadiene plus 2,5-dihydrofuran was 57%, 80%, and 83%, respectively. In addition, different product distributions were observed; the ratio of 1,3-butadiene and 2,5-dihydrofuran products was 5.3:1, 9.0:1, and 15.6:1, respectively (Scheme 6). Upon prolonging the reaction from 1.5 h to 3 h, the total yield of 1,3-butadiene plus 2,5-dihydrofuran increased to 68% and 84% for **2** and *trans*-[(Py_4)ReO_2]PF_6, while a lower 72% was observed for the perrhenate catalyzed reaction. The latter decrease could be the result of secondary reactions of the rather reactive butadiene product under the reactions conditions in the presence of [PyH][ReO_4]. The ratio of 1,3-butadiene and 2,5-dihydrofuran products slightly changed to 5.2:1, 9.5:1, and 11:1 for the prolonged reactions.

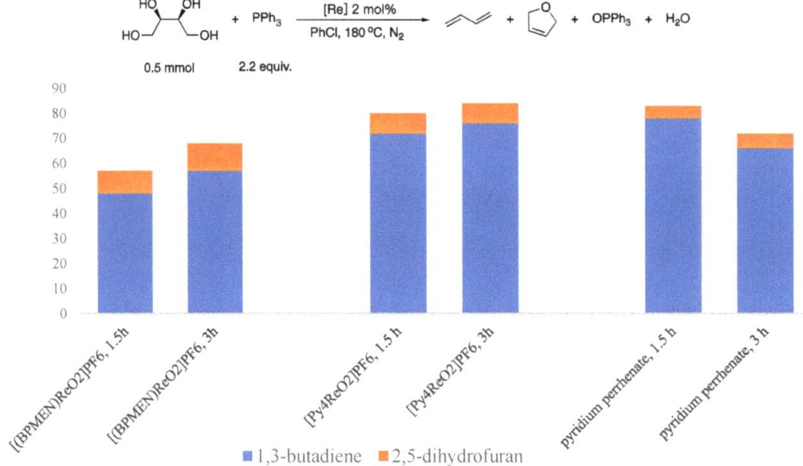

Scheme 6. DODH of erythritol using *cis*-[(BPMEN)ReO_2]PF_6 (**2**), *trans*-[(Py_4)ReO_2]PF_6 (**4**), and pyridinium perrhenate as catalyst.

Overall, the product distribution was quite consistent for these three catalysts. The differences observed in product formation and distribution in these experiments do not provide solid proof that perrhenate is the (only) active species in "ReO_2" catalyzed DODH reactions. In addition, the absence of *cis*-2-butene-1,4-diol, *trans*-2-butene-1,4-diol, and 3-butene-1,2-diol byproducts indicated that these three reactions do not proceed through the same active species and/or do not form the same Re-based side products as Cp^*ReO_3 and $Cp^{ttt}ReO_3$ in the PPh_3/PhCl system. On the other hand, as mentioned above, in all cases of using a secondary alcohol as both solvent and reductant, 2,5-dihydrofuran was the only detected byproduct, just like for our experiments using the PPh_3/PhCl conditions; indicating that for the dioxo-rhenium and perrhenate systems the same or similar active species could actually be involved as for the Re-trioxo systems.

Since neither the time-course profile nor the product distribution studies could provide full evidence that perrhenate is the active species in dioxo-rhenium-catalyzed DODH reaction, we have sought for other means of probing the nature of the active species in these reactions. On basis of the time-course profile the DODH reaction catalyzed by **2** (Scheme 5a), the activation/formation of the active species could proceed in a two-step manner. After a small amount of olefin product had formed, a second phase in product formation was observed before the rate of product formation really took on. Accordingly, we considered the possibility that the active species might be generated through the involvement of the initial rhenium complexes and one of the products formed in the DODH reaction, i.e., next to olefin also water and triphenylphosphine oxide ($OPPh_3$).

We first turned our attention to the possible role of water on the speciation of rhenium in the DODH reactions. It is known that in the present of water methyltrioxorhenium (MTO) will form a gold-colored rhenium oligomer $\{H_{0.5}[(CH_3)_{0.92}ReO_3]\}$ in high yield (ca. 70%) at 70 °C, in addition to the formation of O_2, $HReO_4$, and methane (ca. 30% in total) [31,32]. Thus, we were curious about the species formed from the dioxo-rhenium complexes and water. Accordingly, *cis*-[(BPMEN)ReO_2]BPh_4 (**6**) was heated in CD_3CN at 180 °C for 2 h in the presence of 50 equiv. of water. ^1H NMR analysis of the reaction after this period did not indicate the presence of non-coordinated BPMEN ligand. Next, this "water-treated" complex, i.e., the resulting reaction mixture, was used as catalyst in the DODH of 1,2-octanediol under our standard reaction conditions. In this case, 86% of 1-octene formed after 2 h at 180 °C and again without the formation of olefin isomers. The time-course profile of this reaction shows that only 3% of 1-octene had formed after 10 min, but that the overall induction period was significantly shorter than using pristine *cis*-[(BPMEN)ReO_2]BPh_4 as catalyst (Figure 3). After the induction period, the two reaction profiles are rather similar, which indicates that the same active species might be involved and the formation of this species might involve a reaction with water at high temperature. Comparing the time-course profile of DODH reactions using complex **6** and "water-treated" **6** as catalyst, the latter one is much faster. In both cases though, no olefin isomers were formed during the reaction and the final yield of 1-octene was the same (95%).

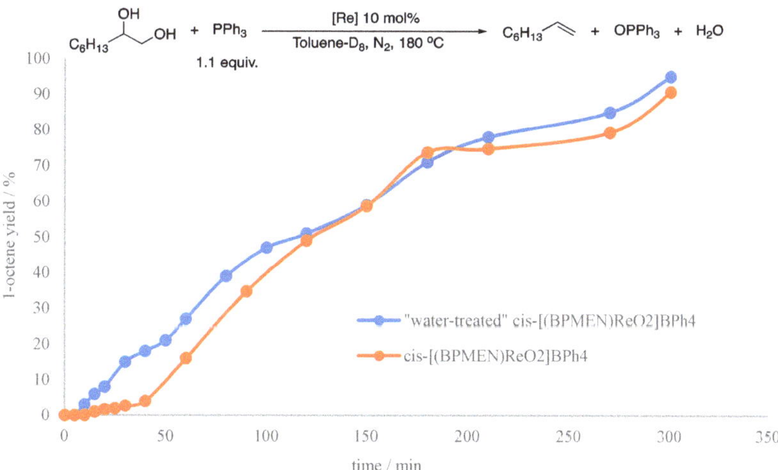

Figure 3. Time-course profile of the DODH reaction of 1,2-octanediol using *cis*-[(BPMEN)ReO_2]BPh_4 and "water-treated" *cis*-[(BPMEN)ReO_2]BPh_4 as catalyst. Reaction conditions: 1,2-octanediol (0.05 mmol), PPh_3 (0.055 mmol, 1.1 equiv.), Re-catalyst (0.005 mmol, 10 mol%), mesitylene (0.05 mmol, 1.0 equiv., internal standard), toluene-D_8 (0.5 mL), 180 °C, N_2.

Next, we compared the IR spectra of **6** before and after treatment with water to that of pyridinium perrhenate (Figure 4). The IR spectrum of pristine *cis*-[(BPMEN)ReO_2]BPh_4 shows a sharp and intense signal at 813 cm^{-1} assigned to Re = O stretching, while the IR of pyridinium perrhenate shows relatively broad and intense signals at 863, 741, and 672 cm^{-1} assigned to Re = O stretching of the perrhenate anion. In the IR spectrum of "water-treated" *cis*-[(BPMEN)ReO_2]BPh_4 the most intense signal in the 500–1000 cm^{-1} range was found at 907 cm^{-1}, which again was assigned to Re = O stretching, albeit at significantly higher wave numbers compared to pristine **6**. Two less intense bands were also observed in this range, at 761 and 706 cm^{-1}. Comparing these three spectra, it is obvious that no *cis*-[(BPMEN)ReO_2]BPh_4 is left after water treatment. The similarities in the 500–1000 cm^{-1} range for pyridinium perrhenate and "water-treated" *cis*-[(BPMEN)ReO_2]BPh_4 suggest that ReO_4^- is formed upon water treatment of **6**. In the range of 1000–2000 cm^{-1}, very different signals were observed, likely due to the presence of different organic ligands.

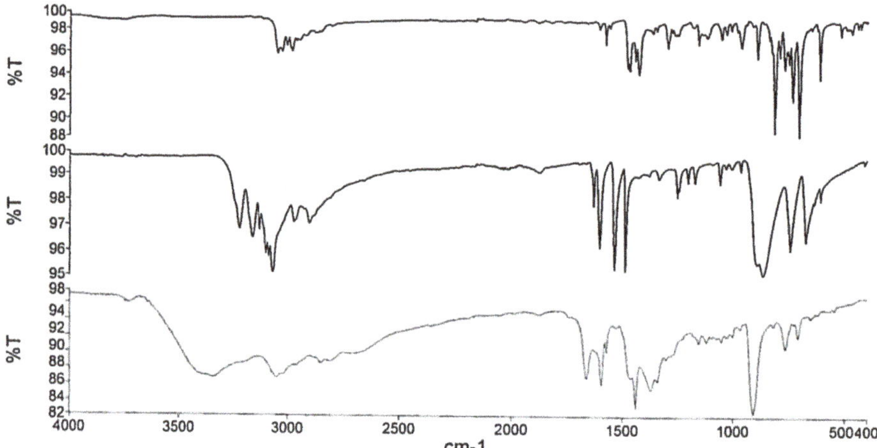

Figure 4. IR spectra of *cis*-[(BPMEN)ReO_2]BPh_4, pyridinium perrhenate, and "water-treated" *cis*-[(BPMEN)ReO_2]BPh_4.

Recently, Marrone and d'Alessandro et al. have reported a study on the active species in rhenium-catalyzed DODH reactions using secondary alcohols as reductant [33]. An induction period was observed for the DODH of glycerol using methyltrioxorhenium (MTO), trioxo-rhenium (ReO_3), rhenium pentachloride ($ReCl_5$), hepta-oxo-dirhenium (Re_2O_7), rhenium triiodide (ReI_3), and $IReO_2(PPh_3)_2$ as catalyst and 2,4-dimethyl-3-pentanol (DMP) as reductant. After removal of the volatile reaction products and the DMP reductant, the residues of these reactions were used as the catalyst in subsequent DODH experiments and no induction periods were observed. IR spectra of these active residues were collected and compared. A few intense and broad signals were detected in the 1800–3750 cm^{-1} range, and these signals could be assigned to C–H and O–H stretching vibrations. In our IR spectra of "water-treated" *cis*-[(BPMEN)ReO_2]BPh_4, these kind of intense and broad signals (3346, 3053, 2852 cm^{-1}) were also observed in this range. In the range of 1000–1750 cm^{-1}, multiple intense signals were detected by Marrone and d'Alessandro while in our case, much less intense signals were observed. Since this range is where C–O stretching and C–C–H, C–O–H bending vibrations are located, these differences indicate that related species may have formed but these are not identical. In both studies, the most intense signals appeared in the 600–1000 cm^{-1} range. In the study of Marrone and d'Alessandro, the most intense signal appears at 920 cm^{-1} for the Re = O stretching vibration in all cases studied, independent from the rhenium sources. In our case this vibration is found at 907 cm^{-1}. A less intense band around 700 cm^{-1} was also observed in all cases by Marrone and d'Alessandro, and assigned to either the out-of-plane bending of O–H bonds or the stretching of the Re–O–Re oxo-bridges. Overall, rather similar spectra were obtained in these two studies. By further investigating the MTO-catalyzed DODH of glycerol, Marrone and d'Alessandro proposed a Re(VII) alkoxide oligomer as the actual active species could form alongside methane from the reaction of MTO and the secondary alcohol reductant. A very important observation from their study is the formation of a catalytically active black solid in all cases. In our present study though, we did not observe the formation of black precipitates.

As described in Marrone and d'Alessandro's work, an active DODH species is formed from the reaction of the Re source and a secondary alcohol. Although in our system, we do not use a secondary alcohol reductant, an active species seems to form from a reaction with water, and obviously a vicinal diol is present as the substrate. A possibility would be that in our case a rhenium-oxide/hydroxide oligomer is formed instead of a Re(VII) alkoxide oligomer as proposed by Marrone and d'Alessandro. The differences seen in the 1000–1750 cm^{-1} range in IR analysis would support this assumption. As mentioned before, IR analysis of the $\{H_{0.5}[(CH_3)_{0.92}ReO_3]\}$ oligomer, the reaction product of MTO and water, shows signals at 912 (vs), 881 (sh), 851 (sh), and 758 cm^{-1} (m). [32] These signals could be assigned to Re = O stretching vibrations. Although not identical, we observed quite similar IR vibrations in the 600–1000 cm^{-1} region (Table 4). For the $\{H_{0.5}[(CH_3)_{0.92}ReO_3]\}$ oligomer, a methyl group is bound to the rhenium center, although not in a 1:1 ratio. In our case, no coordinated ligand was observed after the water treatment at 180 °C. This comparison lends further credit to the proposed formation of a rhenium-oxide/hydroxide oligomer as the active species in DODH catalysis.

Table 4. Comparison of IR signals in 600–1000 cm^{-1} region of "water-treated" *cis*-$[(BPMEN)ReO_2]BPh_4$, $\{H_{0.5}[(CH_3)_{0.92}ReO_3]\}$ [32], and the Re(VII) alkoxide oligomer [33].

"Water-Treated" *cis*-$[(BPMEN)ReO_2]BPh_4$	$\{H_{0.5}[(CH_3)_{0.92}ReO_3]\}$	Re(VII) Alkoxide Oligomer
907 (vs)	912 (vs)	920 (vs)
	881 (sh)	
	851 (sh)	
761 (m)	758 (m)	
706 (w)		around 700 (m)

2.5. Substrate Scope

Previously, *trans*-$[(Py_4)ReO_2]PF_6$ (10 mol%) was used by Nicholas et al. as catalyst in the DODH of 1,2-decanediol using Zn as reductant and benzene as solvent to give 67% of 1-decene with full-substrate conversion at 150 °C in 24 h. In our system, 1,2-octanediol was quantitively converted to 1-octene at 180 °C in 3 h using only 2 mol% of *trans*-$[(Py_4)ReO_2]PF_6$. Remarkably, upon increase of the catalyst loading under our conditions to 10 mol%, the reaction was done in 15 min at 180 °C. For the Cp-based trioxorhenium-catalyzed DODH systems, much longer reaction times are needed to realize the full conversion of 1,2-octanediol. As mentioned before, no isomers were observed under our reaction conditions, while small amount of isomers were observed under the same conditions for Cp-based trioxorhenium catalysts [27]. Furthermore, for the deoxydehydration of erythritol (*vide supra*), *trans*-$[(Py_4)ReO_2]PF_6$ was quite a competitive catalyst (Scheme 6). This initial catalytic capacity test of *trans*-$[(Py_4)ReO_2]PF_6$ encouraged us to apply this precatalyst to other substrates (Table 5). When the aromatic vicinal diol 1-phenylethane-1,2-diol was used as substrate, a quantitative styrene yield was obtained (Table 5, entry 1). For the DODH of 1,4-anhydroerythritol, 97% of 2,5-dihydrofuran was formed (Table 5, entry 2); this 2,5-dihydrofuran yield is competitive to the one obtained from the MTO/3-pentanol system, which provides 95% of 2,5-dihydrofuran [30]. Glycerol, a biomass-derived triol, was also investigated in our reaction system and gave a quantitative amount of allyl alcohol (entry 3). Although quantitative allyl alcohol formation is also obtained by the $Cp^{tt}ReO_3$/3-octanol system [23], the air-stable property of *trans*-$[(Py_4)ReO_2]PF_6$ makes this catalyst system more attractive. *trans*-$[(Py_4)ReO_2]PF_6$ also catalyzes the DODH of biomass-derived L-(+)-tartaric acid and mucic acid. When using L-(+)-tartaric acid as substrate, there was no product detected using PPh_3 as reductant and PhCl as solvent. This is most likely due to the poor substrate solubility in PhCl. Using 3-pentanol as both solvent and reductant gave 84% of the corresponding olefins as a mixture of fumaric acid and fumarate esters. Along a similar vein, a total yield of muconic acid and muconates of 65% was obtained using mucic acid as substrate (Table 5, entry 5). In the case of $Cp^{tt}ReO_3$-catalyzed DODH of mucic acid, a slightly higher product yield of 71% was found, albeit at 5 mol% catalyst loading. Overall, *trans*-$[(Py_4)ReO_2]PF_6$ is a good pre-catalyst for the DODH of diols and biomass-derived polyols.

Table 5. *trans*-[$(Py_4)ReO_2$]PF_6 catalyzed deoxydehydration [a].

Entry	Substrate	Product	Yield	Reaction Conditions
1	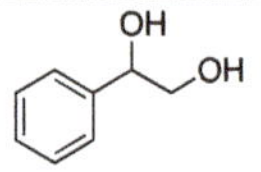	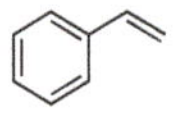	>99%	[Re] 2mol%, PPh_3 (1.1 equiv.) PhCl (0.1M), 180 °C, 3 h, N_2
2	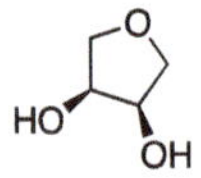	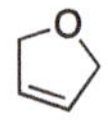	97%	[Re] 2mol%, PPh_3 (1.1 equiv.) PhCl (0.1M), 180 °C, 3 h, N_2
3	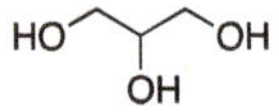	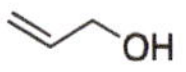	>99%	[Re] 2mol%, PPh_3 (1.1 equiv.) PhCl (0.1M), 180 °C, 3 h, N_2
4	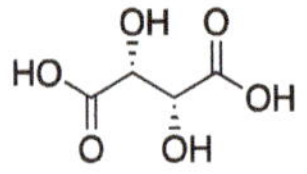	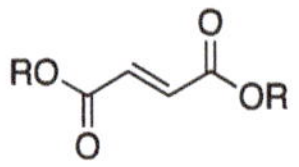	84%	[Re] 2mol%, 3-pentanol (0.1M) 180 °C, 3 h, N_2
5	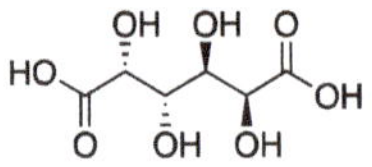	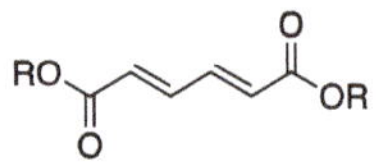	65%	[Re] 2mol%, 3-pentanol (0.1M) 180 °C, 3 h, N_2

[a] Olefin yield was determined by 1H NMR using mesitylene (1.0 equiv., 0.5 mmol) as an internal standard.

3. Materials and Methods

All chemicals including solvents were degassed by either freeze-pump-thaw cycles or degasification under vacuum. Triphenylphosphine was crystallized from ethanol and dried under vacuum. Iododioxobis(triphenylphosphine)rhenium ($IReO_2(PPh_3)_2$) was purchased from Sigma-Aldrich and used without further purification. *cis*-[(*S,S*-BPBP)ReO_2]PF_6 [26], BPMEN [34], $BP^{mdm}MEN$ [34], and *trans*-[$(Py_4)ReO_2$]PF_6 [25] were prepared according to known literature procedures. Unless otherwise stated, all other commercial chemicals were used without further purification. NMR spectra were recorded on a Varian VNMRS400 (400 MHz) instrument at 298 K. ESI-MS spectra were recorded using a Waters LCT Premier XE instrument. IR spectra were recorded with a Perkin–Elmer Spectrum One FTIR spectrometer. All olefinic products are known compounds and were calibrated against mesitylene for quantification.

4. Conclusions

Two new tetradentate N-donor ligand-supported ReO_2^+ complexes, *cis*-[(BPMEN)ReO_2]PF_6 (**2**) and *cis*-[($B^{mdm}PMEN$)ReO_2]PF_6 (**3**), were synthesized and fully characterized in this chapter. These complexes, along with the complex *cis*-[(*S,S*-BPBP)ReO_2]PF_6 (**1**) earlier reported by Che, were found to be active in DODH reactions of diols, suggesting that Re(V) dioxo complexes are able to be involved in DODH catalysis. However, an induction period was observed for the DODH reaction of 1,2-octanediol using these N_2Py_2 supported dioxo-rhenium complexes. Under the same reaction conditions, no induction period was observed for *trans*-[$(Py_4)ReO_2$]PF_6 and a much shorter induction period was observed for pyridinium perrhenate. ESI-MS analysis of the reaction mixtures of DODH reactions catalyzed by *cis*-[(BPMEN)ReO_2]PF_6 showed the formation of protonated ligand and the ReO_4^- anion during catalysis. These combined observations suggested that the Re(V) dioxo complexes might act as precatalysts in DODH catalysis. Treatment of *cis*-[(BPMEN)ReO_2]BPh_4 with an excess amount of water generated a species that is also active in DODH catalysis and that displayed a much shorter induction period than the pristine complex. A similar reaction rate and final 1-octene yield indicated that pristine and "water-treated" *cis*-[(BPMEN)ReO_2]BPh_4 generate the same active DODH species. By comparing the IR spectra of various complexes before and after catalysis and a number of Re-compounds from the literature, we believe that ligand dissociation takes place after the first DODH reaction in which a Re-trioxo species is formed, to subsequently form a rhenium-oxide/hydroxide oligomer that is capable of catalytic turn-over. The remarkably high catalytic activity of *trans*-[$(Py_4)ReO_2$]PF_6 can be explained by the more facile dissociation of pyridine from the Re center compared to the tetradentate N_2Py_2 ligands.

Last but not the least, several vicinal diols and biomass-derived polyols were applied to the simple *trans*-[$(Py_4)ReO_2$]PF_6 catalyst in combination either PPh_3 or 3-pentanol as reductant. Excellent to quantitative olefin yields were obtained for vicinal diol substrates. In the case of bio-based polyols, good olefin yield was obtained when 3-pentanol was used as solvent and reductant.

Supplementary Materials: The following are available online at http://www.mdpi.com/2073-4344/10/7/754/s1, the synthesis and characterization of *cis*-[(BPMEN)ReO_2]PF_6 (**2**), *cis*-[(B^{mdm}PMEN)ReO_2]PF_6 (**3**), and *cis*-[(BPMEN)ReO_2]BPh_4 (**6**); the general procedure of catalysis and the ^{1}H NMR data for olefin products.

Author Contributions: Conceptualization, J.L. and R.J.M.K.G.; methodology, J.L.; software, M.L.; validation, J.L., M.L. and R.J.M.K.G.; formal analysis, J.L.; investigation, J.L.; resources, R.J.M.K.G.; data curation, J.L.; writing—original draft preparation, J.L.; writing—review and editing, J.L., M.L. and R.J.M.K.G.; visualization, J.L.; supervision, R.J.M.K.G.; project administration, R.J.M.K.G.; funding acquisition, R.J.M.K.G. All authors have read and agreed to the published version of the manuscript.

Funding: This research was funded by the Chinese Scholarship Council, Utrecht University and the Netherlands Organization for Scientific Research (NWO).

Conflicts of Interest: The authors declare no conflict of interest.

References

1. Bender, T.A.; Dabrowski, J.A.; Gagné, M.R. Homogeneous catalysis for the production of low-volume, high-value chemicals from biomass. *Nat. Rev. Chem.* **2018**, *2*, 35–46. [CrossRef]
2. Korstanje, T.J.; De Waard, E.F.; Jastrzebski, J.T.; Klein Gebbink, R.J.M. Rhenium-catalyzed dehydration of nonbenzylic and terpene alcohols to olefins. *ACS Catal.* **2012**, *2*, 2173–2181. [CrossRef]
3. Schlaf, M. Selective deoxygenation of sugar polyols to α,ω-diols and other oxygen content reduced materials—A new challenge to homogeneous ionic hydrogenation and hydrogenolysis catalysis. *Dalt. Trans.* **2006**, *39*, 4645–4653. [CrossRef]
4. Adduci, L.L.; McLaughlin, M.P.; Bender, T.A.; Becker, J.J.; Gagné, M.R. Metal-free deoxygenation of carbohydrates. *Angew. Chem. Int. Ed.* **2014**, *53*, 1646–1649. [CrossRef] [PubMed]
5. Raju, S.; Moret, M.-E.; Klein Gebbink, R.J.M. Rhenium-catalyzed dehydration and deoxydehydration of alcohols and polyols: Opportunities for the formation of olefins from biomass. *ACS Catal.* **2015**, *5*, 281–300. [CrossRef]
6. Kaluža, L.; Karban, J.; Gulková, D. Activity and selectivity of Co(Ni)Mo sulfides supported on MgO, Al_2O_3, ZrO_2, TiO_2, MCM-41 and activated carbon in parallel hydrodeoxygenation of octanoic acid and hydrodesulfurization of 1-benzothiophene. *Reac. Kinet. Mech. Catal.* **2019**, *127*, 887–902. [CrossRef]
7. De Vicente Poutás, L.C.; Castiñeira Reis, M.; Sanz, R.; López, C.S.; Faza, O.N. A radical mechanism for the vanadium-catalyzed deoxydehydration of glycols. *Inorg. Chem.* **2016**, *55*, 11372–11382. [CrossRef] [PubMed]
8. Galindo, A. DFT studies on the mechanism of the vanadium-catalyzed deoxydehydration of diols. *Inorg. Chem.* **2016**, *55*, 2284–2289. [CrossRef]
9. Petersen, A.R.; Nielsen, L.B.; Dethlefsen, J.R.; Fristrup, P. Vanadium-catalyzed deoxydehydration of glycerol without an external reductant. *ChemCatChem* **2018**, *10*, 769–778. [CrossRef]
10. Jiang, Y.Y.; Jiang, J.L.; Fu, Y. Mechanism of vanadium-catalyzed deoxydehydration of vicinal diols: Spin-crossover-involved processes. *Organometallics* **2016**, *35*, 3388–3396. [CrossRef]
11. Hills, L.; Moyano, R.; Montilla, F.; Pastor, A.; Galindo, A.; Álvarez, E.; Marchetti, F.; Pettinari, C. Dioxomolybdenum(VI) complexes with acylpyrazolonate ligands: Synthesis, structures, and catalytic properties. *Eur. J. Inorg. Chem.* **2013**, *19*, 3352–3361. [CrossRef]
12. Dethlefsen, J.R.; Lupp, D.; Oh, B.-C.; Fristrup, P. Molybdenum-catalyzed deoxydehydration of vicinal diols. *ChemSusChem* **2014**, *7*, 425–428. [CrossRef] [PubMed]
13. Lupp, D.; Christensen, N.J.; Dethlefsen, J.R.; Fristrup, P. DFT study of the molybdenum-catalyzed deoxydehydration of vicinal diols. *Chem.-A Eur. J.* **2015**, *21*, 3435–3442. [CrossRef]
14. Dethlefsen, J.R.; Lupp, D.; Teshome, A.; Nielsen, L.B.; Fristrup, P. Molybdenum-catalyzed conversion of diols and biomass-derived polyols to alkenes using isopropyl alcohol as reductant and solvent. *ACS Catal.* **2015**, *5*, 3638–3647. [CrossRef]

15. Beckerle, K.; Sauer, A.; Spaniol, T.P.; Okuda, J. Bis(phenolato)molybdenum complexes as catalyst precursors for the deoxydehydration of biomass-derived polyols. *Polyhedron* **2016**, *116*, 105–110. [CrossRef]
16. Stalpaert, M.; De Vos, D. Stabilizing effect of bulky β-diketones on homogeneous Mo catalysts for deoxydehydration. *ACS Sustain. Chem. Eng.* **2018**, *6*, 12197–12204. [CrossRef]
17. Navarro, C.A.; John, A. Deoxydehydration using a commercial catalyst and readily available reductant. *Inorg. Chem. Commun.* **2019**, *99*, 145–148. [CrossRef]
18. Tran, R.; Kilyanek, S.M. Deoxydehydration of polyols catalyzed by a molybdenum dioxo-complex supported by a dianionic ONO pincer ligand. *Dalt. Trans.* **2019**, *48*, 16304–16311. [CrossRef]
19. Dethlefsen, J.R.; Fristrup, P. Rhenium-catalyzed deoxydehydration of diols and polyols. *ChemSusChem* **2015**, *8*, 767–775. [CrossRef]
20. Cook, G.K.; Andrews, M.A. Toward nonoxidative routes to oxygenated organics: Stereospecific deoxydehydration of diols and polyols to alkenes and allylic alcohols catalyzed by the metal oxo complex $(C_5Me_5)ReO_3$. *J. Am. Chem. Soc.* **1996**, *118*, 9448–9449. [CrossRef]
21. Raju, S.; Jastrzebski, J.T.B.H.; Lutz, M.; Klein Gebbink, R.J.M. Catalytic deoxydehydration of diols to olefins by using a bulky cyclopentadiene-based trioxorhenium catalyst. *ChemSusChem* **2013**, *6*, 1673–1680. [CrossRef] [PubMed]
22. Li, J.; Lutz, M.; Otte, M.; Klein Gebbink, R.J.M. A Cp^{tt}-Based Trioxo-Rhenium catalyst for the Deoxydehydration of Diols and Polyols. *ChemCatChem* **2018**, *10*, 4755–4760. [CrossRef] [PubMed]
23. Kilyanek, S.M.; Denike, K.A.; Kilyanek, S.M. Deoxydehydration of vicinal diols by homogeneous catalysts: A mechanistic overview. *R. Soc. Open Sci.* **2019**, *6*, 1–16.
24. Liu, S.; Senocak, A.; Smeltz, J.L.; Yang, L.; Wegenhart, B.; Yi, J.; Kenttämaa, H.I.; Ison, E.A.; Abu-Omar, M.M. Mechanism of MTO-catalyzed deoxydehydration of diols to alkenes using sacrificial alcohols. *Organometallics* **2013**, *32*, 3210–3219. [CrossRef]
25. Michael McClain, J.; Nicholas, K.M. Elemental reductants for the deoxydehydration of glycols. *ACS Catal.* **2014**, *4*, 2109–2112. [CrossRef]
26. Ng, V.Y.-M.; Tse, C.-W.; Guan, X.; Chang, X.; Yang, C.; Low, K.-H.; Lee, H.K.; Huang, J.-S.; Che, C.-M. *cis*-dioxorhenium(V/VI) complexes supported by neutral tetradentate N_4 ligands. Synthesis, characterization, and spectroscopy. *Inorg. Chem.* **2017**, *56*, 15066–15080. [CrossRef]
27. Raju, S.; Van Slagmaat, C.A.M.R.; Li, J.; Lutz, M.; Jastrzebski, J.T.B.H.; Moret, M.-E.; Klein Gebbink, R.J.M. Synthesis of cyclopentadienyl-based trioxo-rhenium complexes and their use as deoxydehydration catalysts. *Organometallics* **2016**, *35*, 2178–2187. [CrossRef]
28. Morris, D.S.; Van Rees, K.; Curcio, M.; Cokoja, M.; Kühn, F.E.; Duarte, F.; Love, J.B. Deoxydehydration of vicinal diols and polyols catalyzed by pyridinium perrhenate salts. *Catal. Sci. Technol.* **2017**, *7*, 5644–5649. [CrossRef]
29. Raju, S. Cyclopentadienyl-Based Trioxo-Rhenium Complexes for the Catalytic Deoxydehydration of Diols and Bio-Based Polyols to Olefins. Ph.D. Thesis, Utrecht University, Utrecht, The Netherlands, 15 June 2015.
30. Shiramizu, M.; Toste, F.D. Deoxygenation of biomass-derived feedstocks: Oxorhenium-catalyzed deoxydehydration of sugars and sugar alcohols. *Angew. Chem. Int. Ed.* **2012**, *51*, 8082–8086. [CrossRef]
31. Herrmann, W.A.; Kühn, F.E. Organorhenium oxides. *Acc. Chem. Res.* **1997**, *30*, 169–180. [CrossRef]
32. Herrmann, W.A.; Fischer, R.W. Multiple bonds between main-group elements and transition metals. 136.[1]"Polymerization" of an organometal oxide: The unusual behavior of methyltrioxorhenium(VII) in water. *J. Am. Chem. Soc.* **1995**, *117*, 3223–3230. [CrossRef] [PubMed]
33. Lupacchini, M.; Mascitti, A.; Canale, V.; Tonucci, L.; Colacino, E.; Passacantando, M.; Marrone, A.; d'Alessandro, N. Deoxydehydration of glycerol in presence of rhenium compounds: Reactivity and mechanistic aspects. *Catal. Sci. Technol.* **2019**, *9*, 3005–3296. [CrossRef]
34. Mukherjee, J.; Balamurugan, V.; Gupta, R.; Mukherjee, R. Synthesis and properties of Fe^{III} and Co^{III} complexes: Structures of $[(L^2)Fe(N_3)_3]$, $[(L^2)Fe_2(\mu\text{-}O_2CMe)_2][ClO_4]_2\cdot 2H_2O$ and $[(L^2)Co_2(\mu\text{-}OH)_2(\mu\text{-}O_2CMe)][ClO_4]_3\cdot MeCN$ [L^2 = methyl[2-(2-pyridyl)ethyl](2-pyridylmethyl)amine]. *Dalton Trans.* **2003**, *2*, 3686–3692. [CrossRef]

Article

Catalytic Cracking of Heavy Crude Oil over Iron-Based Catalyst Obtained from Galvanic Industry Wastes

Estefanía Villamarin-Barriga [1], Jéssica Canacuán [1], Pablo Londoño-Larrea [1], Hugo Solís [1], Andrés De La Rosa [1], Juan F. Saldarriaga [2] and Carolina Montero [1,*]

[1] Chemical Engineering Faculty, Universidad Central del Ecuador, Ritter s/n & Bolivia, Quito 17-01-3972, Ecuador; jevillamarin@uce.edu.ec (E.V.-B.); jessicacanacuan@gmail.com (J.C.); palondono@uce.edu.ec (P.L.-L.); hfsolis@uce.edu.ec (H.S.); adelarosa@uce.edu.ec (A.D.L.R.)

[2] Department of Civil and Environmental Engineering, Universidad de los Andes, Carrera 1Este #19A-40, Bogotá 111711, Colombia; jf.saldarriaga@uniandes.edu.co

* Correspondence: cdmontero@uce.edu.ec; Tel.: +593-22544361

Received: 2 June 2020; Accepted: 17 June 2020; Published: 3 July 2020

Abstract: Sewage sludge from the galvanic industry represents a problem to the environment, due to its high metal content that makes it a hazardous waste and must be treated or disposed of properly. This study aimed to evaluate the sludge from three galvanic industries and determine its possible use as catalysts for the synthesis of materials. Catalyst was obtained from a thermal process based on dried between 100–120 °C and calcination of sludges between 400 to 700 °C. The physical–chemical properties of the catalyst were analyzed by several techniques as physisorption of N_2 and chemisorption of CO of the material. Catalytic activity was analyzed by thermogravimetric analysis of a thermo-catalytic decomposition of crude oil. The best conditions for catalyst synthesis were calcination between 400 and 500 °C, the temperature of reduction between 750 and 850 °C for 15 min. The catalytic material had mainly Fe as active phase and the specific surface between 17.68–96.15 $m^2 \cdot g^{-1}$, the catalysts promote around 6% more weight-loss of crude oil in the thermal decomposition compared with assays without the catalyst. The results show that the residual sludge of galvanic industries after thermal treatment can be used as catalytic materials due to the easiness of synthesis procedures required, the low E-factor obtained and the recycling of industrial waste promoted.

Keywords: sludge; catalytic material; galvanic industry; waste valorization

1. Introduction

The generation of waste worldwide for 2016 was estimated at 2.01 billion tons, while the world generation is expected to reach 3.30 billion $t \cdot day^{-1}$ by 2100 [1]. The global trend shows that industrial waste generation is almost 18 times higher than municipal waste. Industrial waste increases significantly as income level increases [1]. Among these is the waste from the galvanic industry, which produces on average in the United States about 150,000 $t \cdot year^{-1}$ [2], similar values are reported by the European Union [3]. Sludge represents approximately 25 m^3 for every million tons of textile wastewater and two-thirds of these sludges have physicochemical properties that can be modified [4,5].

The recovery of galvanic baths is beneficial in the mechanical industry to avoid corrosion problems in materials [6]. Nevertheless, it generates hazardous industrial waste as the residual sludge of the process baths, mainly from rinsing in the stages of stripping and degrease [7].

Since galvanic sludge wastes are hazardous due to their chemical composition, they need final treatment or disposal in a safe landfill. Worldwide the implemented skills are mainly: disposal in soils, thermal treatments as pyrolysis/incineration [8–11], encapsulation [12], as a component of other solid materials [3,7,13–15]. Similarly, some techniques as inertization of heavy metals present in this type of sludge have been studied [2]. However, these treatments represent additional costs for galvanic companies.

The sludge of the galvanic industry has been mainly treated by thermochemical processes such as pyrolysis and incineration. The byproducts of this process have been used as adsorbents [5,16]. Due to the diverse metallic content of this residual industrial sludge, there is the possibility of recovering these residues for the synthesis of catalysts [15,17–20]. Sludge obtained from waste from textile industries, aluminum, galvanic and tannery can be converted on a catalyst for the oxidation of the propane, getting conversions of more than 95% [21,22].

Catalysts prepared from the sludge of the process of ferrite were tried for CO conversion, the results indicate that the Cu-ferrite catalyst can convert CO to CO_2 at an inlet CO concentration of 4000 ppm and a space velocity of 6000 h^{-1} were held at 140 °C [23]. The selective catalytic NO reduction was studied using metal catalyst doped with carbon from the residual sludge of ferrite, at 300 °C conversion >99.7% of NO was reported [24]. The sludge waste from wastewater treatment in the textile industries has metals like Fe and Cr were used as catalytic materials during their reducing phase for the decomposition of hydrocarbons [25]. Fe–char catalyst from tank cleaning oily sludge for the catalytic cracking of oily sludge at 800 °C, the oil conversion efficiency reached is around 95.8% [26]. Carbon–silica derived from SiC–Si sludge has been proven as support for Fe catalysts; better results are shown when Fe was loaded by chemical vapor infiltration than incipient wetness impregnation [27]. Metallic iron from the dyeing sludge ash was probed as a catalyst for biomass gasification. It showed similar behavior that of the commercially available iron–chrome-based catalyst for the same equivalent total amount of Fe_2O_3 [28].

This work has aimed to evaluate residual sludge from the galvanic industry as catalysts for which this work raises the synthesis of catalytic materials applied in the reactions of cracking of heavy crude oil.

2. Results and Discussion

2.1. Sludge and Catalyst Characterization

The thermal treatment (calcination) provided the catalytic characteristics at the residual sludge as the drying of the internal water, volatile substances. The metallic content in the sludge was oxidized in this process. Elemental analysis was done to evaluate the organic material eliminated by calcination.

The results of the elemental analysis and physical properties of the dried sludges and catalytic material obtained are shown in Table 1. With a prior drying process can be observed that the optimal drying temperature is 120 °C for L1 and L2 and 100 °C for L3, the differences in the drying temperature can be due at the presence of different compounds and additives on the sludges [29–31].

There was observed that thermal treatment promotes changes in the physical–chemical composition of raw sludge. A higher quantity of sulfur was detected on C3, probably due to the use of sulfuric acid and other sulfur compounds in the galvanic process, with the increase of temperature, these compounds could be converted in oxides [32]. On the C1 and C2 catalyst, the sulfur content was constant. From L2 and its derived catalytic material C2, the content of carbon is higher than the others. It could be due to carbonates and surfactants used on the stripping process [33]. Carbon and sulfur values are low compared with the reported by other authors in this type of sludge, finding carbon content between 25% to 40% and for sulfur between 4% and 22% [5,13,16].

Table 1 shows the effects of calcination temperature on the reduction of surface area and pore volume of catalytic materials. The decrease of surface area can be associated with the collapse of the pore structure [21,34]. The catalyst with the highest surface area was C1. For calcination temperature of 400 °C, the three catalysts showed more upper surface area than the other temperatures, but the high organic content can decrease the catalytic activity. Due to the low surface area, the samples calcined at 700 °C have not been used on the catalytic activity test because the physisorption capability can be disfavored.

Table 1. Elemental analysis and physical properties of dried residual sludges (L1–L3) and catalytic material synthesized (C1–C3).

Sample	T, °C	% N	% C	% H	% S	Surface Area, $m^2 \cdot g^{-1}$	Pore Volume, $cm^3 \cdot g^{-1}$	E-Factor
L1	120	0.10	2.36	2.05	0.94	nm	nm	nm
C1	400	0.02	0.65	0.35	0.39	96.15	0.12	0.1
	500	0.05	0.44	0.19	0.38	71.37	0.10	
	700	0.03	0.16	0.08	0.35	14.98	0.02	
L2	120	0.15	6.48	0.52	0.20	nm	nm	nm
C2	400	0.02	5.48	0.18	0.22	73.46	0.10	0.1
	500	0.03	5.48	0.14	0.23	63.36	0.09	
	700	0.01	4.77	0.08	0.20	27.56	0.03	
L3	100	1.11	5.86	1.35	2.94	nm	nm	nm
C3	400	0.04	1.74	0.08	3.71	29.72	0.03	0.3
	500	0.03	0.40	0.08	4.23	17.68	0.02	
	700	0.02	0.20	0.06	4.42	4.53	0.01	

nm: not measured.

To determine the sustainability of this process, the Environmental Factor (E) was calculated [35–37]. E-factor correlates the actual amount of waste produced in the process with the desired product (E = mass of waste/mass of product). In this case, the desired product was the catalyst. The ideal E-factor is zero. The values E-factor are in Table 1; there were between 0.1 and 0.3. This process is sustainable because it can valorize industrial waste minimizing the waste produced in the process [35]. The galvanic companies gave the information that raw sludges had metallic content based on iron and zinc. The MSDS and technical data sheets of the streams on the galvanic processes verified it.

Figure 1 shows the FTIR analysis that allows the identification of functional groups for both sludge and catalytic materials. On L1 and C1 OH groups were observed, probably due to Zn(OH) (1407, 1478 and 1630 cm^{-1}) and other bands in 448 cm^{-1} (Fe_2O_3) and 584 cm^{-1} (Fe_3O_4) [38]. For L2 and C2 were identified bands on 2864 and 875 cm^{-1} correlated with FeOO bond, on 464 cm^{-1} for Fe_2O_3 and 799 cm^{-1} for FeO. The presence of ZnO (606, 712 and 571 cm^{-1}) and Zn(OH) (1801 cm^{-1}) also were detected [39]. Additionally, the band on 448 cm^{-1} identified on L3 and C3 can be associated with $FeSO_4$ [40]. In the sludges, it was observed other peaks at 2923 and 3373 cm^{-1}, related to OH bonds due to the presence of the Fe–Zn hydroxyl groups on the surface [41], these bonds decrease its intensity while the calcination temperature increases.

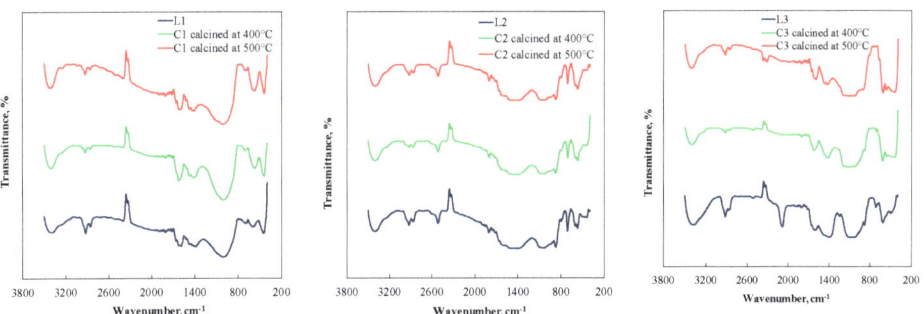

Figure 1. Fourier transform infrared spectroscopy (FTIR) spectra of galvanic sludge and catalytic material.

2.2. Catalyst Reducibility Analysis

The reducibility capability of the synthesized catalysts was studied using the reduction at program temperature (TPR) with hydrogen as a reduction agent, H_2–TPR profiles are shown in Figure 2. C1 and C2 catalysts show three regions for reduction: the first region around 300–550 °C, the second and the prominent area around 550–800 °C, while the third region located between 800–900 °C. In the case of C3 shows only one prominent peak at 650 °C.

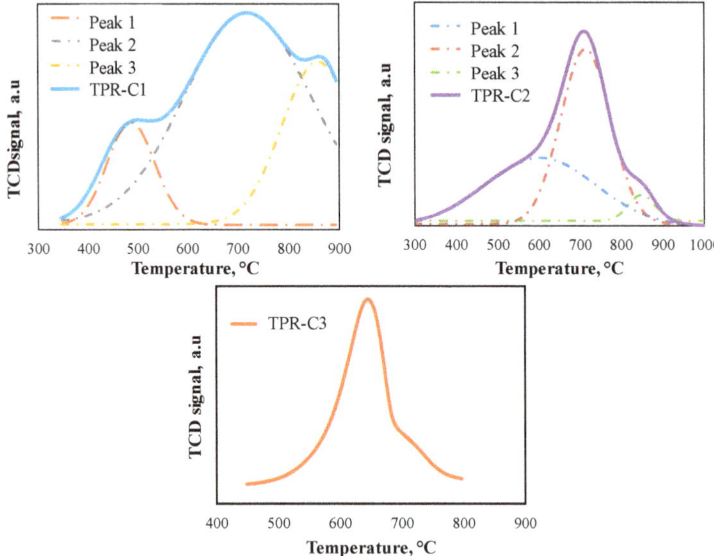

Figure 2. Temperature-programmed reduction (TPR) profile for C1, C2 and C3 calcined at 500 °C.

Similar behavior has been shown for other Fe based catalysts, for Fe/Al_2O_3 catalyst. Other authors have shown the same three areas that were evidenced in this work. The first one between 250–460 °C that can be described as a transformation towards Fe_2O_3, a second region that is between 480–800 °C that the authors associate with Fe_3O_4 formation. Moreover, the last area that is between 830–950 °C, which is related to the Fe formation [42,43].

Likewise, in this work, the second region again is dominant, but concluded that the area could be ascribable at Fe or Zn sulfides. The first region only is present when the Fe sulfides are doped with Zn and the third region increases at the same time, more content of zinc was loaded. Concerning catalyst C3, a particular behavior was found, since the presence of Zn significantly displaced the temperature towards a higher temperature for C1 and C2, this behavior is similar to other studies [44]. Therefore, the reduction peak of ZnO to a metallic state is around 700 °C, but interactions with sulfur compounds can change the reduction temperature at about 650 °C [45]. With this analysis is inferred that the C1–C2 mainly had iron phases while C3 has Fe–Zn.

Figure 3 shows a study of the time influence on the reducibility of the catalyst. The catalysts were analyzed at three different times, 15, 30 and 45 min. The reduction was made at 750 °C for C1, at 850 °C for C2 and 800 °C for C3 to compare the reducibility at different temperatures. In this test, the same behavior of the previous analysis is evidenced, in which the three catalysts present the same peaks (Figure 2). On the three catalysts was observed the complete reduction of the metallic phase at 15 min, longer reduction times did not show significantly higher reducibility. These results are important to be able to extrapolate at some point, the process at an industrial scale.

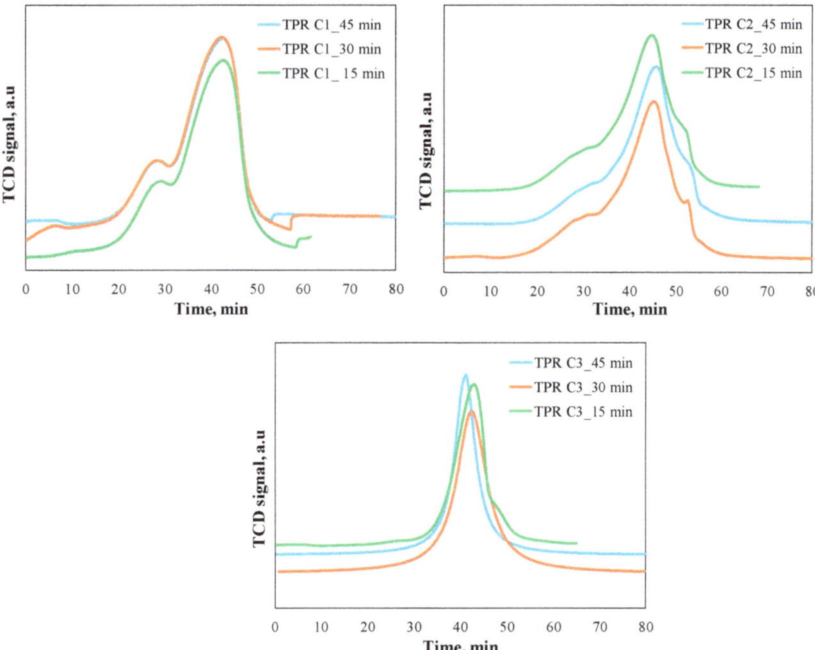

Figure 3. TPR profiles for C1, C2 and C3 at different reduction times on stream.

2.3. Metallic and Textural Properties of the Catalytic Material

The calcined catalyst and the raw sludge were not analyzed through X-ray diffraction (XRD). A similar metallic composition was found in the TPR analysis and while for catalysis application, the synthesized material was reduced. In the case of the reduced catalyst, an XRD analysis was performed. Peak identification was completed with the platform AtomWork: Inorganic Materials Database from NIMS Materials Database (National Institute for Materials Science, Tsukuba, Ibaraki, Japan) [46].

XRD patterns are shown in Figure 4, C1 and C2 had similar patterns, but in the C3, more quantity of peaks had appeared. In the three catalysts were observed mainly peaks corresponding to iron structures and other structures as Zn were defined by FTIR [47–53].

Sulfide structures as FeS (2θ = 26.8, 28.5, 32.85°) were identified in three catalysts, but in C3, ZnS (2θ = 30.5°) were identified too. This compound agreed with the sulfur content quantified in the elemental analysis; for this reason, the peaks in C3 were better defined.

ZnO structures were present in all the catalysts (2θ = 31.7, 34.28°), but C3 had more presence of this structure peaks than the other catalysts (2θ = 47.49, 56.52, 62.8, 67.9 and 72.7°), ZnO_2 (2θ = 41.23°) on C2 and C3, but Zn0 peak (2θ = 36.23°) were seen only in C3. The principal presence in the Zn structures in C3 could be associated with the TPR analysis, in which a single main peak was evident, due to the presence of these structures. In the case of the other two catalysts, the behavior was different because it presented three reduction regions.

Different iron structures were identified in the three catalysts analyzed. Fe_3O_4 (2θ = 35.5°) were found in C1, Fe_2O_3 (2θ = 39.67°) were identified in C2 and FeO (2θ = 50.50, 51.5°), while for C3 the three iron compounds Fe_3O_4 (2θ = 35.5°), Fe_2O_3 (2θ = 39.67°) and FeO (2θ = 50.50, 51.5°) were evidenced. The main peak of FeO (2θ = 44.6, 65.1°) was prominent in C1 and C2, while in the case of C3 was decomposed in two lower peaks. Of the three catalysts, C2 had a lower intensity compared to the other two catalysts.

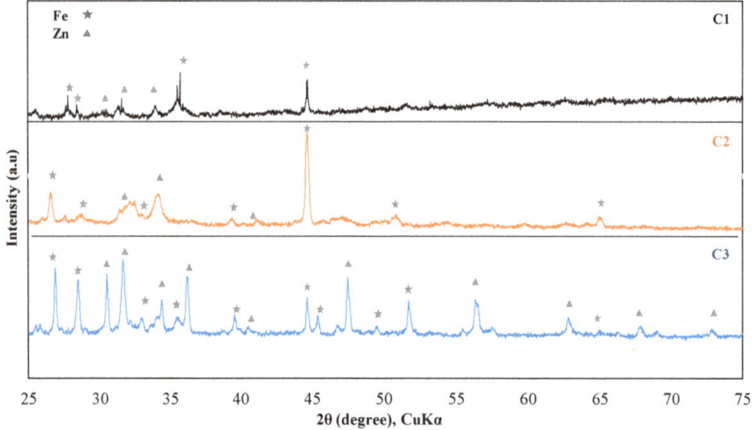

Figure 4. XRD patterns of reduced catalysts C1, C2 and C3.

Scanning electron microscopy-energy dispersive spectrometry (SEM-EDS) confirmed the presence of these metallic phases on the catalytic material. On the surface of C1 were quantified 41.65% Fe and 16.91% Zn, C2 has 19.56% Fe and 7.33% Zn, while C3 has 7.22% Fe and 33.26% Zn. The results were concordant with XRD element identification. The surface morphology of the catalysts by SEM microscopy is shown in Figure 5.

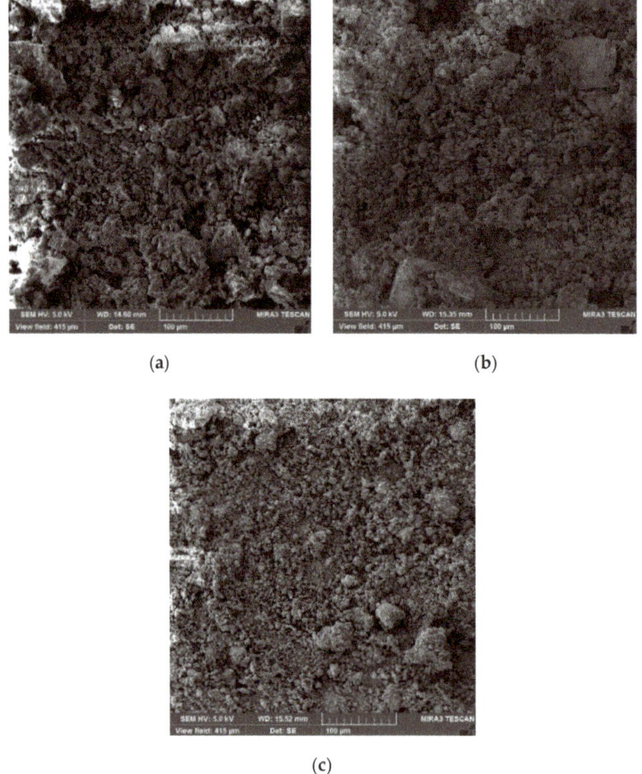

Figure 5. SEM images of C1 (**a**), C2 (**b**) and C3 (**c**) calcined at 550 °C (view at 500×).

No homogeneous particles can be observed in C1–C2, while for C3, it can see some regularity size. Similarly, no crystal-defined structures of catalytic material were found in any of the catalysts evaluated. It may be due to the low homogeneity of the raw sludge. Likewise, differences in morphology between C1 and C3 can be seen, which may be related to the differences in the evaluated surface areas, 96.15 and 17.68 $m^2 \cdot g^{-1}$, respectively (Table 1). To obtain irregular particles with good crystallinity is needed high reaction temperature of 1050 °C [54], but this temperature can cause sinterization for catalyst purposes.

2.4. Catalytic Evaluation of the Synthesized Materials

The chemisorption capacity of the catalytic material was analyzed with CO pulses (Figure 6). This information is important to identify catalytic applications correlated at the CO conversion. For catalysts C1 and C2, it was observed that both could adsorb CO. It is evident that in both the saturation of the surface is in the third pulse, around 10 min. In addition, the adsorption rate is similar for C1 and C2. Therefore, these could be used in CO reactions.

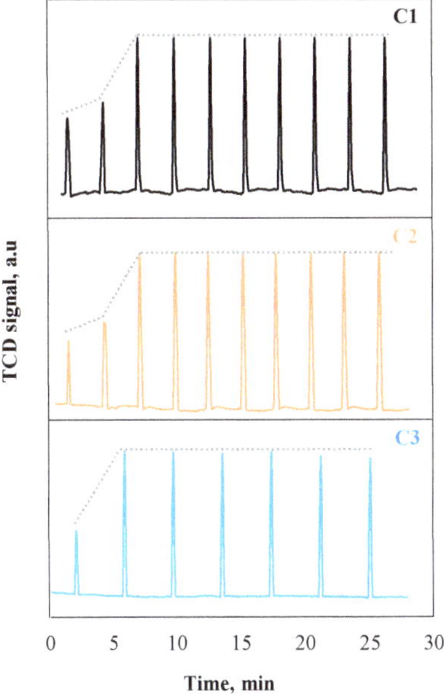

Figure 6. Thermal conductivity detector (TCD) response to CO pulse chemisorption by C1, C2 and C3. C1 and C2 saturation reached after 2 peaks (10 min), C3 saturation reached after 1 peak (3 min).

A different behavior was found for C3. The first time it was not possible to detect any signal. The CO flux was increased, observing a rapid saturation on the second peak around 3 min, a shorter time than in the other two catalysts. This behavior could be attributed to the Zn content compared to the other two that the main compound is iron. In addition, another factor is its low specific surface.

Therefore, this catalyst, C3, does not have characteristics to be used as a catalyst in CO processes.

The catalytic activity of the synthesized materials was analyzed in a reaction to the thermal decomposition of heavy crude oil. The reaction was carried out by thermogravimetric analysis (TGA) in an N_2 atmosphere. This technique has been used previously to characterize the heavy crude oils [55–57].

The loss of mass is considered as indicative of catalytic decomposition, where higher mass-loss on crude oil was regarded as better catalytic activity. These assays were compared with reactions without catalyst and a reaction performed with a commercial catalyst (FCC type, 265 $m^2 \cdot g^{-1}$) at the same space–time (0.15 g catalyst g^{-1} crude oil). The results are in Table 2; all the catalysts promote the weight-loss while the temperature increases as expected.

Table 2. Effect of the catalyst on the weight loss of crude oil on the thermal decomposition reaction.

Temperature of Reaction, °C	Sample	Weight Loss, %	Kinetic Expression	R^2	Rate of Mass Loss (avg.), mg·min^{-1}
400	Crude oil without catalyst	54.30	$\frac{dm}{dt} = 0.0015\ m_A^{0.5}$	0.90	0.57
	Crude oil + C1	59.28	$\frac{dm}{dt} = 0.0017\ m_A^{0.5}$	0.92	0.61
	Crude oil + commercial cat.	56.00	$\frac{dm}{dt} = 0.0020\ m_A^{0.5}$	0.95	0.65
450	Crude oil without catalyst	73.20	$\frac{dm}{dt} = 0.0016\ m_A^{0.5}$	0.91	0.64
	Crude oil + C2	84.50	$\frac{dm}{dt} = 0.0018\ m_A^{0.5}$	0.90	0.71
	Crude oil + C3	82.10	$\frac{dm}{dt} = 0.0019\ m_A^{0.5}$	0.92	0.69
	Crude oil + commercial cat.	77.95	$\frac{dm}{dt} = 0.0019\ m_A^{0.5}$	0.94	0.81

dm, differential mass; dt, differential time; mA, mass at time (t), mg.

The three catalysts promote around 5% more mass-loss than the experiment without catalyst and some higher mass-loss compared with a commercial catalyst (C2–C3). For C2 and C3, the mass-loss was enhanced at temperatures higher than 400 °C. The catalytic effect was analyzed by mass-loss data, and it was used to define a kinetic model. The kinetic analysis was done applying the integral method of linearization of kinetic expression 2 $m^{0.5}$ vs. time. The kinetic order of 0.5 provided the best fit with the experimental data according to the determination coefficient (R^2).

The average rate of mass loss was calculated for the reactions. The experimental data were replaced in the kinetic expression for each time. The numerical value at $\left(\frac{dm}{dt}\right)$ was obtained for the extension of reaction.

Kinetic expressions showed that the rate of thermal decomposition promoted by catalytic material derived from sludge had similar values than the commercial catalyst, and an evident higher rate than the test without a catalyst, mainly in C2 and C3 to the reaction at 450 °C. For a technical limitation due to the highly exothermic reaction presented by crude oil itself, it was not able to perform thermal decomposition experiments above 450 °C.

These results permit concluded that the catalyst can be used in cracking applications to replace commercial catalysts.

3. Materials and Methods

The samples of sludges were taken from three galvanic companies with similar technologies located in Quito, Ecuador (Table 3). For company 1 the sludge was named L1; for company 2, L2 and company 3, L3. The catalytic materials derived from the sludges were identified as C1, C2 and C3.

Table 3. Nomenclature used for the different sludges evaluated.

Company	Sludge	Catalytic Material
Company 1	L1	C1
Company 2	L2	C2
Company 3	L3	C3

All the analyses of characterization were performed in triplicate to guarantee the best possible results.

3.1. Catalytic Preparation

The residual sludges were dried as is described on ASTM-D2216 in a drying oven (Nabertherm-TR60, Bahnhofstr, Lilienthal, Germany) for 4 h. The dried sludges were sieved between 150–180 μm with a Tyler's-sieve series.

The residual sludges were calcined between 400–700 °C for 4 h in a muffle furnace (Thermo Scientific- Thermolyne, Waltham, MA, USA). This process was done to remove the organic residues and to oxidize the metallic phase on the sludges. After the thermal process, the residual sludges were considered as catalytic material.

The sludges and the catalytic material were analyzed by elemental analysis based on the ASTM-D5373 to identify the organic material in an elemental analyzer (Elementar-Vario Macro Cube, Langenselbold, Germany). Around 10 mg of the sample was exposed to oxidative decomposition at 1150 °C and subsequent reduction at 850 °C, to quantify carbon, hydrogen, nitrogen and sulfur (CHNS) composition. Before the experiment, the samples were prepared by digestion microwave system (Milestone-ETHOS UP, Sorisole, BG, Italy).

Fourier-transformed infrared spectroscopy (FTIR) (PerkinElmer-Spectrum Two spectrometer, Waltham, MA, USA) was used to identify the bonds related to organic and inorganic compounds on the sludges. The dried samples were pulverized in an agate mortar and then mixed with KBr (PerkinElmer, Waltham, MA, USA), in a weight ratio 1:100. Later, a pellet was formed in a press and then read on FTIR.

3.2. Catalyst Characterization

The surface area and pore volume of the catalytic material were determined by N_2 adsorption–desorption in a surface area analyzer (Horiba-SA 9600, minami-ku, Kyoto, Japan). The equipment uses the flowing gas method to acquire gas adsorption and desorption curves, and the surface area was by the single-point BET method. A sample of 0.15 g was loaded in a *U*-tube and degasified for 2 h at 300 °C to clean the surface. Later, the N_2 adsorption–desorption was done using liquid N_2 (Enox S.A, Quito, Pichincha, Ecuador).

Temperature-programmed reduction (TPR) was done to study the reducibility of the catalysts in an automated chemisorption analyzer (Micromeritics-AutoChem II, 2920, Norcross, GA, USA). The catalyst samples (200 mg) were reduced at the heating rate of 10 $°C{\cdot}min^{-1}$ up 1000 °C under a flow (50 $mL{\cdot}min^{-1}$) of 10% H_2/Ar (99.999%, INDURA, Quito, Pichincha, Ecuador). The consumption of hydrogen was monitored with a thermal conductivity detector (TCD). Experiments were performed at 15, 30 and 45 min to define the effect to the time on the total reduction.

X-ray diffraction was done (Malvern Panalytical-Empyrean, Malvern Worcestershire, UK) with CuKα radiation source and Scanning electron microscopy SEM/EDS in a (TESCAN-MIRA3 FEG SEM, Brno, Kohoutovice, Czech Republic) were used to study the metallic content on the reduced catalyst. These assays were done in an external laboratory CENCINAT ESPE Laboratory (Sangolquí, Pichincha, Ecuador).

3.3. Catalytic Material Evaluation

The CO chemisorption was studied with CO pulses technique (Micromeritics-AutoChem II, 2920, Norcross, GA, USA). Before the experiments, the catalysts were reduced, as described previously. Chemisorption experiments were done with 200 mg of the samples at 50 °C [58] exposed at the catalyst at 10% CO/He (99,999%, INDURA, Quito, Pichincha, Ecuador). The total adsorption amount of CO was detected by TCD. The gas uptake was measured from a sequence of small pulses until saturation was obtained.

The catalytic activity of the synthesized materials was evaluated with a thermal decomposition reaction with a crude oil sample (18.9 °API, PETROECUADOR, Esmeraldas, Esmeraldas, Ecuador). A thermogravimetric analysis in a thermo-balance (Mettler Toledo-TGA1 SF/1100, Columbus, OH, USA), was realized in a reaction atmosphere of N_2 (30 mL·min^{-1}, 99,999%, INDURA, Quito, Pichincha, Ecuador) to determine the loss of mass promoted for the catalytic activity. In this assay, a space–time of 0.15 g catalyst g^{-1} crude oil was used. The studies were performed at 400 °C for C1 and 450 °C for C2–C3 with a rate of heating to 10 °C·min^{-1}. The reaction had 1 h to time on stream and the mass loss was recorded every 35 s.

Before the study, the catalytic material was reduced in a 10% H2/Ar. Assays without catalyst and FCC commercial catalyst (GRACE-ResidCrackeR, Columbia, MD, USA) were done to compare the catalytic effect on the crude oil. The reaction rate of thermo-catalytic decomposition was determined through the integral method of data analysis considering the loss of mass vs. time of reaction.

4. Conclusions

Due to the metal content—mainly iron in their composition—sludge from wastewater treatment from galvanic industries can be used as catalytic material in the thermal decomposition of hydrocarbons—after a thermal treatment.

The best catalytic activity was evidenced at the sludges calcinated at 400 °C for C1 and 500 °C for C2–C3. The catalysts had specific surface areas of 96.15, 63.36 and 17.68 m^2·g^{-1}, respectively. For the reduction of the catalyst, the best condition of time was 15 min and 750, 850 and 800 °C for C1, C2 and C3, respectively. The prepared catalyst evidenced at CO chemisorption capacity, those with the best performance were catalysts C1 and C2.

The catalytic material enhanced the rate of reaction on the thermal decomposition of crude oil promotes more weight-loss in a thermo-gravimetric analysis compared to the reactions without the catalyst.

The synthesis presented here is a good and sustainable alternative to commercial (heavy oil cracking) catalysts due to the easiness of synthesis procedures required, the low E-factor obtained and the recycling of industrial waste promoted.

Author Contributions: Conceptualization, C.M., H.S. and A.D.L.R.; methodology, C.M., H.S. and A.D.L.R.; validation, C.M., H.S. and A.D.L.R.; formal analysis, E.V.-B., J.C., P.L.-L. and C.M.; investigation, E.V.-B., J.C., P.L.-L. and C.M.; resources, C.M. and A.D.L.R.; data curation E.V.-B., J.C. and P.L.L.; writing—original draft preparation, C.M., E.V.-B., A.D.L.R., P.L.-L. and J.F.S.; writing—review and editing, C.M., H.S. and J.F.S.; visualization, C.M., H.S. and A.D.L.R.; supervision, C.M.; project administration, C.M.; funding acquisition, C.M. and A.D.L.R. All authors have read and agreed to the published version of the manuscript.

Funding: This work was carried with the financial support of the Central University of Ecuador (UCE)—Advanced Projects Program 2017, Project N° 23.

Acknowledgments: The authors are grateful to the Chemical Engineering Faculty (UCE) for the lab facilities and J. Alvear from Laboratory of Catalyst (FIQ-UCE) for her technical assistance. Juan F. Saldarriaga thanks to Dept. Civil and Environmental Engineering of the Universidad de los Andes.

Conflicts of Interest: The authors declare no conflict of interest.

References

1. Kaza, S.; Yao, L.C.; Bhada-Tata, P.; Van Woerden, F. *What a Waste 2.0: A Global Snapshot of Solid Waste Management to 2050*; Urban Development; World Bank: Washington, DC, USA, 2018.
2. Mymrin, V.; Borgo, S.C.; Alekseev, K.; Avanci, M.A.; Rolim, P.H.B.; Argenda, M.A.; Klitzke, W.; Gonçalves, A.J.; Catai, R.E. Galvanic Cr-Zn and spent foundry sand waste application as valuable components of sustainable ceramics to prevent environment pollution. *Int. J. Adv. Manuf. Technol.* **2020**, *107*, 1239–1250. [CrossRef]
3. Pérez-Villarejo, L.; Martínez-Martínez, S.; Carrasco-Hurtado, B.; Eliche-Quesada, D.; Ureña-Nieto, C.; Sánchez-Soto, P.J. Valorization and inertization of galvanic sludge waste in clay bricks. *Appl. Clay Sci.* **2015**, *105–106*, 89–99. [CrossRef]
4. Huang, M.; Chen, L.; Chen, D.; Zhou, S. Characteristics and aluminum reuse of textile sludge incineration residues after acidification. *J. Environ. Sci.* **2011**, *23*, 1999–2004. [CrossRef]
5. Sohaimi, K.S.A.; Ngadi, N.; Mat, H.; Inuwa, I.M.; Wong, S. Synthesis, characterization and application of textile sludge biochars for oil removal. *J. Environ. Chem. Eng.* **2017**, *5*, 1415–1422. [CrossRef]

6. Cagno, E.; Trucco, P. Cleaner technology transfer in the Italian galvanic industry: Economic and know-how issues. *J. Clean. Prod.* **2008**, *16*, S32–S36. [CrossRef]
7. Stepanov, S.; Morozov, N.; Morozova, N.; Ayupov, D.; Makarov, D.; Baishev, D. Efficiency of Use of Galvanic Sludge in Cement Systems. *Procedia Eng.* **2016**, *165*, 1112–1117. [CrossRef]
8. Svoboda, K.; Baxter, D.; Martinec, J. Nitrous oxide emissions from waste incineration. *Chem. Pap.* **2006**, *60*, 78–90. [CrossRef]
9. Werle, S. A reburning process using sewage sludge-derived syngas. *Chem. Pap.* **2012**, *66*, 99–107. [CrossRef]
10. Rossini, G.; Bernardes, A.M. Galvanic sludge metals recovery by pyrometallurgical and hydrometallurgical treatment. *J. Hazard. Mater.* **2006**, *131*, 210–216. [CrossRef]
11. Cichowicz, R.; Stelęgowski, A. Effect of thermal sludge processing on selected components of air quality in the vicinity of a wastewater treatment plant. *Chem. Pap.* **2019**, *73*, 843–849. [CrossRef]
12. Castañeda Bocanegra, J.J.; Espejo Mora, E.; Cubillos González, G.I. Encapsulation in ceramic material of the metals Cr, Ni, and Cu contained in galvanic sludge via the solidification/stabilization method. *J. Environ. Chem. Eng.* **2017**, *5*, 3834–3843. [CrossRef]
13. Felisberto, R.; Santos, M.C.; Arcaro, S.; Basegio, T.M.; Bergmann, C.P. Assessment of environmental compatibility of glass–ceramic materials obtained from galvanic sludge and soda–lime glass residue. *Process Saf. Environ. Prot.* **2018**, *120*, 72–78. [CrossRef]
14. Luz, C.A.; Rocha, J.C.; Cheriaf, M.; Pera, J. Valorization of galvanic sludge in sulfoaluminate cement. *Constr. Build. Mater.* **2009**, *23*, 595–601. [CrossRef]
15. Bednarik, V.; Vondruska, M.; Koutny, M. Stabilization/solidification of galvanic sludges by asphalt emulsions. *J. Hazard. Mater.* **2005**, *122*, 139–145. [CrossRef] [PubMed]
16. Wong, S.; Yac'cob, N.A.N.; Ngadi, N.; Hassan, O.; Inuwa, I.M. From pollutant to solution of wastewater pollution: Synthesis of activated carbon from textile sludge for dye adsorption. *Chin. J. Chem. Eng.* **2018**, *26*, 870–878. [CrossRef]
17. Amaral, F.A.D.; dos Santos, V.S.; Bernardes, A.M. Metals recovery from galvanic sludge by sulfate roasting and thiosulfate leaching. *Miner. Eng.* **2014**, *60*, 1–7. [CrossRef]
18. Huyen, P.T.; Dang, T.D.; Tung, M.T.; Huyen, N.T.T.; Green, T.A.; Roy, S. Electrochemical copper recovery from galvanic sludge. *Hydrometallurgy* **2016**, *164*, 295–303. [CrossRef]
19. Jandová, J.; Štefanová, T.; Niemczyková, R. Recovery of Cu-concentrates from waste galvanic copper sludges. *Hydrometallurgy* **2000**, *57*, 77–84. [CrossRef]
20. Silva, J.E.; Paiva, A.P.; Soares, D.; Labrincha, A.; Castro, F. Solvent extraction applied to the recovery of heavy metals from galvanic sludge. *J. Hazard. Mater.* **2005**, *120*, 113–118. [CrossRef]
21. Klose, F.; Scholz, P.; Kreisel, G.; Ondruschka, B.; Kneise, R.; Knopf, U. Catalysts from waste materials. *Appl. Catal. B Environ.* **2000**, *28*, 209–221. [CrossRef]
22. Sushil, S.; Scholz, P.; Pollok, K.; Ondruschka, B.; Batra, V.S. Application of industrial waste based catalysts for total oxidation of propane. *Chem. Eng. J.* **2011**, *166*, 568–578. [CrossRef]
23. Lou, J.-C.; Chang, C.-K. Catalytic Oxidation of CO Over a Catalyst Produced in the Ferrite Process. *Environ. Eng. Sci.* **2006**, *23*, 1024–1032. [CrossRef]
24. Zhang, J.; Zhang, J.; Xu, Y.; Su, H.; Li, X.; Zhou, J.Z.; Qian, G.; Li, L.; Xu, Z.P. Efficient Selective Catalytic Reduction of NO by Novel Carbon-doped Metal Catalysts Made from Electroplating Sludge. *Environ. Sci. Technol.* **2014**, *48*, 11497–11503. [CrossRef] [PubMed]
25. Montero, C.; Castañeda, K.M.; Suntasig, Y.M.O.; Oña, D.R.F.; De La Rosa, A. Catalyst Based on Sludge Derived from Wastewater Treatment of Textile Industry. *Chem. Eng. Trans.* **2018**, *70*, 931–936. [CrossRef]
26. Lin, B.; Huang, Q.; Yang, Y.; Chi, Y. Preparation of Fe-char catalyst from tank cleaning oily sludge for the catalytic cracking of oily sludge. *J. Anal. Appl. Pyrolysis* **2019**, *139*, 308–318. [CrossRef]
27. Lee, M.S.; Park, K.Y.; Park, H.K.; Kang, T.W.; Jang, H.D.; Han, S.S.; Jeon, J.-K. Prospective application of carbon-silica derived from SiC-Si sludge as a support for Fe catalysts. *Korean J. Chem. Eng.* **2017**, *34*, 100–104. [CrossRef]
28. Nam, S.-B.; Park, Y.-S.; Yun, Y.-S.; Gu, J.-H.; Sung, H.-J.; Horio, M. Catalytic application of metallic iron from the dyeing sludge ash for benzene steam reforming reaction in tar emitted from biomass gasification. *Korean J. Chem. Eng.* **2016**, *33*, 465–472. [CrossRef]
29. Zhu, F.; Jiang, H.; Zhang, Z.; Zhao, L.; Wang, J.; Hu, J.; Zhang, H. Research on Drying Effect of Different Additives on Sewage Sludge. *Procedia Environ. Sci.* **2012**, *16*, 357–362. [CrossRef]

30. Ronda, A.; Gómez-Barea, A.; Haro, P.; de Almeida, V.F.; Salinero, J. Elements partitioning during thermal conversion of sewage sludge. *Fuel Process. Technol.* **2019**, *186*, 156–166. [CrossRef]
31. Guangyin, Z.; Youcai, Z. Chapter Three—Sewage Sludge Solidification/Stabilization and Drying/Incineration Process. In *Pollution Control and Resource Recovery for Sewage Sludge*; Guangyin, Z., Youcai, Z., Eds.; Butterworth-Heinemann: Oxford, UK, 2017; pp. 101–160, ISBN 978-0-12-811639-5.
32. Kuzin, E.N.; Chernyshev, P.I.; Vizen, N.S.; Krutchinina, N.E. The Purification of the Galvanic Industry Wastewater of Chromium(VI) Compounds Using Titanium(III) Chloride. *Russ. J. Gen. Chem.* **2018**, *88*, 2954–2957. [CrossRef]
33. Kliopova, I.; Staniškis, J. Optimization of Galvanic Wastewater Treatment Processes. In *Modern Tools and Methods of Water Treatment for Improving Living Standards*; Omelchenko, A., Pivovarov, A.A., Swindall, W.J., Eds.; Springer: Dordrecht, the Netherlands, 2005; pp. 197–208.
34. Torres-Luna, J.A.; Carriazo, J.G.; Sanabria-González, N.R. Calcination Temperature Effect on structural and textural properties of Fe(iii)-TiO_2. *Rev. Fac. Cienc. Básicas* **2014**, *10*, 186–195. [CrossRef]
35. Sheldon, R.A. Fundamentals of green chemistry: Efficiency in reaction design. *Chem. Soc. Rev.* **2012**, *41*, 1437–1451. [CrossRef] [PubMed]
36. Dicks, A.P.; Hent, A. The E Factor and Process Mass Intensity. In *Green Chemistry Metrics: A Guide to Determining and Evaluating Process Greenness*; SpringerBriefs in Molecular Science; Dicks, A.P., Hent, A., Eds.; Springer International Publishing: Cham, Switzerland, 2015; pp. 45–67, ISBN 978-3-319-10500-0.
37. Tieves, F.; Tonin, F.; Fernández-Fueyo, E.; Robbins, J.M.; Bommarius, B.; Bommarius, A.S.; Alcalde, M.; Hollmann, F. Energising the E-factor: The E+-factor. *Tetrahedron* **2019**, *75*, 1311–1314. [CrossRef]
38. Sahoo, S.K.; Agarwal, K.; Singh, A.K.; Polke, B.G.; Raha, K.C. Characterization of γ- and α-Fe_2O_3 nano powders synthesized by emulsion precipitation-calcination route and rheological behaviour of α-Fe_2O_3. *Int. J. Eng. Sci. Technol.* **2010**, 2. [CrossRef]
39. Ortego, J.D.; Barroeta, Y.; Cartledge, F.K.; Akhter, H. Leaching effects on silicate polymerization. An FTIR and silicon-29 NMR study of lead and zinc in portland cement. *Environ. Sci. Technol.* **1991**, *25*, 1171–1174. [CrossRef]
40. Paterson, E. The Iron Oxides. Structure, Properties, Reactions, Occurrences and Uses. *Clay Min.* **2006**, *34*, 209–210. [CrossRef]
41. Iqbal, A.; Jacob, J.; Mahmood, A.; Mehboob, K.; Mahmood, K.; Ali, A.; Bukhari, T.H.; Adrees, M.; Ibrahim, M.; Ahmad, M. Synthesis and characterization of Zn–Mn–Fe nano oxide composites for the degradation of reactive yellow 15 dye. *Phys. B Condens. Matter* **2020**, *588*, 412210. [CrossRef]
42. Fakeeha, A.; Khan, W.; Ibrahim, A.; Al-Otaibi, R.; Alfatesh, A.; Soliman, M.; Abasaeed, A. Alumina supported iron catalyst for hydrogen production: Calcination study. *Int. J. Adv. Chem. Eng. Biol. Sci.* **2015**, *2*, 139–141. [CrossRef]
43. Li, H.; Liu, J.; Li, J.; Hu, Y.; Wang, W.; Yuan, D.; Wang, Y.; Yang, T.; Li, L.; Sun, H.; et al. Promotion of the Inactive Iron Sulfide to an Efficient Hydrodesulfurization Catalyst. *ACS Catal.* **2017**, *7*, 4805–4816. [CrossRef]
44. Liang, K.; Zhang, C.; Xiang, H.; Yang, Y.; Li, Y. Effects of modified SiO_2 on H_2 and CO adsorption and hydrogenation of iron-based catalysts. *J. Fuel Chem. Technol.* **2019**, *47*, 769–779. [CrossRef]
45. Song, H.; Cui, H.; Li, F. The effect of Zn–Fe modified S_2O_8 2−/ZrO_2–Al_2O_3 catalyst for n-pentane hydroisomerization. *Res. Chem. Intermed.* **2016**, *42*, 3029–3038. [CrossRef]
46. National Institute for Materials Science NIMS Materials Database (MatNavi). Available online: https://mits.nims.go.jp/index_en.html (accessed on 12 May 2020).
47. Glavee, G.N.; Klabunde, K.J.; Sorensen, C.M.; Hadjipanayis, G.C. Chemistry of Borohydride Reduction of Iron(II) and Iron(III) Ions in Aqueous and Nonaqueous Media. Formation of Nanoscale Fe, FeB, and Fe_2B Powders. *Inorg. Chem.* **1995**, *34*, 28–35. [CrossRef]
48. Legodi, M.A.; de Waal, D. The preparation of magnetite, goethite, hematite and maghemite of pigment quality from mill scale iron waste. *Dyes Pigments* **2007**, *74*, 161–168. [CrossRef]
49. Picasso, G.; Sun Kou, R.; Gómez, G.; Hermoza, E.; López, A.; Pina, M.P.; Herguido, J. Nanosized catalyst based on Fe Oxide for combustion of n-hexane. *Rev. Soc. Quím. Perú* **2009**, *75*, 163–176.
50. Caballero, D.; Mass, J.; Landinez, D. Optical and Structural Characterization of Zn_2 TiO_4 capped with Mg. *Tumbaga* **2011**, *6*, 165–172.
51. Kumar, H.; Rani, R. Structural and Optical Characterization of ZnO Nanoparticles Synthesized by Microemulsion Route. *Int. Lett. Chem. Phys. Astron.* **2013**, *14*, 26–36. [CrossRef]

52. Dou, J.; Li, X.; Tahmasebi, A.; Xu, J.; Yu, J. Desulfurization of coke oven gas using char-supported Fe-Zn-Mo catalysts: Mechanisms and thermodynamics. *Korean J. Chem. Eng.* **2015**, *32*, 2227–2235. [CrossRef]
53. Yan, Z.; Kang, Y.; Li, D.; Liu, Y.C. Catalytic oxidation of sulfur dioxide over α-Fe_2O_3/SiO_2 catalyst promoted with Co and Ce oxides. *Korean J. Chem. Eng.* **2020**, *37*, 623–632. [CrossRef]
54. Yin, X.; Yue, M.; Lu, Q.; Liu, M.; Wang, F.; Qiu, Y.; Liu, W.; Zuo, T.; Zha, S.; Li, X.; et al. An Efficient Process for Recycling Nd–Fe–B Sludge as High-Performance Sintered Magnets. *Engineering* **2020**, *6*, 165–172. [CrossRef]
55. Kök, M.V.; Varfolomeev, M.A.; Nurgaliev, D.K. Crude oil characterization using TGA-DTA, TGA-FTIR and TGA-MS techniques. *J. Pet. Sci. Eng.* **2017**, *154*, 537–542. [CrossRef]
56. Simo, S.M.; Naman, S.A.; Ahmed, K.R.; Faritovich, A.A. Evaluation of Two Kurdistan-Iraq Crude Oil (T-21A, PF2) by Derivatographic Method. *Int. Res. J. Pure Appl. Chem.* **2020**, 38–46. [CrossRef]
57. Park, Y.C.; Paek, J.-Y.; Bae, D.-H.; Shun, D. Study of pyrolysis kinetics of Alberta oil sand by thermogravimetric analysis. *Korean J. Chem. Eng.* **2009**, *26*, 1608–1612. [CrossRef]
58. Liu, S.; Ren, J.; Zhu, S.; Zhang, H.; Lv, E.; Xu, J.; Li, Y.-W. Synthesis and characterization of the Fe-substituted ZSM-22 zeolite catalyst with high n-dodecane isomerization performance. *J. Catal.* **2015**, *330*, 485–496. [CrossRef]

Article

A Highly Efficient Monolayer Pt Nanoparticle Catalyst Prepared on a Glass Fiber Surface

Teruyoshi Sasaki [1,*], Yusuke Horino [2,3], Tadashi Ohtake [2,4], Kazufumi Ogawa [2] and Yoshifumi Suzaki [2]

1 Graduate School of Engineering, Kagawa University, Takamatsu, Kagawa 761-0396, Japan
2 Faculty of Engineering and Design, Kagawa University, Takamatsu, Kagawa 761-0396, Japan; tomohalida72@gmail.com (Y.H.); tadashi.ohtake@gmail.com (T.O.); kaogawa@pe.kagawa-u.ac.jp (K.O.); suzaki@eng.kagawa-u.ac.jp (Y.S.)
3 Kurashiki Kako Co., Ltd., Kurashiki, Okayama 712-8555, Japan
4 Rika Research Institute G.K., Hamamachi, Daito Shi, Osaka 574-0041, Japan
* Correspondence: s18d551@stu.kagawa-u.ac.jp

Received: 9 April 2020; Accepted: 23 April 2020; Published: 25 April 2020

Abstract: Over the past few years, various nanoparticle-supported precious metal-based catalysts have been investigated to reduce the emission of harmful substances from automobiles. Generally, precious metal nanoparticle-based exhaust gas catalysts are prepared using the impregnation method. However, these catalysts suffer from the low catalytic activity of the precious metal nanoparticles involved. Therefore, in this study, we developed a novel method for preparing highly efficient glass fiber-supported Pt nanoparticle catalysts. We uniformly deposited a single layer of platinum particles on the support surface using a chemically adsorbed monomolecular film. The octane combustion performance of the resulting catalyst was compared with that of a commercial catalyst. The precious metal loading ratio of the proposed catalyst was approximately seven times that of the commercial catalyst. Approximately one-twelfth of the mass of the proposed catalyst exhibited a performance comparable to that of the commercial catalyst. Thus, the synthesis method used herein can be used to reduce the weight, size, and manufacturing cost of exhaust gas purification devices used in cars.

Keywords: nanoparticle; exhaust gas catalyst; precious metal; chemically adsorbed monomolecular film

---

1. Introduction

Carbon monoxide, hydrocarbons, nitrogen oxides (NO_x), and particulate matter such as soot are harmful air pollutants emitted from automobiles. Vehicle emission regulations are becoming increasingly stringent. To reduce the emission of harmful pollutants from automobiles and meet the demand for energy conservation, researchers have attempted to improve the performance of precious metal catalysts used in automobile exhaust gas purification devices and fuel cells. Conventionally, these catalysts are supported on a porous oxide such as silicon dioxide (silica: SiO_2) or aluminum oxide (alumina: Al_2O_3) or a substrate with a large specific surface area such as activated carbon. Supported precious metal nanoparticle catalysts based on metals such as Pt, Rh, and Pd have been widely investigated [1–4]. Such precious metal nanoparticle-based catalysts are typically prepared using the impregnation method [5]. However, the precious metal nanoparticles in the catalysts prepared using this method aggregate at high temperatures, and hence show poor dispersion. Moreover, it is difficult to control the nanoparticle diameter using this method. This results in a significant reduction in the effective surface area and catalytic activity of the nanoparticles. Moreover, precious metals (particularly Pt group metals) are highly expensive, and hence should ideally be used in limited quantities. Motivated by these challenges, herein, we developed a novel method for preparing highly efficient glass fiber-supported Pt nanoparticle catalysts. We deposited Pt nanoparticles uniformly on a chemically adsorbed monomolecular film formed on the surface of a glass fiber support. The nanoparticles were densely packed in a single layer to avoid wastage. The monomolecular film could be removed by heat treatment. The objective of this study was to develop a highly efficient Pt catalyst with a large effective surface area that could maintain high stability and catalytic activity even under high temperature environments such as engine exhaust devices.

2. Results and Discussion

2.1. Observation of Pt Catalyst

Figure 1 shows the field emission scanning electron microscopy (FE-SEM) images of the Pt catalyst surface. Table 1 lists the precious metal loading rates (precious metal particle weight/catalyst weight × 100) of the Pt catalyst prepared herein and the commercial catalyst.

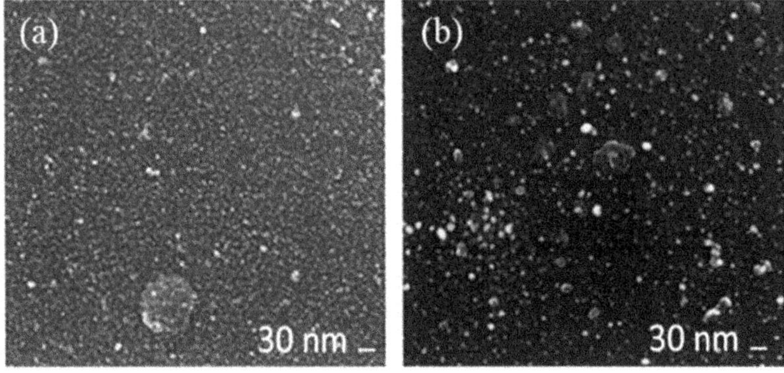

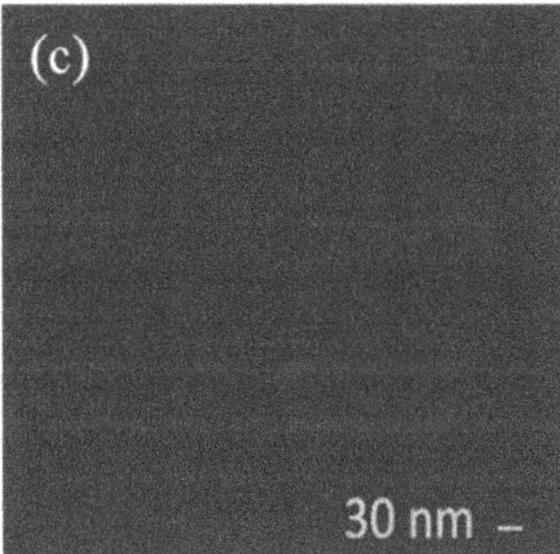

Figure 1. FE-SEM images of the Pt catalyst on the glass fiber substrate: (**a**) before curing, (**b**) after curing, and (**c**) on the glass fiber substrate without the 3-mercaptopropyltrimethoxysilane (MPTS) monolayer after the Pt nanoparticle treatment.

Table 1. Precious metal loading rates (precious metal particle weight/catalyst weight × 100) of the catalysts.

-	Catalyst Weight (g)	Precious Metal Particle Weight (g)	Loading Rate (wt%)
Pt catalyst	0.047	0.0060	12.7
Commercial catalyst sample	0.9722	0.0165	1.7

Figure 1a,b reveal that Pt nanoparticles (on the scale of 10 nm) were dispersed on a single layer on the surface of the glass fiber substrate. Moreover, the Pt particles exhibited minimal aggregation because of sintering at 350 °C (Figure 1b). However, the Pt particles could not be grafted when the 3-mercaptopropyltrimethoxysilane (MPTS) monolayer was formed (Figure 1c). This suggests that using a chemically adsorbed monolayer with a terminal thiol (-SH) is an effective route for forming a single layer of Pt particles with a large specific surface area. As can be observed from Table 1, the loading ratio of the Pt catalyst prepared in this study was approximately seven times larger than that of the commercially available catalyst.

2.2. Thermal Characterization during Catalyst Preparation

Figure 2 shows the thermogravimetry–differential scanning calorimetry (TG–DSC) results of the Pt catalyst. The catalyst showed weight loss as the temperature approached 200 °C. Hence, a DSC peak was observed. This peak can be attributed to the volatilization of MPTS, whose boiling point is approximately 215 °C. The results show that the weight of the MPTS lost by volatilization decreased and that the chemisorbed monomolecular film was sufficiently removed by heating the catalyst at 350 °C.

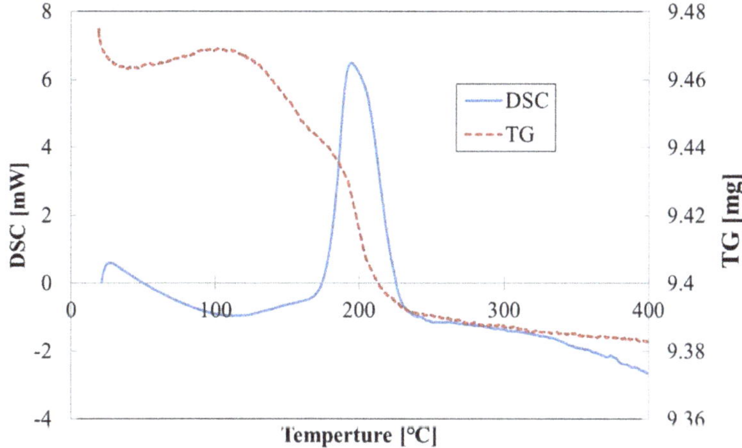

Figure 2. Thermogravimetry–differential scanning calorimetry (TG–DSC) curves of the Pt catalyst.

2.3. Effective Surface Area of the Catalyst

Figure 3 shows the X-ray diffraction (XRD) results of the Pt catalyst before and after the heat treatment. The catalyst showed a Pt peak at 2θ = approximately 40°. After heat treatment, the intensity of this peak increased, whereas the full width at half maximum decreased. This suggests that the particles aggregated, and the particle diameter increased after the heat treatment. This also suggests that the specific surface area of the particles decreased after the heat treatment. However, as the peak obtained after the heat treatment was not extremely sharp, it can be stated that the specific surface area required for the catalytic reaction was sufficient.

It was assumed that the particles shown in Figure 1 a were approximately 5 nm in diameter. In addition, the surface area and volume per particle were approximately 3.14×10^{-16} m^2 and 5.24×10^{-19} cm^3, respectively. Since the density of platinum was 21.45 g/cm^3, the weight and number of units per particle were 1.12×10^{-17} g and 8.90×10^{16}, respectively. Moreover, the BET surface area was approximately 27.96 m^2/g. In contrast, Figure 3 shows that the BET surface area decreased to approximately 14.02 m^2/g because the half width before and after heating had nearly doubled in size. Furthermore, as described above, since the catalytic effect was considered to be saturated at 300 °C or higher, it was found that a sufficient catalytic effect was exhibited by the BET surface area of the platinum particles after the heat treatment. In the referenced literature, the average particle size of platinum was 62 nm. The BET surface area calculated from this was approximately 4.37 mm^2/g. From this, it is considered that the platinum catalyst developed in this study has a large BET surface area and exhibits high performance [6].

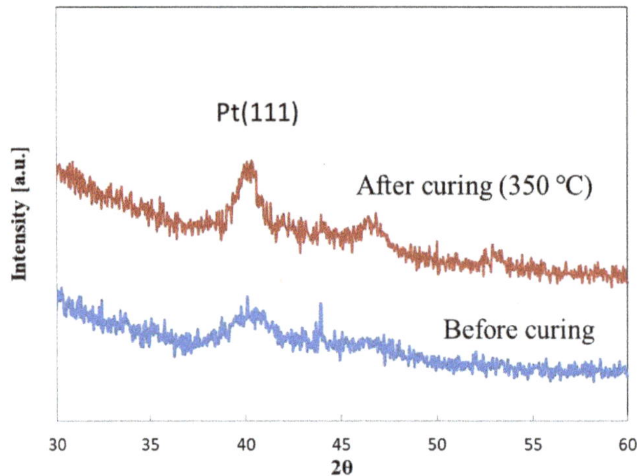

Figure 3. XRD patterns of the Pt catalyst before and after the heat treatment.

2.4. Catalyst Performance Evaluation

Figures 4 and 5 show the amount of water generated and the octane combustion rates of the catalysts as a function of temperature, respectively. In the absence of a catalyst, the amount of water produced by the combustion of octane remained nearly constant at temperatures up to 300 °C. However, the amount of water generated increased substantially with an increase in temperature to 400 °C. This can be attributed to octane self-burning. Volatile organic compounds (VOCs) typically burn completely to form water and carbon dioxide with an increase in temperature from approximately 750 to 850 °C (direct combustion method) [7]. Thus, at 400 °C, octane acted as a VOC and underwent combustion even in the absence of the catalyst, thus producing a large amount of water. At 200 °C, the amount of water produced in the presence of the catalysts (both the Pt and commercial catalyst) was higher than that produced in the absence of the catalysts. Furthermore, at temperatures ≥300 °C, the catalysts showed no significant change in the amount of water produced. This indicates that the performance of the catalysts could be evaluated at temperatures ≥200 °C. However, the performance of the catalysts saturated at temperatures higher than 300 °C. These results indicate that the performance of the Pt catalysts prepared in this study was comparable to that of the commercial catalyst. As can be observed from Table 2, the Pt catalyst showed a higher octane combustion rate per gram of the catalyst and precious metal particles than the commercial catalyst.

Table 2. Precious metal nanoparticle utilization efficiency of the catalysts.

Heating Temperature	Combustion Rate of Octane per g of Catalyst (%)		Octane Combustion Rate per g of Precious Metal Particles (%)	
-	Commercial Catalyst	Pt Catalyst	Commercial Catalyst	Pt Catalyst
50	2.2	56.2	97.8	466.3
100	5.1	49.9	232.4	414.2
200	18.3	195.3	825.5	1620.3
300	40.9	460.7	1848.5	3822.5
400	43.5	536.4	1964.7	4450.6

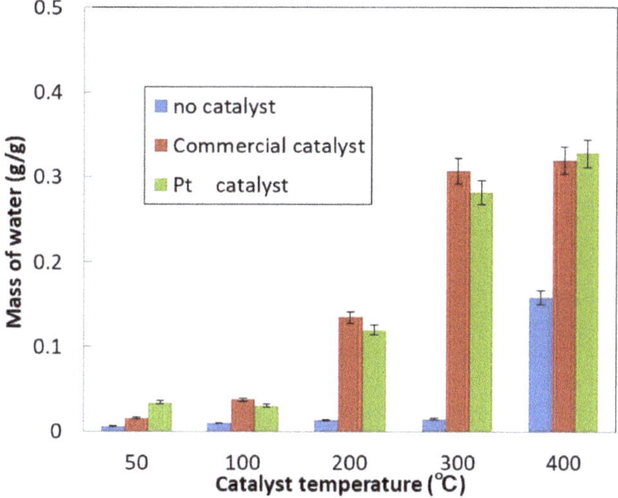

Figure 4. Amount of water produced by the combustion of octane as a function of the catalyst temperature.

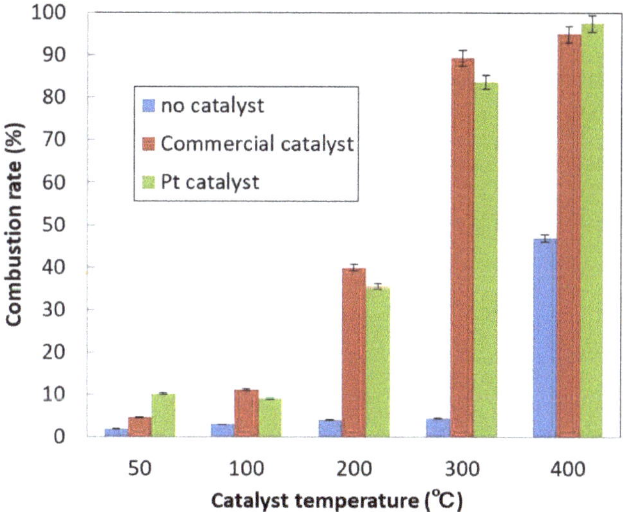

Figure 5. Combustion rate of octane for the catalysts as a function of the catalyst temperature.

At temperatures of 200 °C or higher, the Pt catalyst showed an octane combustion rate that was approximately 12 times higher per gram than that of the commercial catalyst. On the contrary, the octane combustion rate per gram of the precious metal particles of the Pt catalyst was approximately two times higher than that of the commercial catalyst. Pt and Pd show the same catalytic performance for the combustion of hydrocarbons [8]. However, the performance of 1/12th of the mass of the Pt catalyst prepared in this study was comparable to that of the commercial catalyst prepared using the conventional method. This indicates that the precious metal utilization efficiency of the Pt catalyst was higher than that of the commercial catalyst. These results suggest that the Pt catalyst developed in this study was approximately two times more efficient than the commercially available catalyst. Therefore, the synthesis method used in this study can be applied to prepare catalysts for exhaust gas purification devices used in cars to reduce their weight, size, and cost. However, further investigation is required to evaluate the efficiency of this method in the presence of coke.

3. Materials and Methods

3.1. Preparation of Platinum Catalyst

The glass fiber substrate (manufactured by Masuda Rika Kogyo Co., Ltd.) (Osaka, Japan) was heated at 350 °C for 1 h using an electric furnace to remove organic matter from its surface. After the heat treatment, the glass fiber was ultrasonically cleaned with ethanol for 5 min. After drying, the material was further irradiated under air atmosphere for 10 min using an excimer light irradiation device (Ushio Inc., UER20-172B) (Tokyo, Japan), and the surface was further oxidized with generated ozone to achieve hydrophilization. After the irradiation, the glass fiber was immersed at room temperature for 2 h in a chemical adsorption solution prepared under the conditions listed in Table 3 [9].

Table 3. Conditions for preparing the adsorption solution.

-	Chemicals	Concentration	Amount
Solvent	Aqua solvent G-21 (Aqua Chemical Co., Ltd., Chuo-ku, Japan)	Undiluted	50 mL
Adsorbent	3-mercaptopropyltrimethoxysilane (MPTS) (dehydrated chloroform dilution)	0.1 M	76.5 μL
Catalyst	Tetra chlorosilane (TCS) (dehydrated chloroform dilution)	0.01 M	2.5 mL

MPTS (Shin-Etsu Chemical Co., Ltd.) (Tokyo, Japan) was used as the chemical adsorbent in this study (Figure 6). After immersion in the chemical adsorption solution, the glass fiber was removed and subjected to ultrasonic washing with ethanol for 5 min to deposit a monomolecular film on its surface. The glass fiber with the monomolecular film was then placed in a sample case and left under air atmosphere for at least 24 h. Finally, the glass fiber was washed with chloroform, acetone, and ethanol, and then dried.

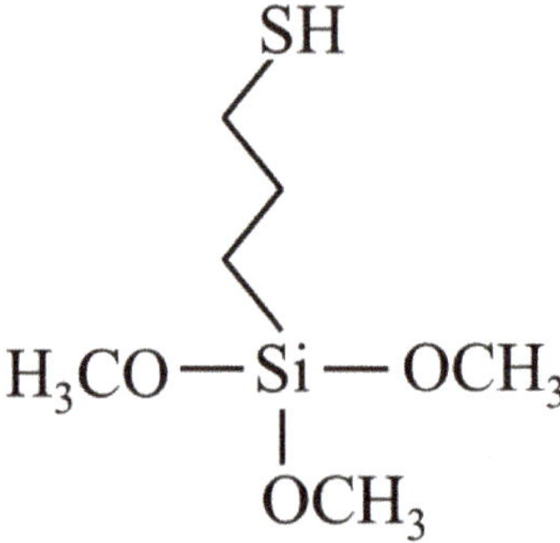

Figure 6. Chemical structure of MPTS.

Figure 7 shows the Fourier-transform infrared-reflection absorption spectroscopy (FTIR-RAS) (Massachusetts, USA) profile of the glass substrate with the MPTS film. The absorption peaks corresponding to the carbon chains of MPTS confirm the chemical adsorption of MPTS onto the glass substrate [10]. In order to graft the substrate with Pt particles, it was immersed for 4 h in a dispersion of Pt nanoparticles (manufactured by Shikoku Keisoku Co.; size distribution of the particles is shown in Figure 8) (Kagawa, Japan) (adsorption solvent = ethylene glycol) prepared to a particle concentration of 2.5 wt% [11].

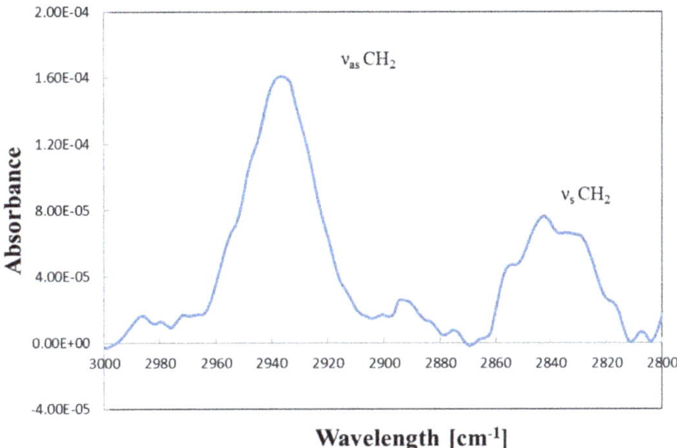

Figure 7. Fourier-transform infrared-reflection absorption spectroscopy (FTIR-RAS) profile of the MPTS monolayer adsorbed on the glass plate.

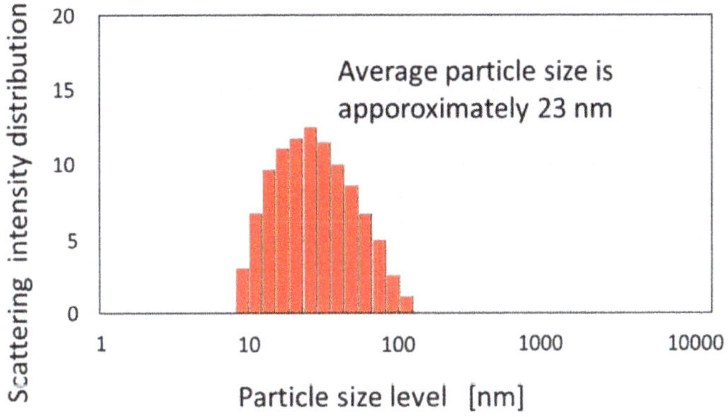

Figure 8. Particle-size distribution of the Pt nanoparticles.

Immersion was achieved by ultrasonically stirring the Pt nanoparticle dispersion using an ultrasonic homogenizer. Then, ultrasonic cleaning was carried out using ethanol to prepare the glass fibers and to graft Pt nanoparticles onto them. Thereafter, the monomolecular film was decomposed and Pt was cured by heating the resulting assembly at 350 °C, using an electric furnace (Tokyo, Japan) under air atmosphere for 30 min. Finally, the catalyst with Pt nanoparticles deposited on the surface of glass fibers was obtained. The schematic of the platinum catalyst preparation process is shown in Figure 9.

The surface morphology of the Pt catalyst was observed before and after the heat treatment by carrying out FE-SEM measurements (Tokyo, Japan) to confirm the grafting of Pt particles. For comparison, the FE-SEM analysis of the glass fibers without the MPTS film was also carried out. In order to measure the weight of the Pt nanoparticles grafted onto the glass fiber surface, the catalyst was added to a hydrofluoric acid aqueous solution (which was handled with the utmost care) [12]. The glass fibers dissolved into this solution and the Pt nanoparticles could be recovered by filtering. The recovered Pt nanoparticles were weighed, and the loading ratio was calculated by measuring the mass of the remaining Pt particles.

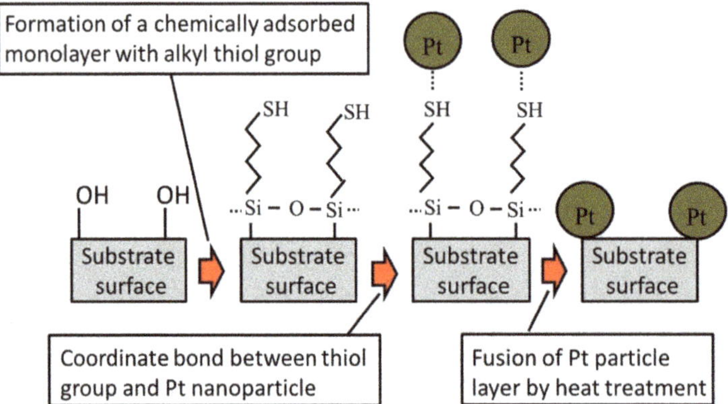

Figure 9. Schematic of the Pt catalyst preparation process.

3.2. Catalyst Performance Evaluation

In order to compare the catalyst loading ratio, a commercially available catalyst (Suzuki Corporation, Exhaust Manifold: Precious metal component: Pd) (Nagano, Japan) was dissolved in hydrofluoric acid, concentrated nitric acid, and an aqueous hydrogen peroxide solution. This metal catalyst was then filtered and weighed to calculate its metal loading ratio. Then, tubing was piped so that the gas flowed from the air inlet to the flask, catalyst section, trap pipe, and draft, as shown in Figure 10. In the catalyst section, the Pt catalyst with a measured mass was packed in a U-shaped pipe, which was placed in a mantle heater. The temperature of the mantle heater was controlled using a voltage regulator and thermostat. Furthermore, octane (C_8H_{18}) (Osaka, Japan) was introduced into the flask beforehand, molecular sieves were placed in the trap tube and weighed, and dry air (humidity approximately 0.2%) was injected at a flow rate of 0.8 L/min for 90 min from the air inlet. After 90 min, the flow of dry air was interrupted, and the mass of the trap tube containing the octane-containing flask and the molecular sieve weight were measured using an electronic balance. These tests were carried out at 50, 100, 200, 300, and 400 °C. The same experiments were carried out for the commercially available catalyst, the Pd catalyst from Suzuki. Furthermore, the TG–DSC curve of the Pt catalyst was obtained before carrying out the heat treatment to determine the difference between the weight of the catalyst in the presence and absence of the chemically adsorbed monomolecular film. The XRD measurements (Kyoto, Japan) of the catalyst were conducted before and after the heat treatment to confirm the change in the state of Pt.

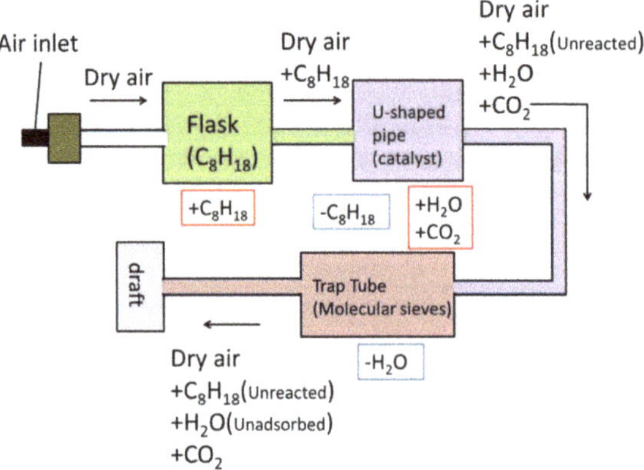

Figure 10. Schematic of the device used for evaluating the catalyst performance.

4. Conclusions

Herein, we proposed a novel method for developing highly efficient Pt catalysts using a reactive chemically adsorbed monomolecular film, Pt nanoparticle dispersion, and glass fiber substrate. The Pt catalyst investigated in this study was prepared by depositing a reactive chemically adsorbed monomolecular film with a thiol group at the molecular terminal on the surface of a hydrophilized glass fiber. The monomolecular film enabled the uniform and dense grafting of Pt nanoparticles on the surface of the glass fiber as a single layer. In addition, the Pt loading ratio of the catalyst was approximately 12.7 wt%. This indicates that the chemically adsorbed monomolecular film on the Pt catalyst could support more platinum than the commercially available catalyst. The catalytic performances of the Pt and commercially available catalysts were evaluated on the basis of their octane combustion rates. Both the catalysts exhibited catalytic performance at temperatures ≥ 200 °C. At 300 °C, the octane combustion rates of the catalysts were 80% or more. Moreover, at 50–400 °C, the octane combustion rates of both the catalysts were approximately the same. Hence, no difference was observed in the catalytic performance of the catalysts. Furthermore, Pt catalysts have been found to exhibit catalytic performance comparable to that of Pd catalysts under lean conditions [6]. In addition, Pt is cheaper than Pd [13]. In this study, we successfully prepared a highly efficient Pt catalyst with superior catalytic performance and lower Pt loading than the commercially available catalyst. The catalyst prepared in this study is a potential candidate for application in car exhaust purification devices as it can reduce the weight, size, and cost of these devices. However, in this study, we evaluated the catalyst performance only on the basis of octane combustion. Hence, further evaluation under a gas atmosphere is also necessary.

Author Contributions: Conceptualization, K.O. and T.O.; Validation, T.S. and Y.H.; Writing—Original Draft Preparation, T.S.; Writing—Review & Editing, K.O. and Y.S.; Supervision, K.O.; Funding Acquisition, K.O. and Y.S. All authors have read and agreed to the published version of the manuscript.

Funding: This work was supported by JSPS KAKENHI, Grant number JP19K05272.

Acknowledgments: We would like to thank Editage (www.Editage.com) for English language editing.

Conflicts of Interest: The authors declare no conflicts of interest.

References

1. Toshima, N.; Wang, Y. Polymer-protected Cu/Pd bimetallic clusters. *Adv. Mater.* **1994**, *6*, 245–247. [CrossRef]
2. Mizukoshi, Y.; Oshima, R.; Maeda, Y.; Nagata, Y. Preparation of Platinum Nanoparticles by Sonochemical Reduction of the Pt(II) Ion. *Langmuir* **1999**, *15*, 2733–2737. [CrossRef]
3. Seino, S.; Kinoshita, T.; Nakagawa, T.; Kojima, T.; Taniguchi, R.; Okuda, S.; Yamamoto, T.A. Radiation induced synthesis of gold/iron-oxide composite nanoparticles using high-energy electron beam. *J. Nanopart. Res.* **2008**, *10*, 1071–1076. [CrossRef]
4. Kageyama, S.; Sugano, Y.; Hamaguchi, Y.; Kugai, J.; Ohkubo, Y.; Seino, S.; Nakagawa, T.; Ichikawa, S.; Yamamoto, T.A. Pt/TiO_2 composite nanoparticles synthesized by electron beam irradiation for preferential CO oxidation. *Mater. Res. Bull.* **2013**, *48*, 1347–1351. [CrossRef]
5. Mohamed, R.M. Characterization and catalytic properties of nano-sized Pt metal catalyst on TiO_2-SiO_2 synthesized by photo-assisted deposition and impregnation methods. *J. Mater. Process. Tech.* **2009**, *209*, 577–583. [CrossRef]
6. Ogawa, M.; Sato, R.; Kikuchi, S.; Iwachido, K. Effects of TWC's PGM Loading Amount on the Exhaust-emission Purification Performance and on the PGM Sintering. *Trans. Soc. Automot. Eng. Jpn.* **2017**, *48*, 1015–1020. [CrossRef]
7. Suda, A.; Sobukawa, H.; Suzuki, T.; Kandori, T.; Ukyo, Y.; Sugiura, M. Synthesis of Ceria Zirconia Solid Solution and Performance as Catalytic Promoter. *Electron. J. R&D Rev. Toyota CRDL* **1998**, *3*, 3–12. Available online: https://www.tytlabs.com/japanese/review/rev333j.html (accessed on 18 March 2018).
8. Akama, H. Study on Effective Removal of Residual Methane in Wasted Gas originated from the Fuel Reforming Gas by Catalytic Combustion. *J. Combust. Soc. Jpn.* **2010**, *52*, 76–85. [CrossRef]

9. Tsuji, I.; Ohkubo, Y.; Ogawa, K. Study on Super-Hydrophobic and Oleophobic Surfaces Prepared by Chemical Adsorption Technique. *Jpn. J. Appl. Phys.* **2008**, *48*, 040205. [CrossRef]
10. Ogawa, K.; Mino, N.; Nakajima, K.; Azuma, Y.; Ohmura, T. Studies of molecular alignments of monolayers deposited by a chemical adsorption technique. *Langmuir* **1991**, *7*, 1473–1477. [CrossRef]
11. Sawada, S.; Masuda, Y.; Zhu, P.; Koumoto, K. Micropatterning of copper on a poly(ethylene terephthalate) substrate modified with a self-assembled monolayer. *Langmuir* **2006**, *22*, 332–337. [CrossRef] [PubMed]
12. Murahashi, S.; Sakakibara, S. Anhydrous hydrogen fluoride as solvent. *J. Synthetic Org. Chem. Jpn.* **1967**, *25*, 1176–1191. [CrossRef]
13. Tanaka Kikinzoku Kogyo Co., Ltd. Latest Precious Metal Market Trends. Available online: https://pro.tanaka.co.jp/library/rate/ (accessed on 15 October 2019).

Article

Kinetic and Mechanistic Study of Rhodamine B Degradation by H_2O_2 and $Cu/Al_2O_3/g\text{-}C_3N_4$ Composite

Chunsun Zhou †, Zhongda Liu †, Lijuan Fang, Yulian Guo, Yanpeng Feng and Miao Yang *

School of Chemistry, Chemical Engineering and Life Sciences, Wuhan University of Technology, 122 Luoshi Road, Wuhan 430070, China; zhouchunsun@whut.edu.cn (C.Z.); liuzhongda@whut.edu.cn (Z.L.); fanglijuan@whut.edu.cn (L.F.); guoyulian@whut.edu.cn (Y.G.); fengyanpeng1996@gmail.com (Y.F.)

* Correspondence: yangmiao@whut.edu.cn; Tel.: +86-1898-617-6465

† Authors contributed equally to this work and are recognized as co-first authors.

Received: 25 January 2020; Accepted: 9 March 2020; Published: 10 March 2020

Abstract: The classic Fenton reaction, which is driven by iron species, has been widely explored for pollutant degradation, but is strictly limited to acidic conditions. In this work, a copper-based Fenton-like catalyst $Cu/Al_2O_3/g\text{-}C_3N_4$ was proposed that achieves high degradation efficiencies for Rhodamine B (Rh B) in a wide range of pH 4.9–11.0. The Cu/Al_2O_3 composite was first prepared via a hydrothermal method followed by a calcination process. The obtained Cu/Al_2O_3 composite was subsequently stabilized on graphitic carbon nitride ($g\text{-}C_3N_4$) by the formation of C–O–Cu bonds. The obtained composites were characterized through FT-IR, XRD, TEM, XPS, and N_2 adsorption/desorption isotherms, and the immobilized Cu^+ was proven to be active sites. The effects of Cu content, $g\text{-}C_3N_4$ content, H_2O_2 concentration, and pH on Rh B degradation were systematically investigated. The effect of the catalyst dose was confirmed with a specific reaction rate constant of $(5.9 \pm 0.07) \times 10^{-9}$ $m{\cdot}s^{-1}$ and the activation energy was calculated to be 71.0 kJ/mol. In 100 min 96.4% of Rh B (initial concentration 20 mg/L, unadjusted pH (4.9)) was removed in the presence of 1 g/L of catalyst and 10 mM of H_2O_2 at 25 °C, with an observed reaction rate constant of 6.47×10^{-4} s^{-1}. High degradation rates are achieved at neutral and alkaline conditions and a low copper leaching (0.55 mg/L) was observed even after four reaction cycles. Hydroxyl radical (HO·) was identified as the reactive oxygen species by using isopropanol as a radical scavenger and by ESR analysis. HPLC-MS revealed that the degradation of Rh B on Cu/Al_2O_3/CN composite involves N-de-ethylation, hydroxylation, de-carboxylation, chromophore cleavage, ring opening, and the mineralization process. Based on the results above, a tentative mechanism for the catalytic performance of the $Cu/Al_2O_3/g\text{-}C_3N_4$ composite was proposed. In summary, the characteristics of high degradation rate constants, low ion leaching, and the excellent applicability in neutral and alkaline conditions prove the $Cu/Al_2O_3/g\text{-}C_3N_4$ composite to be a superior Fenton-like catalyst compared to many conventional ones.

Keywords: $Cu/Al_2O_3/g\text{-}C_3N_4$; Fenton-like; H_2O_2; hydroxyl radical; Rhodamine B

1. Introduction

With the rapid development of industry, persistent organic pollutants in water have attracted widespread attention due to their persistence, bioaccumulation, and high toxicity [1,2]. So far, advanced oxidation processes (AOPs) have been found to be one of the most promising methods to treat persistent organic pollutants in water [3–7]. Due to the in-situ formed highly reactive and non-selective hydroxyl radicals (HO·) during the process, AOPs are capable of mineralizing almost all organic compounds to CO_2, H_2O, and small organic compounds [8–11].

As a typical AOP, Fenton reaction is efficient for HO· production, but still faces some limitations, such as the strict acidic pH range (pH < 4) [12,13], formation of iron sludge [14,15], and high cost for catalyst recycling [16,17]. A number of non-ferrous metals, such as copper [18], manganese [19], and titanium [20], have been developed as alternatives. In particular, Cu^+ reacts with H_2O_2 in a similar manner to Fe^{2+}, but with a much higher reaction rate constant (as shown in Equations (1) and (2)). In contrast to Fe^{3+} that forms the insoluble $[Fe(H_2O)_6]^{3+}$ complex at pH > 5, Cu^{2+} forms the aquo complex $[Cu(H_2O)_6]^{2+}$ that predominates at neutral conditions, making Cu^+ own a wider pH range for application [21]. Additionally, it is known that Cu^{2+} could form certain complexes with organic degradation intermediates, which could react with H_2O_2 to generate more HO· [22].

$$Fe^{2+} + H_2O_2 \rightarrow Fe^{3+} + HO\cdot + OH^- \quad (k = 63 - 76\ M^{-1}\cdot s^{-1}) \tag{1}$$

$$Cu^{+} + H_2O_2 \rightarrow Cu^{2+} + HO\cdot + OH^- \quad (k = 1.0 \times 10^4\ M^{-1}\cdot s^{-1}) \tag{2}$$

However, Cu^+ ions are prone to disproportionation in acidic aqueous solution and can be easily oxidized by dissolved O_2, thereby limiting the application in the aqueous environment [23]. One common strategy for the preparation of copper-based catalysts is the immobilization of copper species on support materials, for example, the immobilization of Cu^+/Cu^{2+}, copper oxide, or the copper-organic complex [24] on various matrixes like metal oxides [25–27], molecular sieve [28,29], and graphitic carbon nitride (g-C_3N_4) [30,31].

With a suitable band gap (2.7 eV) and high response to visible light, g-C_3N_4 has been intensively explored as a photocatalyst for energy and environmental applications [3,32–34]. Apart from this, g-C_3N_4 could also be applied as a Fenton-like catalyst in the absence of light irradiation through the combination with other materials [35,36]. g-C_3N_4 has a 2D planar structure in which tri-s-triazine units are connected by tertiary amines. Typical π-conjugated graphitic planes are formed via the sp^2 hybridization of carbon and nitrogen atoms [37], which brings a large specific surface area, thus providing more reaction sites for heterogeneous reaction. Besides, g-C_3N_4 shows strong affinity for H_2O_2 and could easily adsorb them to the surface, thereby providing larger chances for their contact with other catalytic materials loaded on g-C_3N_4 [34]. Additionally, numerous external groups ($-NH_2$, $-NH$, $-N$, and $-OH$) on the surface may serve as strong Lewis base sites for nanoparticle deposition or metal inclusion [38]. The incorporation of metallic elements into the g-C_3N_4 matrix could induce the production of delocalized electrons, which promotes the catalytic reactions [36,39]. As a typical example, Xu et al. constructed the Cu-Al_2O_3-g-C_3N_4 system, in which a small amount of g-C_3N_4 was used to coordinate with Cu ions to induce the formation of an electron-rich Cu centre and decrease the electron density of the π-electron conjugated system through cation-π interactions [40].

Inspired by the above-mentioned research, a similar but distinguishing efficient Cu/Al_2O_3/g-C_3N_4 composite was proposed in this work, with which H_2O_2 activating was promoted in a different approach. Copper species were bonded to the Al_2O_3 framework to act as the catalytic component and g-C_3N_4 was used in large amounts to act as support for the Cu/Al_2O_3 composite. The decomposition of H_2O_2 was promoted by the strong adsorption of H_2O_2 on the g-C_3N_4 matrix. The obtained composites were characterized through Fourier transform infrared (FT-IR) spectroscopy, X-ray diffraction (XRD), transmission electron microscopy (TEM), X-ray photoelectron spectroscopy (XPS), and N_2 adsorption/desorption isotherms. The optimal synthetic parameters and experimental conditions, including the Cu content, g-C_3N_4 content, H_2O_2 concentration, and pH value were determined. Besides, the effect of the catalyst dose and temperature were confirmed by calculating specific reaction rate constant and activation energy, and the durability and low leaching were evaluated in recycling experiment. In addition, the reactive oxygen species generated in the present system were identified by scavenging experiments and electron spin resonance (ESR) analysis, while the degradation products of Rh B were identified by high performance liquid chromatography-mass spectrometry (HPLC-MS) analysis. Based on the results, the catalytic mechanism of Rh B degradation on $Cu/Al_2O_3/g\text{-}C_3N_4$ composite was proposed.

2. Results and Discussion

2.1. Structural Characterization of Composites

Cu_{12}/Al_2O_3 and $Cu_{12}/Al_2O_3/CN_{1.3}$ (the naming rules of catalysts are indicated in experimental section) were selected as the typical Cu/Al_2O_3 and $Cu/Al_2O_3/CN$ composite to compare their structures with Al_2O_3 and CN. The FT-IR spectra of Al_2O_3, Cu/Al_2O_3, CN, and $Cu/Al_2O_3/CN$ samples were recorded to distinguish the functional groups. As shown in Figure 1, the two peaks at 1513 cm^{-1} and 1636 cm^{-1} in the FT-IR spectra of Al_2O_3 and Cu/Al_2O_3 are attributed to the C=C vibrations and C=O vibrations, which are derived from glucose added in the preparation process [40]. As observed for CN, the sharp peak at 810 cm^{-1} is ascribed to the breathing mode of tri-s-triazine units, and the peaks in the region of 1200–1600 cm^{-1} are assigned to the stretching vibration of the CN heterocycle [31]. In addition, the peak at 3173 cm^{-1} corresponds to the N–H stretching vibration [36]. All these characteristic peaks could be observed in the $Cu/Al_2O_3/CN$ composite, indicating that the chemical structure of CN was not affected by the introduction of Cu/Al_2O_3.

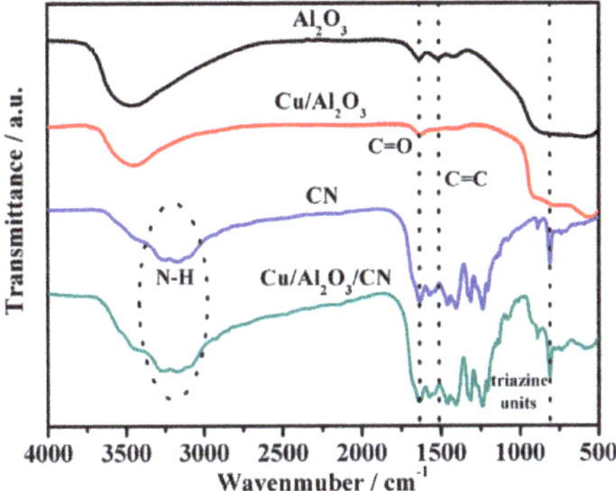

Figure 1. FT-IR spectra of Al_2O_3, Cu/Al_2O_3, CN, and $Cu/Al_2O_3/CN$ samples.

The X-ray diffraction patterns of Al_2O_3, Cu/Al_2O_3, CN, and $Cu/Al_2O_3/CN$ samples were investigated to determine the crystalline structures, as presented in Figure 2. It is clearly observed that the major crystalline structure of Al_2O_3 is the γ-alumina phase (JCPDS No. 10-0425) [41]. In comparison with pure Al_2O_3, new peaks are founded at 35.2°, 38.5°, 48.8°, 53.4°, 58.2°, 66.2°, and 68.1° in the XRD pattern of Cu/Al_2O_3, which represent the (002), (111), (20-2), (020), (202), (31-1), and (220) planes of the copper oxide phase (JCPDS No. 48-1548), respectively [40]. The peak at 61.5° is referred to the diffraction of the (220) plane of cuprous oxide phase (JCPDS No. 65-3288) [42]. These results imply that the copper species were successfully bonded to the Al_2O_3 framework. From the pattern of CN, two main diffraction peaks could be found at 13.0° and 26.9°, which correspond to the interlayer stacking (100) plane of tri-s-triazine units and the (002) plane arising from the interlayer stacking of aromatic systems (JCPDS No. 87-1526) [43]. All the diffraction peaks of Cu/Al_2O_3 and CN could be found in the pattern of $Cu/Al_2O_3/CN$ composite, indicating the successful synthesis of the $Cu/Al_2O_3/CN$ composite.

The microstructures of Al_2O_3, Cu/Al_2O_3, CN, and $Cu/Al_2O_3/CN$ samples were analyzed by TEM. The results are shown in Figure 3. As can be seen in Figure 3A,B, the lattice spacing was confirmed to be 0.29 nm, which corresponds to the (220) lattice planes of γ-alumina [44], and was consistent with the results obtained by XRD. In Figure 3C, it can be seen that the copper bonded alumina framework was rod like. The lattice spacing in Figure 3D was confirmed to be 0.24 nm, which corresponds to the (111) lattice planes of copper oxide confirmed by XRD [45]. In Figure 3E, it is clear that CN shows a two-dimensional structure with a wide pore size distribution [35]. According to Figure 3F, it can be observed that the rod like Cu/Al_2O_3 was embedded in the CN matrix, indicating the successful combination of Cu/Al_2O_3 and CN.

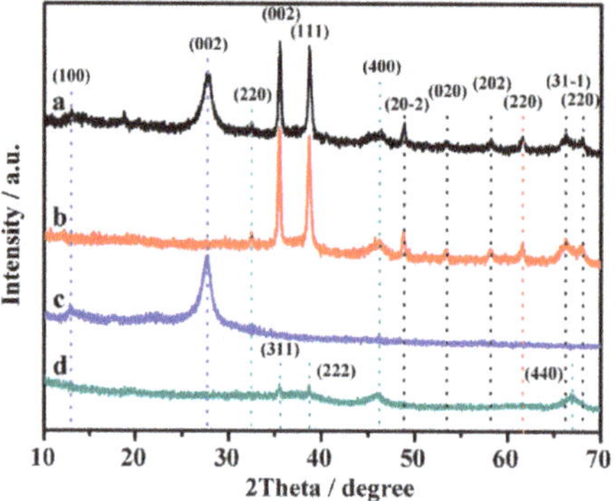

Figure 2. XRD patterns of the synthesized samples (**a**) $Cu/Al_2O_3/CN$, (**b**) Cu/Al_2O_3, (**c**) CN, and (**d**) Al_2O_3.

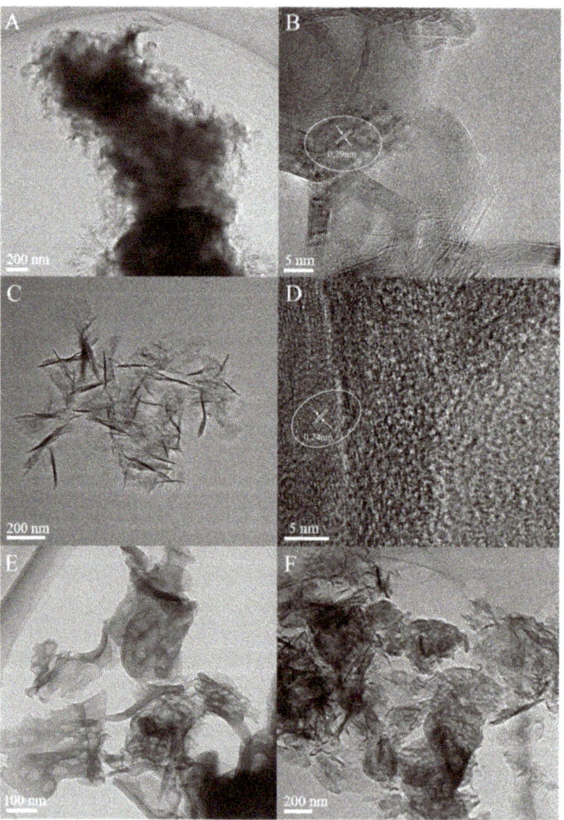

Figure 3. TEM images of (**A**,**B**) Al_2O_3, (**C**,**D**) Cu/Al_2O_3, (**E**) CN, and (**F**) $Cu/Al_2O_3/CN$.

The XPS spectra of the $Cu/Al_2O_3/CN$ composite before and after the reaction were obtained to investigate its surface chemical states and elemental composition. The results are shown in Figure 4 and Table 1. The peaks of the C 1s spectrum at 281.5 eV and 284.6 eV were assigned to C–H and C–C, respectively. Notably, the peak at 285.1 eV was attributed to C–O–H or C–O–metal, which could confirm that Cu/Al_2O_3 was stabilized on the CN matrix by the C–O–Cu bonds [36,40]. For N 1s, the peaks at 395.1 eV, 395.7 eV, and 397.1 eV were ascribed to triazine rings C–N–C, tertiary nitrogen N–$(C)_3$, and sp^2-type C=N bond, indicating that the structure of CN was not changed [36]. The XPS spectrum of Cu 2p in Figure 4C displayed three main peaks at 932.0 eV, 934.0 eV, and 941.5 eV, which correspond to Cu^+, Cu^{2+}, and satellite peaks of copper species, respectively [40,41]. As the binding energy of Cu^+ and Cu^0 are very close, it is difficult to distinguish between Cu^+ and Cu^0 by the XPS feature of Cu $2p_{3/2}$. The Cu LMM peak is normally applied to distinguish between the Cu^0 and Cu^+ according to previous studies [46,47]. However, due to the relatively low copper content in the sample, the presence of Cu^+ cannot be judged by the Cu LMM peak. The results of the XRD patterns could be used instead to confirm the presence of Cu^+. Two peaks of the Al 2p spectrum were observed at 74.1 eV and 75.0 eV in Figure 4D, which were attributed to Al–O–Al and Al–O–Cu, indicating that copper species were bonded to Al_2O_3 framework [46]. The Cu 2p XPS spectrum of the used $Cu/Al_2O_3/CN$ composite is shown in Figure 4E. The peaks at 931.7 eV, 933.6 eV, and 941.5 eV still represent Cu^+, Cu^{2+}, and satellite peaks of copper species, respectively. The Cu^+ to Cu^{2+} ratio on the $Cu/Al_2O_3/CN$ composite was 2.34 before reaction and decreased to 1.42 after reaction, indicating that copper species were the active sites and the conversion of Cu^+ to Cu^{2+} was involved in the reaction [48].

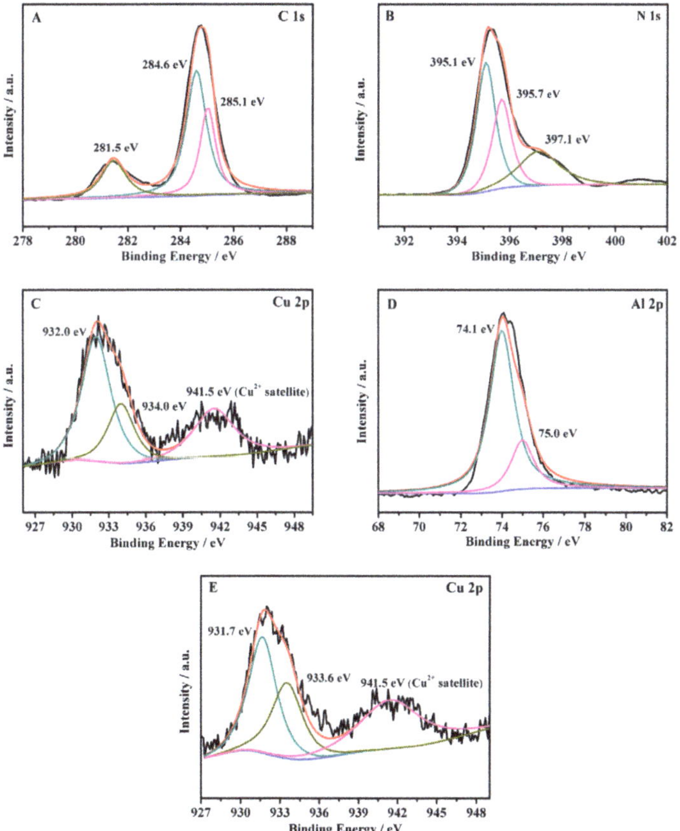

Figure 4. (**A**) C 1s, (**B**) N 1s, (**C**) Cu 2p, and (**D**) Al 2p XPS spectra of $Cu/Al_2O_3/CN$ composite before reaction, and (**E**) Cu 2p spectrum of the $Cu/Al_2O_3/CN$ composite after reaction.

Table 1. XPS results of the $Cu/Al_2O_3/CN$ composite.

	BE/eV	Chemical Bonds		BE/eV	Chemical Bonds
C 1s	281.5	C–H	N 1s	395.1	C–N–C
	284.6	C–C		395.7	N–(C)3
	285.1	C–O		397.1	C=N
Cu 2p	932.0/931.7	Cu^+	Al 2p	74.1	Al–O–Al
	934.0/933.6	Cu^{2+}		75.0	Al–O–Cu

BE denotes the binding energy.

The N_2 adsorption/desorption isotherm and the pore size distribution of the Al_2O_3, Cu/Al_2O_3, CN, and $Cu/Al_2O_3/CN$ samples are shown in Figure 5. The specific surface area of Al_2O_3 reaches 133.1 ± 0.5 m^2/g with a pore width of 6.7 nm, and the introduction of copper species leads to a decline to 119.5 ± 1.0 m^2/g and 5.3 nm, respectively. All the CN containing samples show typical IV isotherms with the H3 hysteresis loop, indicating the existence of a typical mesoporous structure with slit-like pores in CN [22]. Additionally, when Cu/Al_2O_3 was stabilized on the CN matrix, the specific surface area increased from 72.2 ± 0.9 m^2/g for CN to 146.6 ± 1.0 m^2/g for $Cu/Al_2O_3/CN$, which is more beneficial for the adsorption of organic pollutants and H_2O_2 on the catalyst surface and proves that the fabrication of the $Cu/Al_2O_3/CN$ composite is not merely a physical mixing process. In summary, a higher specific surface area and smaller pore width are achieved in the $Cu/Al_2O_3/CN$ composite.

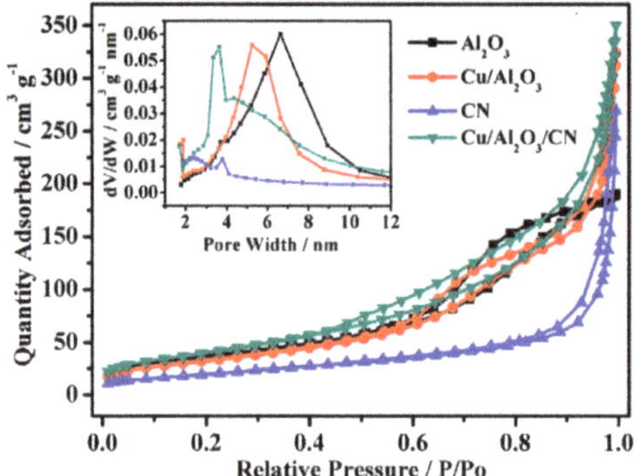

Figure 5. N_2 adsorption/desorption isotherm and the pore size distribution (insert) of the Al_2O_3, Cu/Al_2O_3, CN, and $Cu/Al_2O_3/CN$.

2.2. Catalytic Performance of Composites

The effect of Cu content, CN content, H_2O_2 concentration, and pH value were investigated to obtain the optimal synthetic parameters and reaction conditions. Besides, the effect of the catalyst dose and temperature was confirmed by calculating the specific reaction rate constant and activation energy (E_a). The durability and ion leaching of the composite were evaluated by the recycling experiments. In addition, the reactive species generated were identified by ESR analysis and radical scavenging experiments, and then the catalytic mechanism of Rh B on $Cu/Al_2O_3/CN$ composite was proposed by confirming intermediates using HPLC-MS.

2.2.1. Effect of Cu Content

In order to obtain the optimal Cu content for catalyst fabrication, a series of Cu_x/Al_2O_3 composites (x = 0, 4, 7, 9, 12, and 15 wt %) were prepared to compare their catalytic performances on the removal of Rh B in the presence of H_2O_2. As shown in Figure 6, in the absence of copper decoration, Al_2O_3 exhibited negligible catalytic capacity for Rh B degradation, with 3.0% Rh B removed in 100 min. The degradation was enhanced by copper doping and the rate progressively increased from x = 0 to x = 12, after which the degradation rate started to decrease. During the synthesis process of Cu/Al_2O_3 composite, copper species were bonded to the Al_2O_3 framework. However, there is an upper limit (x = 12) for the amount of copper species that could enter the Al_2O_3 framework, so part of the copper species would be present as the extra framework if excess copper was introduced [47]. The extra framework copper species may impede the contact between H_2O_2 and framework copper species, thereby limiting the decomposition of H_2O_2 [22,41]. Therefore, the optimal Cu content was confirmed to be x = 12 and Cu_{12}/Al_2O_3 was selected as optimal components for further studies.

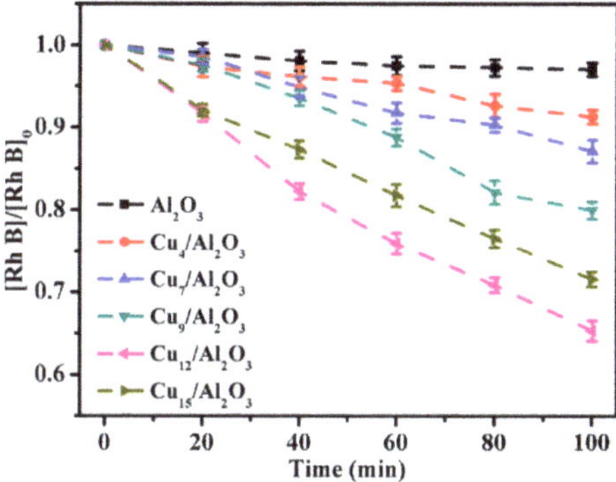

Figure 6. [Rh B]/[Rh B]$_0$ as a function of time in the presence of 1 g/L Cu_x/Al_2O_3 composite (x = 0, 4, 7, 9, 12, and 15 wt %). Reaction conditions: [Rh B]$_0$ = 20 mg/L, $[H_2O_2]_0$ = 10 mM, V = 100 mL, T = 25 °C, pH 4.9 (unadjusted).

2.2.2. Effect of CN Content

A fixed amount (0.1 g) of the Cu_{12}/Al_2O_3 composite was loaded on different amounts of CN to determine the optimal CN content. The degradation of Rh B in the presence of $Cu_{12}/Al_2O_3/CN_y$ (y = 0, 0.7, 1.3, 2.0, and 2.7 wt %) is presented in Figure 7. As shown, the Cu_{12}/Al_2O_3 composite shows acceptable catalytic property for Rh B degradation, removing 34.6% Rh B in 100 min. Additionally, it is apparent that the introduction of CN leads to a corresponding enhancement of the degradation rate. With CN incorporated, part of $Al_2O_3/Cu_{0.9}$ was stabilized on CN sheets by the formation of C–O–Cu bonds. Aqueous H_2O_2 are easily bonded to the surface of CN [34], providing larger chances for its contact with immobilized Cu^+, which brought a significant improvement in catalytic property. Consequently, the degradation ratio within 100 min increased from 34.6% to 96.4% in the CN range of y = 0 to y = 1.3. However, excess CN incorporation from y = 1.3 to y = 2.7 did not further improve the degradation ratio. In the presence of adequate CN (y = 1.3), Cu_{12}/Al_2O_3 is fully loaded on the CN interlayers, resulting in the largest degradation ratio exhibited in Figure 7. When the amount of CN was largely in excess to Cu_{12}/Al_2O_3, the mass ratio of Cu_{12}/Al_2O_3 in $Cu/Al_2O_3/CN$ decreased accordingly, thereby limiting the further improvement of catalytic property. Therefore, the optimal CN content was confirmed as y = 1.3 and $Cu_{12}/Al_2O_3/CN_{1.3}$ was selected as the typical composite for further studies. The $Cu/Al_2O_3/CN$ composite in the following text refers to the $Cu_{12}/Al_2O_3/CN_{1.3}$ composite unless addition descriptions were used.

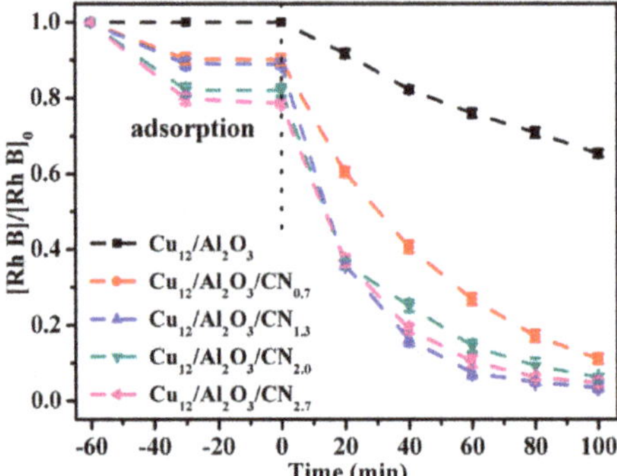

Figure 7. [Rh B]/[Rh B]$_0$ as a function of time in the presence of 1 g/L $Cu_{12}/Al_2O_3/CN_y$ composite (y = 0, 0.7, 1.3, 2.0, and 2.7 wt %). Reaction conditions: [Rh B]$_0$ = 20 mg/L, $[H_2O_2]_0$ = 10 mM, V = 100 mL, T = 25 °C, pH 4.9 (unadjusted).

2.2.3. Synergistic Effect

To verify the synergistic effect of Cu_{12}/Al_2O_3 and CN on the degradation of Rh B, four samples were prepared and their catalytic properties were compared in the same condition. The results are presented in Figure 8. It can be seen that both Cu_{12}/Al_2O_3 and CN show a weak catalytic property. Notably, CN exhibited a fine adsorption capacity for Rh B, which is attributed to its large specific surface area and multiple groups on the surface. By physically mixing Cu_{12}/Al_2O_3 and CN, the adsorption capacity for Rh B decreased but a higher degradation rate was observed. This may be attributed to the fact that immobilized copper species form an adsorption competition with Rh B on CN sheets, leading to a portion of adsorptive sites occupied by immobilized copper, which lowers the adsorption of Rh B but accelerates its degradation. In the $Cu/Al_2O_3/CN$ composite, much stronger bonds were formed between Cu and C–O–H as C–O–Cu, leading to a much lower adsorption capacity and a synergistically enhanced degradation property for Rh B.

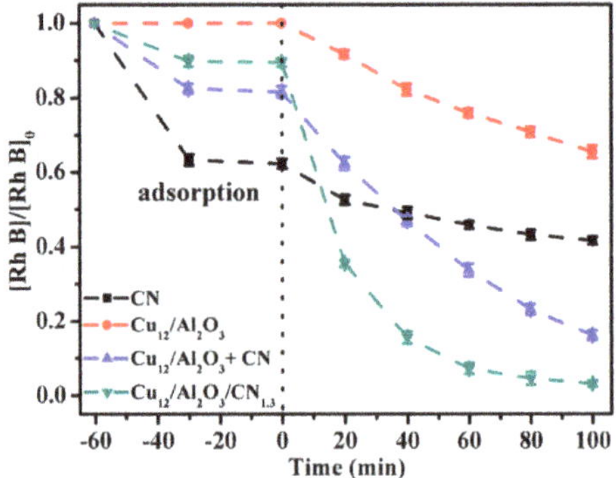

Figure 8. [Rh B]/[Rh B]$_0$ as a function of time in the presence of 1 g/L different samples. Reaction conditions: [Rh B]$_0$ = 20 mg/L, $[H_2O_2]_0$ = 10 mM, V = 100 mL, T = 25 °C, pH 4.9 (unadjusted).

2.2.4. Effect of pH

The homogeneous Fenton reaction is strictly limited to acidic conditions (pH < 4), which limits its application in neutral and alkaline conditions [4,12]. Therefore, the effect of pH is an important criterion for the evaluation of catalytic performance of catalysts, especially the ones that involve potential dissolved metal ion species. To evaluate the practical applicability of the $Cu/Al_2O_3/CN$

composite in various pH conditions, a series of experiments were carried out ranging from pH 3.0 to pH 11.0. The initial pH value of 20 mg/L Rh B is 4.9, so this value (unadjusted pH) was used as a replacement of pH 5.0. The Rh B solutions were adjusted to predetermined pH values with the addition of 1 M HCl or NaOH. In all the experiments, the initial pH values underwent a slight increase with the extent of 1.0 ± 0.3 units, which could be attributed to the formation of OH^- in the H_2O_2 decomposition process. The degradation of Rh B is shown in Figure 9. It is evident that rapid degradations of Rh B were achieved at pH values higher than pKa of Rh B (3.7), and a lower pH value of 3.0 led to a remarkable repression. This is mainly attributed to the different existence forms of Rh B at different pH conditions. At pH values higher than 3.7, the carboxyl group of cationic RhB^+ is deprotonated and the dye is transformed into zwitterionic $RhB^{\pm}$ with a higher hydrophilic character that is prone to reacting with HO· radical [49]. Additionally, the scavenging effect of HO· radical by H^+ becomes stronger at low pH values [50,51], and H_2O_2 would become more stable under strong acid conditions to form oxonium $[H_3O_2]^+$, which inhibits its reaction with active species to generate HO· radical in the presence of a large amount of H^+ [48]. According to the Eh-pH diagram of Cu-H_2O [52,53], CuO and Cu_2O remain stable under alkaline conditions, which is beneficial for the catalytic degradation of Rh B. Despite the fact that more $O_2\cdot^-$ radicals are produced in alkaline conditions that consumes HO· radical and H_2O_2, the degradation rates did not decline in higher pH conditions and even remarkably increased at pH 11.0, which may be attributed to the change of hydrophobic property of the dye and its interaction with negatively charged catalyst surface in strong alkaline conditions [49]. In summary, a remarkable degradation rate could be achieved in neutral and alkaline conditions, which is superior to classic Fenton reaction. Since there is no drastic difference in the removal of Rh B in the range of pH 4.9–9.0, Rh B solution with the unadjusted pH value of 4.9 was used for subsequent experiments.

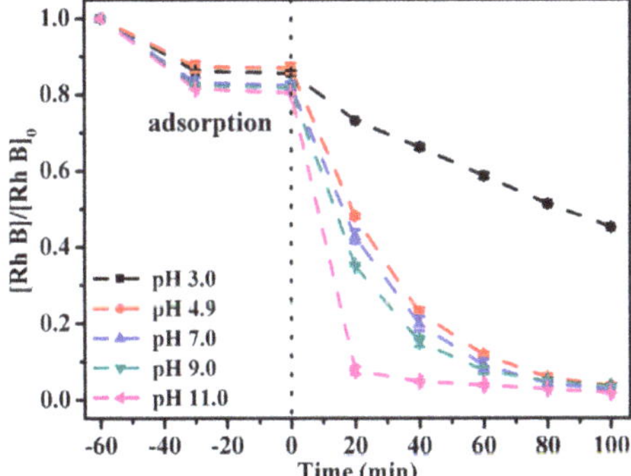

Figure 9. [Rh B]/$[Rh B]_0$ as a function of time with the different pH conditions in the presence of 1 g/L of the Cu/Al_2O_3/CN composite. Reaction conditions: $[Rh B]_0$ = 20 mg/L, $[H_2O_2]_0$ = 10 mM, V = 100 mL, T = 25 °C.

2.2.5. Effect of H_2O_2 Concentration

Since H_2O_2 is the main source of HO· in the Fenton-like reaction, the effect of H_2O_2 on the degradation of Rh B in the presence of Cu/Al_2O_3/CN composite was investigated. As calculated by Equation (3), the stoichiometric amount of H_2O_2 to achieve total mineralization of 20 mg/L Rh B was calculated to be 3.0 mM, and the effect of H_2O_2 concentration was investigated in the range of 1–10 mM.

$$C_{28}H_{31}ClN_2O_3 + 73\ H_2O_2 \rightarrow 28\ CO_2 + 87\ H_2O + HCl + 2\ HNO_3 \quad (3)$$

The evolution of H_2O_2 with an initial concentration of 10 mM was taken as an example for kinetics analysis, the result of which is presented in Figure S1A. The residual concentration of H_2O_2 evolves linearly with reaction time, demonstrating the zero-order reaction kinetics during the reaction, which is attributed to the fact that H_2O_2 was present in large excess compared to Rh B. Only 11.9% of H_2O_2 was consumed in this experiment, which is far less significant than the Rh B decline that reached up to 96.4%. Therefore, more emphasis was stressed on the evolution of Rh B in this work. The normalized concentration of Rh B ([Rh B]/[Rh B]$_0$) and its logarithm are plotted as a function of the reaction time in Figure S1B,C for kinetic analysis. The plot of ln([Rh B]/[Rh B]$_0$) versus reaction time was linearly fitted with an R^2 of 0.9913, which demonstrated that the Rh B decline followed a pseudo first-order kinetics, as confirmed by some photocatalytic and Fenton-like systems [48,54]. The negative slope of the fitted curve represents the observed reaction rate constant k_{obs}, which was 5.14×10^{-4} s^{-1} for this experiment. The decline of Rh B with different initial H_2O_2 concentrations was recorded in Figure 10 together with the corresponding reaction rate constants. As expected, the degradation rate increased with increasing H_2O_2 concentration. The initial H_2O_2 concentration of 10 mM led to an efficient degradation with a rate constant of 5.14×10^{-4} s^{-1}, which is a satisfactory value for subsequent activity evaluations. Certainly, larger reaction rates could be achieved with higher H_2O_2 concentrations, but considering the fact that H_2O_2 also acts as an HO· scavenger (as shown in Equation (4)) that lowers the efficiency of using H_2O_2, exorbitant concentrations were not used [4]. The emphasis is not to seek for an 'optimal' H_2O_2 concentration, so $[H_2O_2]_0$ was fixed at 10 mM for other experiments in this work.

$$H_2O_2 + HO\cdot \rightarrow HO_2\cdot + H_2O \quad (4)$$

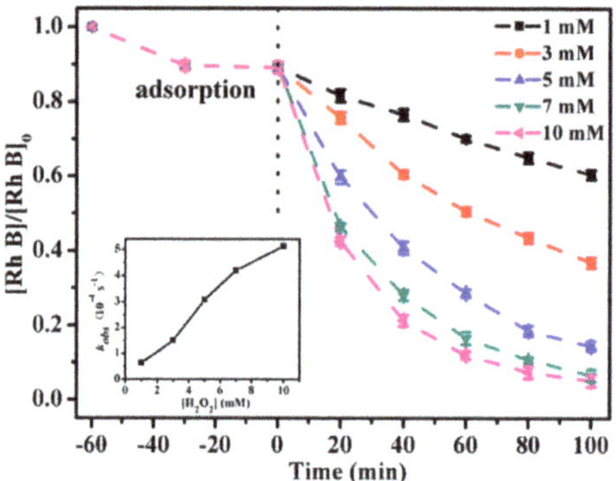

Figure 10. [Rh B]/[Rh B]$_0$ as a function of time and the corresponding reaction rate constants (insert) with different H_2O_2 concentrations in the presence of 1 g/L of the Cu/Al_2O_3/CN composite at room temperature. Reaction conditions: [Rh B]$_0$ = 20 mg/L, V = 100 mL, pH 4.9 (unadjusted).

2.2.6. The Effect of the Catalyst Dose

The value of k_{obs} describes the rate that a reaction happens, and this value is dependent on the dose of the catalyst, which means a larger amount of the catalyst leads to a larger reaction rate constant. In this work, different amounts of catalyst (0.4–1.0 g/L) were used to catalyze Rh B degradation to check the effect of the catalyst dose on the reaction rate. Considering the facts that surface reaction dominates in heterogeneous systems and surface area plays a significant role, it would be unfair to use mass concentration for intersystem comparisons, which does not give consideration to the particle size and shape. Thus, in this work, the catalyst dose is represented by surface area concentration (SA/V), which is the product of mass concentration (g/L) and specific surface area (m^2/g) and has a unit of m^{-1}. The concept of surface area concentration was first proposed by Matheson and Tratnyek as a general independent variable to correlate with the contaminants removal rate constants, as it incorporates most of the effects of grain size and shape [55], and now it is widely used in many heterogeneous systems for kinetic study [35,56–59]. One example for the unit conversion is exhibited in Equation (5), with a mass concentration of 1 g/L and the specific surface area of 146.6 m^2/g for the $Cu/Al_2O_3/CN$ composite.

$$\frac{SA}{V} = 1\,g/L \times 146.6\,m^2/g = 146.6\,m^2/L = 146.6\,m^2/(10^{-3}\,m^3) = 1.466 \times 10^5\,m^2/m^3 = 1.466 \times 10^5\,m^{-1} \quad (5)$$

Figure 11A depicts ln([Rh B]/[Rh B]$_0$) as a function of time with different surface area concentrations (SA/V) of catalyst at 25 °C. All the ln([Rh B]/[Rh B]$_0$) plots were linearly fitted with reaction time, reconfirming the pseudo first-order kinetics in all cases. The key parameters of the fitted curves are listed in Table 2, and k_{obs} is plotted as a function of SA/V in Figure 11B. As shown, a linear fitted curve was achieved (R^2 = 0.9948), demonstrating that k_{obs} was linearly related to SA/V within certain limits, which is in accordance with previous reports [35,60,61]. The slope of this fitted line represents the specific reaction rate constant k_{SA}, which is the normalization of k_{obs} to SA/V (i.e., k_{SA} = k_{obs} normalized to SA/V) [62]. The units for k_{obs} and SA/V are s^{-1} and m^{-1}, respectively, so the unit for k_{SA} should be $m \cdot s^{-1}$. The specific reaction rate constant k_{SA} could be regarded as a general descriptor of reactivity of heterogeneous catalysts [55,62], and is calculated to be (5.9 ± 0.07) × 10^{-9} $m \cdot s^{-1}$ in this work. Judging from the k_{obs} data obtained in this work, much more rapid degradation of Rh B is achieved in the presence of the proposed catalyst compared with many other conventional ones like CuO nanowires, CuO nanoparticles, CuO nanoflowers, and commercial CuO [63,64].

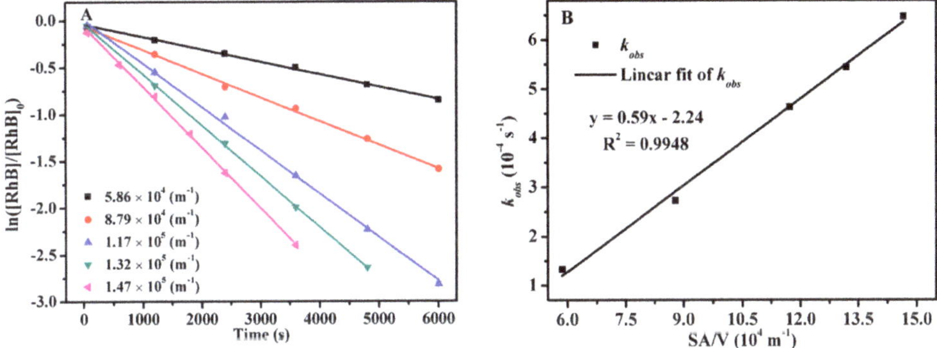

Figure 11. (**A**) ln([Rh B]/[Rh B]$_0$) as a function of reaction time with different surface area concentrations (5.86–14.66 × 10^4 m^{-1}). Reaction conditions: [Rh B]$_0$ = 20 mg/L, $[H_2O_2]_0$ = 10 mM, V = 100 mL, T = 25 °C, pH 4.9 (unadjusted). (**B**) k_{obs} as a function of SA/V. The k_{obs} values were obtained from Figure 11A.

Table 2. The key parameters of the fitted curves in Figure 11A.

Catalyst Dose (g/L)	SA/V (10^5 m^{-1})	k_{obs} (10^{-4} s^{-1})	Standard Deviation (10^{-5} s^{-1})	R^2 (%)
0.4	0.586	1.33	0.317	99.72
0.6	0.879	2.73	0.592	99.73
0.8	1.172	4.63	0.859	99.83
0.9	1.319	5.44	0.409	99.98
1.0	1.466	6.47	1.042	99.87

2.2.7. Activation Energy

Generally, the activation energy (E_a) of the reaction could be calculated by the Arrhenius equation: $k_{obs} = k_1 \cdot e^{-E_a/RT}$, which can be integrated as

$$ln(k_{obs}) = \frac{-E_a}{RT} + ln(k_1) \quad (6)$$

where E_a, R, and T represent the activation energy for the reaction, the gas constant, and the absolute temperature, respectively [35]. To estimate the activation energy of the reaction, a series of experiments were conducted at different temperatures from 20 to 40 °C. The ln([Rh B]/[Rh B]$_0$) was plotted as a function of time in Figure 12A to obtain k_{obs} at different temperatures. The key parameters of the fitted curves were listed in Table 3, and ln(k_{obs}) was plotted as a function of 1/T (K^{-1}) in Figure 12B to obtain E_a.

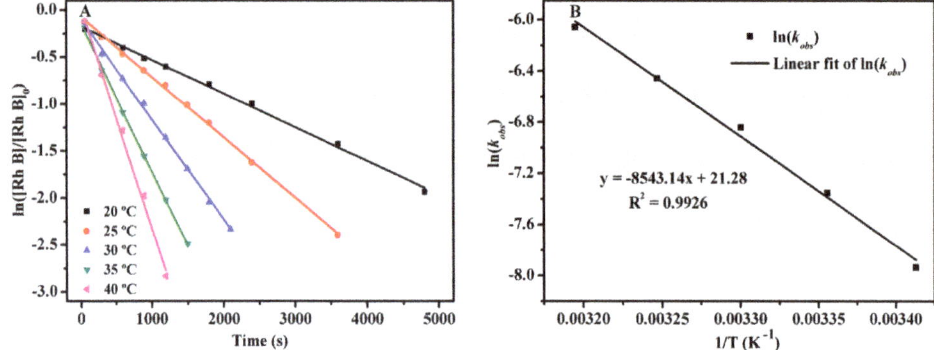

Figure 12. (**A**) ln([Rh B]/[Rh B]$_0$) as a function of reaction time at different temperatures (20–40 °C) in the presence of 1 g/L of the Cu/Al_2O_3/CN composite. Reaction conditions: [Rh B]$_0$ = 20 mg/L, [H_2O_2]$_0$ = 10 mM, V = 100 mL, pH 4.9 (unadjusted). (**B**) ln(k_{obs}) as a function of 1/T. The k_{obs} values are obtained from Figure 12A.

Table 3. The key parameters of the fitted curves in Figure 12A.

Temperature (°C)	k_{obs} (10^{-3} s^{-1})	Standard Deviation (10^{-5} s^{-1})	R^2 (%)
20	0.358	0.689	99.70
25	0.647	1.042	99.87
30	1.07	2.184	99.71
35	1.57	2.081	99.91
40	2.34	8.081	99.53

As can be seen from Figure 12A, with the increase of temperature, the degradation rate of Rh B increased significantly. In addition, all the ln([Rh B]/[Rh B]$_0$) plots were linearly fitted with reaction time, reconfirming the pseudo first-order kinetics at different temperatures. The ln(k_{obs}) is linearly fitted with 1/T with an R^2 of 0.9926 as shown in Figure 12B. According to the Arrhenius equation, the slope of this fitting curve equals to $-E_a$/R, and the activation energy E_a was calculated to be 71.0 kJ/mol. This value was lower than a series of other similar reactions, such as the degradation of Congo red by Fenton reagent (85.9 kJ/mol), the degradation of phenol by diamond supported gold nanoparticles (90.0 kJ/mol), and the iron oxide coating Fenton-like catalyst (96.9 kJ/mol) [65–67].

2.2.8. Recycling Experiments

In order to evaluate the stability of the Cu/Al_2O_3/CN composite, recycling experiments were performed. After each cycle, the suspension was filtered, washed, and dried to obtain the remaining composite. The recovered composite was subsequently subjected to a second cycle under the same reaction conditions. The leaching of copper species was measured at the end of each reaction cycle. The results are presented in Figure 13 and Table 4. Despite that the reaction rate constant decreased to some extent, efficient removal of Rh B was still achieved with a degradation ratio of 89.7% in 50 min after four cycles with a k_{obs} value of 0.77 × 10^{-3} s^{-1}, indicating that the catalytic property of the composite was efficient and long-lasting. The slight decline in catalytic performance may be attributed to the loss of Cu species in catalyst. The leaching of copper species increased from 0.16 mg/L in the first cycle to 0.55 mg/L in the fourth cycle, which shows an increasing trend. This could be regarded as a shortcoming of the proposed catalyst, and it also provides a direction for further improvement. It is remarkable that, despite the fact that copper leaching increases with cycle numbers, only 0.55 mg/L copper species were released to the solution even after four reaction cycles, i.e., copper leaching was remained at a very low level, which is far below the European standard of effluent (2 mg/L) [48].

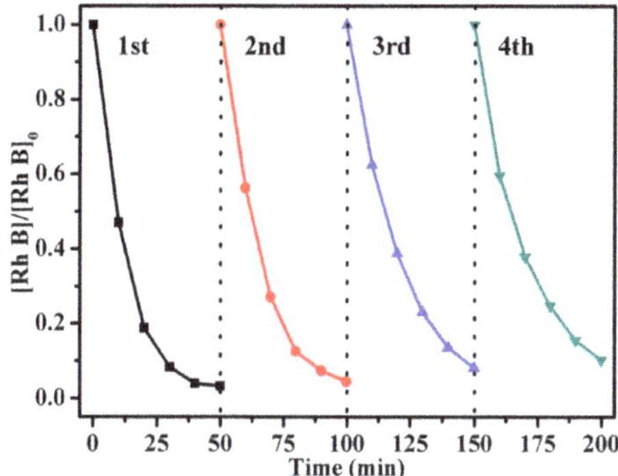

Figure 13. Recycling performance of the $Cu/Al_2O_3/CN$ composite for Rh B degradation in the presence of 1 g/L of the $Cu/Al_2O_3/CN$ composite. Reaction conditions: $[Rh\ B]_0$ = 10 mg/L, $[H_2O_2]_0$ = 10 mM, V = 100 mL, 25 °C, pH 4.9 (unadjusted).

Table 4. k_{obs} and Cu leaching in the recycling experiments.

Recycling Cycles	1st	2nd	3rd	4th
Removal ratio	96.8%	95.5%	91.8%	89.7%
k_{obs} (10^{-3} s^{-1})	1.37	1.12	0.83	0.77
Cu leaching (mg/L)	0.16	0.37	0.43	0.55

In some systems the dissolved metal species could form certain complexes that are able to catalyze the decomposition of H_2O_2, thereby persistently generating HO· [36,68]. To determine the effect of dissolved copper species on Rh B degradation, the suspension containing $Cu/Al_2O_3/CN$ composite was filtered at 40 min to remove catalyst particles, and the residual Rh B in the filtrate was monitored over time. As shown in Figure S2, negligible Rh B degradation was observed in the filtrate, indicating that the dissolved copper species had little effect on Rh B degradation, so only surface reaction dominates in the system containing the $Cu/Al_2O_3/CN$ composite. With a tiny loss of copper species, the $Cu/Al_2O_3/CN$ composite remains highly reactive and could efficiently catalyze the degradation of Rh B. The main features of the proposed $Cu/Al_2O_3/CN$ composite and some other copper-based Fenton-like catalysts on Rh B degradation are listed in Table S1 [27,40,48,63,64,69–73]. The superiority of the $Cu/Al_2O_3/CN$ composite in this work is supported by its high degradation rate constant, low copper leaching, and excellent applicability in neutral and alkaline conditions.

2.3. Catalytic Mechanism

In order to ascertain the Rh B degradation mechanism by the $Cu/Al_2O_3/CN$ composite, both the reactive species and degradation products were identified in the Rh B degradation process. The reactive species were identified by a radical scavenger and ESR analysis, and HPLC-MS analysis was used to separate the degradation products and to identify each component. A tentative mechanism was proposed based on the abovementioned results.

2.3.1. Scavenging Experiments

Isopropanol (IPA) was employed as the HO· scavenger due to its rapid reaction rate with HO· (1.9×10^{-9} $M^{-1}\cdot s^{-1}$) [38]. As shown in Figure 14A, by increasing the IPA concentration from 10 to 100 mM, the scavenging effect became more significant, which indicates that HO· radical is the primary oxidation species in the catalytic reaction. This conclusion is also supported by the ESR experiment in which different radicals were trapped by DMPO to form different adducts that have different spectra [74,75]. As shown in Figure 14B, no ESR signal was observed in the absence of H_2O_2. With the addition of H_2O_2, a four-line ESR signal was observed with the relative intensity of 1:2:2:1, which is the characteristic spectrum of the DMPO-HO· adduct. This result reconfirmed the existence of HO· as main reactive species. Additionally, the characteristic spectrum of DMPO-$O_2\cdot^-$ with a relative intensity of 1:1:1:1 was also observed in Figure 14C. $O_2\cdot^-$ may arise from the reaction between HO· and H_2O_2 as shown in Equation (4). Notably, the intensity of the DMPO-HO· signal reached maximum at 10 min, while the intensity of DMPO-$O_2\cdot^-$ signal continuously increased over time, which supports that the observed $O_2\cdot^-$ radicals were generated from HO· radicals.

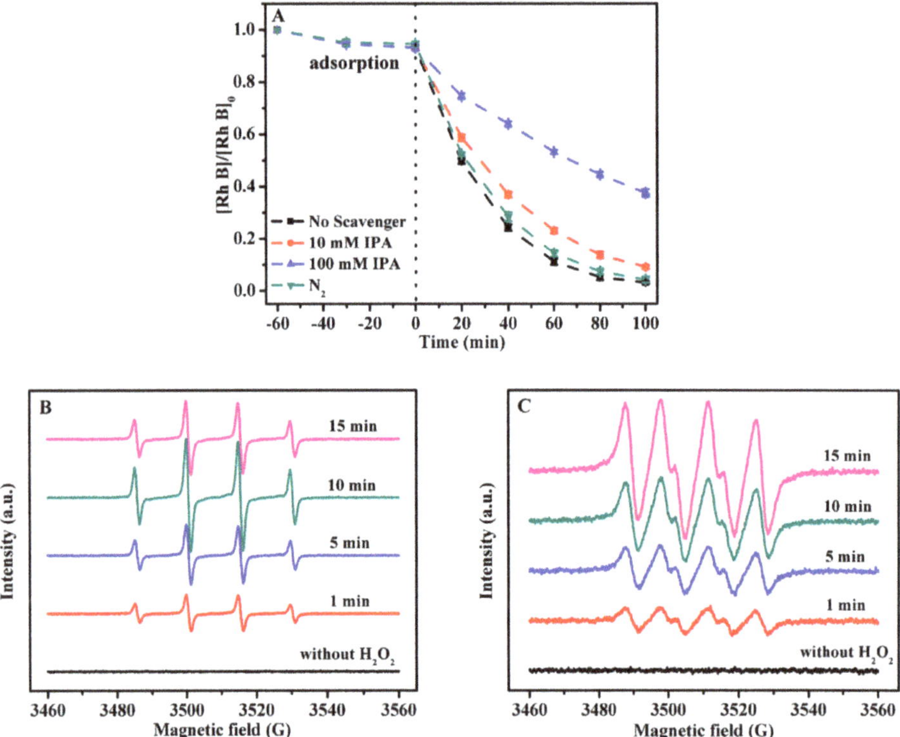

Figure 14. (**A**) [Rh B]/[Rh B]$_0$ as a function of time in the presence of different concentrations of IPA and N_2 bubbling. ESR spectra of (**B**) DMPO-HO· adducts in aqueous solution and (**C**) DMPO-$O_2\cdot^-$ adducts in methanol solution recorded with the Cu/Al_2O_3/CN composite.

Moreover, the effect of dissolved oxygen on Rh B degradation was checked by N_2 bubbling. The Rh B solution was purged by N_2 for 30 min before the reaction to dissipate dissolved oxygen. As shown in Figure 14A, the presence and absence of dissolved oxygen exerted no significant influences on Rh B degradation, indicating that oxygen was not the source of HO·, and the HO· radicals mainly arose from H_2O_2. In summary, HO· was identified as the major reactive species generated in the reaction, and the $O_2\cdot^-$ was generated via the reaction of HO· and H_2O_2.

2.3.2. HPLC-MS Analysis

Since HO· was confirmed as the main contributing reactive species, the $Cu/Al_2O_3/CN$ system exhibited a universal degradation peroperty for other organic pollutants as expected. As shown in Figure S3, the degradation ratios of methyl orange (MO), Rh B, and methylene blue (MB) in 100 min were 55.3%, 96.4%, and 93.1%, respectively. Particularly, the degradation products of Rh B were identified by HPLC-MS analysis.

The HPLC chromatograms of samples extracted at different reaction times were recorded in Figure S4. The two main peaks at 14.7 and 16.9 min corresponded to N-de-ethylated intermediate and Rh B, respectively, which were confirmed by MS identification [76]. It is clear that the peak of Rh B decreased with time, which was accompanied with the formation and subsequent elimination of other peaks. This implies that some transformation products were formed in the degradation process, and these produced compounds were further degraded into smaller products.

Based on the results above and previous reports [76–79], the possible degradation pathways of Rh B are proposed in Figure 15, which could be mainly summarized as N-de-ethylation, chromophore cleavage, ring opening and mineralization process. The MS spectra of each product were shown in Figure S5. A compound with a *m/z* value of 443 was identified to be Rh B, and the intermediates with *m/z* values of 415 and 387 were identified as N, N-diethyl-N′-ethyl rhodamine (DER), and N, N-diethyl rhodamine (DR), respectively. The intermediates with the *m/z* value of 387 could also be N-ethyl-N′-ethyl rhodamine (EER), which is the isomer of DR and has an ethyl on each side. These intermediates prove that in the first stage Rh B was degraded through a stepwise N-de-ethylation pathway, which finally forms an intermediate with an *m/z* value of 318 [79]. At the same time, the de-carboxylation process occurs and forms the product with a *m/z* value of 274 [80], and the central carbon of Rh B was also attacked by HO· radicals to form the hydroxylated product (*m/z* = 459) via the additive reaction [78]. There were two unidentified peaks with m/z values of 453 and 475, which could be regarded as the $[M + H]^+$ and $[M + Na]^+$ adducts of an unknown intermediate. To the best of our knowledge, no such pseudomolecule ion with m/z = 453 was previously reported in Rh B degradation, and we deemed that this compound might not exert a significant influence on the whole degradation pathway. Thus, it is marked unidentified in the present work. In the next stage, non-selective HO· directly attacks the center carbon of these N-de-ethylated intermediates, leading to the cleavage of the conjugated xanthene structure [81]. The intermediates with *m/z* values of 340, 245, 230, and 223 were identified from this process [82–85]. After the chromophore cleavage process, the ring opening process occurs and forms smaller molecules. The ring opening products such as tricarballylic acid, 7-oxooctanoic acid, and propionic acid with m/z values of 177, 158, and 74 were identified [81]. In the final stage, these small molecules were possibly mineralized to CO_2, H_2O, NO_3^-, NH_4^+, etc.

Based on the results and analysis above, a tentative mechanism for the catalytic performance of the $Cu/Al_2O_3/CN$ composite was proposed. Copper species were bonded to the Al_2O_3 framework, which facilitates the conversion of H_2O_2 to HO· by Fenton-like reaction and broadens the pH range for application. CN provides a large surface area and multiple functional groups to load Cu/Al_2O_3 composite by the formation of C–O–Cu bonds. In the presence of the $Cu/Al_2O_3/CN$ composite, aqueous H_2O_2 could be easily adsorbed to the surface of CN, increasing its chance to contact with immobilized copper species to generate HO·. The catalytic process is surface predominant and the tiny leaching of copper species leads to an efficient and long-lasting catalytic property. $O_2 \cdot^-$ is generated via the reaction of HO· and H_2O_2, and the highly reactive HO· attacks Rh B molecules, leading to the degradation of Rh B into smaller molecules.

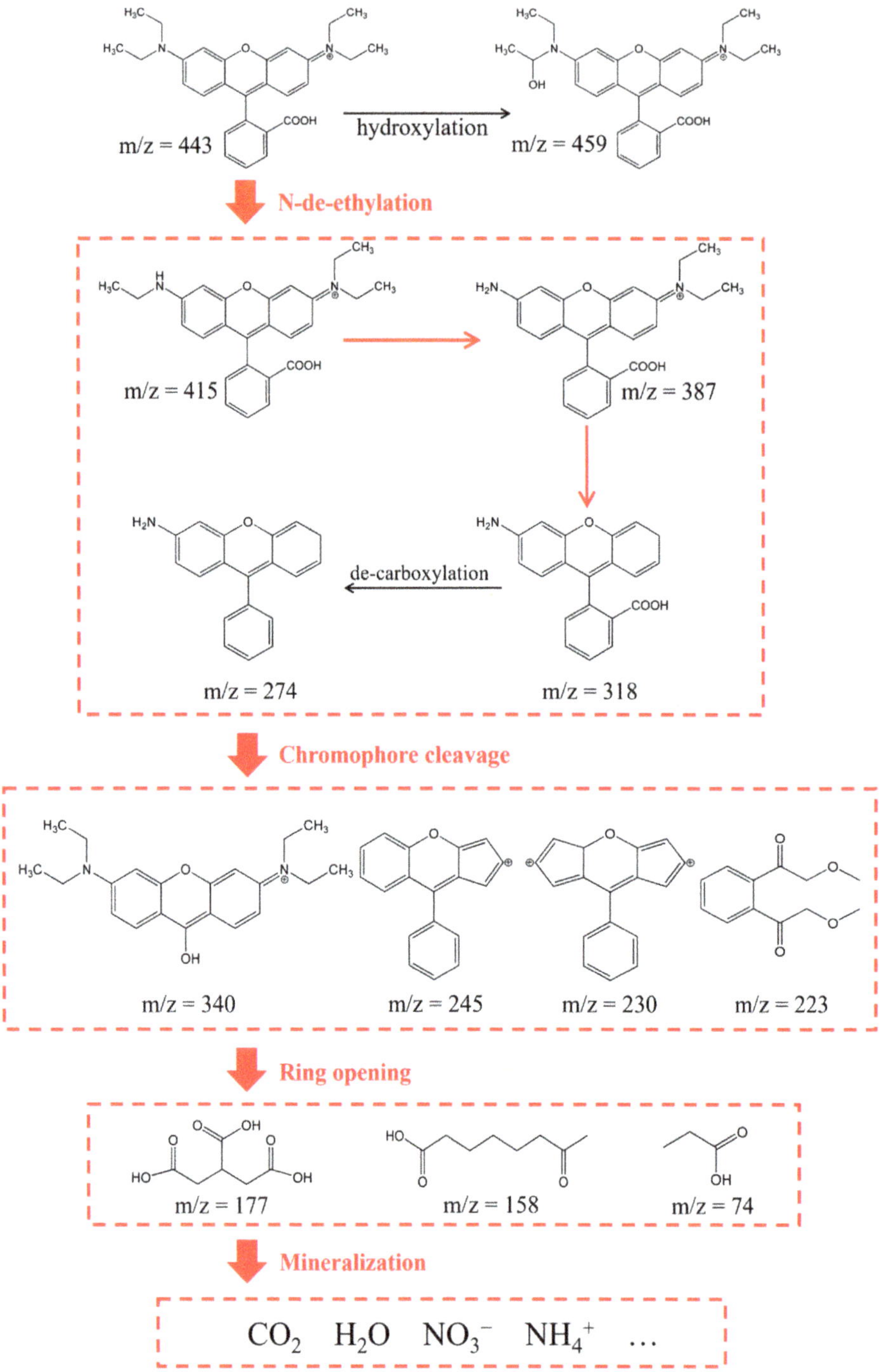

Figure 15. Illustration of the proposed Rh B degradation pathways.

3. Experimental

3.1. Materials

The chemicals used in the present work included aluminum nitrate nonahydrate ($Al(NO_3)_3{\cdot}9H_2O$, CAS[7784-27-2], 99%), copper(II) nitrate trihydrate ($Cu(NO_3)_3{\cdot}3H_2O$, CAS[10031-43-3], 99%), D(+)–glucose monohydrate ($C_6H_{12}O_6{\cdot}H_2O$, CAS[14431-43-7], 99%), urea (CH_4N_2O, CAS[57-13-6], 99%), Rhodamine B (Rh B, $C_{28}H_{31}ClN_2O_3$, CAS[81-88-9], 99.7%), methylene blue trihydrate (MB, $C_{16}H_{16}ClN_3S{\cdot}3H_2O$, CAS[7720-79-3], 82%), methyl orange (MO, $C_{14}H_{14}N_3NaO_3S$, CAS[547-58-0], 98.5%), hydrogen peroxide (H_2O_2, CAS[7722-84-1], 30 wt %), potassium iodide (KI, CAS[7681-11-0], 99%), acetic acid ($C_2H_4O_2$, CAS[64-19-7], 99.5%), sodium acetate anhydrous (CH_3COONa, CAS[127-09-3], 99%), isopropanol (IPA, C_3H_8O, CAS[67-63-0], 99.7%) (Sinopharm, Beijing, China), and ammonium molybdate (di) ($(NH_4)_2Mo_2O_7$, CAS[27546-07-2], Mo 56.5%, Macklin, Shanghai, China). All chemicals were used without further purification. All the solutions were prepared in deionized water.

3.2. Preparation of Composites

3.2.1. Preparation of Cu/Al_2O_3

Cu/Al_2O_3 composite was prepared via a hydrothermal method followed by a calcination process [40]. Typically, 7.5 g of $Al(NO_3)_3{\cdot}9H_2O$, 5.0 g of $C_6H_{12}O_6$, 0.9 g of $Cu(NO_3)_2{\cdot}3H_2O$, and 60 mL of ultrapure water were added to a 100 mL breaker, and stirred at 25 °C for 1 h. The solution was transferred into a 100 mL Teflon-lined autoclave, then heated to 180 °C and kept for 20 h in an oven. After suction filtering, and water washing to neutral, the obtained product was added to an alumina crucible with a cover and then heated to 550 °C with a ramping rate of 5 °C/min in a muffle furnace and hold at 550 °C for 4 h. After naturally cooling down to room temperature, the product Cu/Al_2O_3 was obtained, and labeled as Cu_{12}/Al_2O_3, where 12 represents the initial mass ratio of $Cu(NO_3)_2{\cdot}3H_2O$ to $Al(NO_3)_3{\cdot}9H_2O$ in percentage. Following the same procedure, a series of Cu_x/Al_2O_3 composites (x = 4, 7, 9, 12, and 15 wt %) were prepared in order to obtain the optimal Cu content.

3.2.2. Preparation of $g\text{-}C_3N_4$

Pure $g\text{-}C_3N_4$ was prepared following the thermal polymerization process reported in a previous study [35]. Typically, 50 g of urea was added to an alumina crucible with a cover and then heated to 550 °C with a ramping rate of 5 °C/min in a muffle furnace and hold at 550 °C for 3 h. After naturally cooling down to room temperature, the product $g\text{-}C_3N_4$ was obtained, and labeled as CN.

3.2.3. Preparation of $Cu/Al_2O_3/CN$

The $Cu/Al_2O_3/CN$ composite was prepared using the samples obtained in Sections 3.2.1 and 3.2.2. Typically, 0.1 g of the Cu_{12}/Al_2O_3 composite and 0.1 g of CN were immersed in 100 mL ultrapure water. The suspension was dispersed by ultrasound for 30 min. After drying at 80 °C and grinding to powder, the product $Cu/Al_2O_3/CN$ composite was obtained, and labeled as $Cu_{12}/Al_2O_3/CN_{1.3}$, where 1.3 represents the initial mass ratio of CN to $Al(NO_3)_3{\cdot}9H_2O$ in percentage. Additionally, a series of $Cu_{12}/Al_2O_3/CN_y$ composites (y = 0.7, 1.3, 2.0, and 2.7 wt %) were prepared by the same synthetic process to obtain the optimal CN content.

3.3. Characterization of Prepared Composites

The Fourier transform infrared (FT-IR) spectra of prepared samples were recorded by a Nicolet iS5 (Thermo Fisher Scientific, Waltham, MA., USA) FT-IR spectrometer with KBr pellets to confirm the functional groups of materials.

X-ray diffraction (XRD) patterns were recorded with D8 advance (Bruker, Karlsruhe, Germany) diffractometer using Bragg–Brentano geometry over the 2θ range of 10–70° with a Cu Kα irradiation (λ = 1.54 Å) to obtain the crystal structure of materials.

The morphology of the synthesized samples was visualized by a JET-2100F (JEOL, Akishima, Japan) transmission electron microscope (TEM).

X-ray photoelectron spectroscopy (XPS) data was recorded at 100 W power with a 20 eV pass energy on an ESCALAB 250Xi instrument (Thermo Fisher Scientific, Waltham, MA., USA).

The Brunauer–Emmett–Teller (B.E.T) method via isothermal adsorption and desorption of high purity nitrogen using a TriStar II 3020 (Micromeritics, Norcross, GA., USA) instrument was used to determine the specific surface area of synthesized samples.

The degradation intermediates were identified by a high performance liquid chromatography-mass spectrometry (HPLC-MS, thermo TSQ Quantum Ultra, Waltham, MA., USA). The samples were chromatographically separated by a 2.1 mm × 150 mm XB C18 column with a particle size of 3 μm (Welch, Shanghai, China) with acetonitrile/0.1% formic acid (8:92, *v*/*v*) as the mobile phase The flow rate was 0.3 mL/min and the injection volume was 20 μL. The eluent was detected by a diode array detector at the wavelength of 555 nm. The total ion current was recorded by quadrupole in a positive mode with electrospray ionization as ion source.

The concentration of the dissolved copper species was measured by a Prodigy 7 (Teledyne Leeman Labs, Hudson, NH., USA) inductively coupled plasma-optical emission spectrometer (ICP-OES).

The electron spin resonance (ESR) signals of radicals spin-trapped by 5,5-dimethyl-pyrroline-N-oxide (DMPO) were detected by a Bruker/A300 spectrometer (Bruker, Karlsruhe, Germany) with or without the addition of H_2O_2 in different air saturated methanol/aqueous dispersions.

The concentration of H_2O_2 and Rh B was measured by the UV-5500PC UV-visible spectrophotometer (METASH, Shanghai, China) and V-5600 spectrophotometer (METASH, Shanghai, China). The samples were weighted using a ME204E microbalance (Mettler Toledo, Shanghai, China). The pH of the solution was measured by Starter 2100 pH meter (Ohaus, Shanghai, China) with an accuracy of ± 0.01 pH units.

3.4. Catalytic Experiments

Rh B was selected to evaluate the catalytic performance of the synthesized samples. All experiments were performed in dark conditions. Typically, 0.1 g of sample was placed into 100 mL of 20 mg/L Rh B solution, and then the suspension was stirred continuously for 1 h prior to the addition of H_2O_2, in order to achieve Rh B adsorption/desorption equilibrium. Afterwards, predetermined amount of H_2O_2 was added in the suspension under constant stirring to trigger the reaction. At given time intervals, 4 mL samples were collected with a syringe and filtered using a Millipore filter (220 nm, Tianjin, China). The concentrations of Rh B and H_2O_2 in the filtrate were measured and plotted as a function of time.

The concentrations of Rh B and H_2O_2 in solutions were measured using a UV-visible spectrophotometer due to its convenience and easy operation. The produced intermediates may interfere the measurements, but the experimental error in the determination of Rh B and H_2O_2 concentration was less than 2%, which is acceptable for subsequent discussions. Generally, the absorbance of Rh B was measured at 555 nm in which Rh B had the maximum absorption. The concentration of H_2O_2 was determined by the Ghormley triiodide method reported in previous work [68]. The I^- could be oxidized by H_2O_2 to form I_3^- in the presence of catalyst $(NH_4)_2Mo_2O_7$ (ADM). The I_3^- has the maximum absorption peak at the wavelength of 350 nm, thus the standard curve of the absorbance of I_3^- as a function of H_2O_2 concentration can be obtained.

Radical scavenger experiments were carried out to investigate the reactive oxygen species in the system. The scavenger IPA was added into the suspension after the adsorption/desorption equilibrium prior to the addition of H_2O_2. Additionally, Rh B solution was purged with N_2 prior to the addition of composite to exclude O_2.

4. Conclusions

An efficient Fenton-like Cu/Al_2O_3/CN composite was proposed in this work. The characterizations show that Cu species were immobilized on the Al_2O_3 framework in the form of Cu^{2+} and Cu^{+}, and the introduction of CN increases its specific surface area and adsorption capacity for Rh B. The Cu/Al_2O_3/CN composite showed an excellent catalytic performance in a wide range of pH (4.9–11.0). The specific reaction rate constant of Rh B degradation was calculated as $(5.9 \pm 0.07) \times 10^{-9}$ m·s^{-1}, and the activation energy was calculated to be 71.0 kJ/mol. The recycling experiment demonstrated its durability for Rh B removal and proved that the degradation reaction was surface donimated, with a negligible leaching of copper species in solution. HO· and $O_2{\cdot}^-$ were both major reactive oxygen species generated in the reaction, and HO· was mainly responsible for Rh B degradation, which involves N-de-ethylation, hydroxylation, de-carboxylation, chromophore cleavage, ring opening, and the mineralization process. This work shows that Cu/Al_2O_3/CN composite is a promising Fenton-like catalyst with high activity and stability for the treatment of water pollution.

Supplementary Materials: The following are available online at http://www.mdpi.com/2073-4344/10/3/317/s1, Figure S1: (**A**) [H_2O_2], (**B**) [Rh B]/[Rh B]$_0$, and (**C**) ln ([Rh B]/ [Rh B]$_0$) as a function of time in the presence of 1 g/L of the Cu/Al_2O_3/CN composite at room temperature. Reaction conditions: [Rh B]$_0$ = 20 mg/L, [H_2O_2]$_0$ = 10 mM, V = 100 mL, pH 4.9 (unadjusted); Figure S2: [Rh B]/[Rh B]$_0$ as a function of time in nonfiltered Cu/Al_2O_3/CN suspension and filtrates obtained at selected time intervals. Reaction conditions: [catalyst] = 1 g/L, [Rh B]$_0$ = 20 mg/L, [H_2O_2]$_0$ = 10 mM, V = 100 mL, 25 °C, pH 4.9 (unadjusted); Figure S3: C/C$_0$ as a function of time with different organic pollutants in the presence of 1 g/L of the Cu/Al_2O_3/CN composite. Reaction conditions: [dye] = 20 mg/L, [H_2O_2]$_0$ = 10 mM, V = 100 mL, 25 °C, pH 4.9 (unadjusted); Figure S4: HPLC chromatograms of samples extracted at different reaction times. Reaction conditions: [catalyst] = 1 g/L, [dye] = 20 mg/L, [H_2O_2]$_0$ = 10 mM, V = 100 mL, 25 °C, pH 4.9 (unadjusted); Figure S5: Mass spectrograms of the degradation products in sample extracted at 60 min in Rh B degradation.; Table S1: Comparison of the catalytic activities of the Cu/Al_2O_3/CN composite with other copper-based Fenton-like catalysts.

Author Contributions: M.Y. conceived the experiments; C.Z. and Z.L. designed and performed the experiments; L.F. and Y.G. analyzed the data; Y.F. contributed analysis tools; C.Z., Z.L. and M.Y. wrote the paper. All authors have read and agreed to the published version of the manuscript.

Funding: This work is supported by National Natural Science Foundation of China (21707108) and Independent Innovation Foundation of Wuhan University of Technology (20411057 and 20410962).

Conflicts of Interest: The authors declare no conflict of interest.

References

1. Wang, J.L.; Wang, S.Z. Removal of pharmaceuticals and personal care products (PPCPs) from wastewater: A review. *J. Environ. Manag.* **2016**, *182*, 620–640. [CrossRef] [PubMed]
2. Pavithra, K.G.; Kumar, P.S.; Jaikumar, V.; Rajan, P.S. Removal of colorants from wastewater: A review on sources and treatment strategies. *J. Ind. Eng. Chem.* **2019**, *75*, 1–19. [CrossRef]
3. Li, H.P.; Liu, J.Y.; Hou, W.G.; Du, N.; Zhang, R.J.; Tao, X.T. Synthesis and characterization of g-C_3N_4/Bi_2MoO_6 heterojunctions with enhanced visible light photocatalytic activity. *Appl. Catal. B* **2014**, *160*, 89–97. [CrossRef]
4. Adityosulindro, S.; Barthe, L.; González-Labrada, K.; Jáuregui Haza, U.J.; Delmas, H.; Julcour, C. Sonolysis and sono-Fenton oxidation for removal of ibuprofen in (waste)water. *Ultrason. Sonochem.* **2017**, *39*, 889–896. [CrossRef] [PubMed]
5. Bai, C.P.; Xiong, X.F.; Gong, W.Q.; Feng, D.X.; Xian, M.; Ge, Z.X.; Xu, N.A. Removal of rhodamine B by ozone-based advanced oxidation process. *Desalination* **2011**, *278*, 84–90.
6. Masomboon, N.; Ratanatamskul, C.; Lu, M.C. Kinetics of 2,6-dimethylaniline oxidation by various Fenton processes. *J. Hazard. Mater.* **2011**, *192*, 347–353. [CrossRef]
7. Patil, S.P.; Bethi, B.; Sonawane, G.H.; Shrivastava, V.S.; Sonawane, S. Efficient adsorption and photocatalytic degradation of Rhodamine B dye over Bi_2O_3-bentonite nanocomposites: A kinetic study. *J. Ind. Eng. Chem.* **2016**, *34*, 356–363. [CrossRef]
8. Madhavan, J.; Grieser, F.; Ashokkumar, M. Combined advanced oxidation processes for the synergistic degradation of ibuprofen in aqueous environments. *J. Hazard. Mater.* **2010**, *178*, 202–208. [CrossRef]
9. Rao, Y.F.; Xue, D.; Pan, H.M.; Feng, J.T.; Li, Y.J. Degradation of ibuprofen by a synergistic UV/Fe(III)/Oxone process. *Chem. Eng. J.* **2016**, *283*, 65–75. [CrossRef]

10. Al-Hamadani, Y.A.J.; Park, C.M.; Assi, L.N.; Chu, K.H.; Hoque, S.; Jang, M.; Yoon, Y.; Ziehl, P. Sonocatalytic removal of ibuprofen and sulfamethoxazole in the presence of different fly ash sources. *Ultrason. Sonochem.* **2017**, *39*, 354–362. [CrossRef]
11. Cheng, M.; Zeng, G.M.; Huang, D.L.; Lai, C.; Xu, P.; Zhang, C.; Liu, Y. Hydroxyl radicals based advanced oxidation processes (AOPs) for remediation of soils contaminated with organic compounds: A review. *Chem. Eng. J.* **2016**, *284*, 582–598. [CrossRef]
12. Babuponnusami, A.; Muthukumar, K. A review on Fenton and improvements to the Fenton process for wastewater treatment. *J. Environ. Chem. Eng.* **2014**, *2*, 557–572. [CrossRef]
13. Saleh, R.; Taufik, A. Degradation of methylene blue and congo-red dyes using Fenton, photo-Fenton, sono-Fenton, and sonophoto-Fenton methods in the presence of iron (II,III) oxide/zinc oxide/graphene (Fe_3O_4/ZnO/graphene) composites. *Sep. Purif. Technol.* **2019**, *210*, 563–573. [CrossRef]
14. Bello, M.M.; Raman, A.A.A.; Asghar, A. A review on approaches for addressing the limitations of Fenton oxidation for recalcitrant wastewater treatment. *Process Saf. Environ. Prot.* **2019**, *126*, 119–140. [CrossRef]
15. Wang, P.; Wang, N.N.; Zheng, T.; Zhang, G.S. A review on Fenton-like processes for organic wastewater treatment. *J. Environ. Chem. Eng.* **2016**, *4*, 762–787. [CrossRef]
16. de Luna, M.D.G.; Colades, J.I.; Su, C.C.; Lu, M.C. Comparison of dimethyl sulfoxide degradation by different Fenton processes. *Chem. Eng. J.* **2013**, *232*, 418–424. [CrossRef]
17. Karpinska, A.; Sokol, A.; Karpinska, J. Studies on the kinetics of carbamazepine degradation in aqueous matrix in the course of modified Fenton's reactions. *J. Pharm. Biomed. Anal.* **2015**, *106*, 46–51. [CrossRef]
18. Yamaguchi, R.; Kurosu, S.; Suzuki, M.; Kawase, Y. Hydroxyl radical generation by zero-valent iron/Cu (ZVI/Cu) bimetallic catalyst in wastewater treatment: Heterogeneous Fenton/Fenton-like reactions by Fenton reagents formed in-situ under oxic conditions. *Chem. Eng. J.* **2018**, *334*, 1537–1549. [CrossRef]
19. Xia, D.H.; Xu, W.J.; Hu, L.L.; He, C.; Leung, D.Y.C.; Wang, W.J.; Wong, P.K. Synergistically catalytic oxidation of toluene over Mn modified g-C_3N_4/ZSM-4 under vacuum UV irradiation. *J. Hazard. Mater.* **2018**, *349*, 91–100. [CrossRef]
20. Su, Y.H.; Chen, P.; Wang, F.L.; Zhang, Q.X.; Chen, T.S.; Wang, Y.F.; Yao, K.; Lv, W.Y.; Liu, G.G. Decoration of TiO_2/g-C_3N_4 Z-scheme by carbon dots as a novel photocatalyst with improved visible-light photocatalytic performance for the degradation of enrofloxacin. *RSC Adv.* **2017**, *7*, 34096–34103. [CrossRef]
21. Nieto-Juarez, J.I.; Pierzchla, K.; Sienkiewicz, A.; Kohn, T. Inactivation of MS_2 coliphage in Fenton and Fenton-like systems: Role of transition metals, hydrogen peroxide and sunlight. *Environ. Sci. Technol.* **2010**, *44*, 3351–3356. [CrossRef] [PubMed]
22. Zhang, L.L.; Xu, D.; Hu, C.; Shi, Y.L. Framework Cu-doped $AlPO_4$ as an effective Fenton-like catalyst for bisphenol A degradation. *Appl. Catal. B* **2017**, *207*, 9–16. [CrossRef]
23. Li, Z.F.; Soroka, I.L.; Min, F.Y.; Jonsson, M. pH-Control as a way to fine-tune the Cu/Cu_2O ratio in radiation induced synthesis of Cu_2O particles. *Dalton Trans.* **2018**, *47*, 16139–16144. [CrossRef] [PubMed]
24. Zhang, L.; Wu, B.D.; Zhang, G.Y.; Gan, Y.H.; Zhang, S.J. Enhanced decomplexation of Cu(II)-EDTA: The role of acetylacetone in Cu-mediated photo-Fenton reactions. *Chem. Eng. J.* **2019**, *358*, 1218–1226. [CrossRef]
25. Xu, J.H.; Wang, W.Z.; Gao, E.P.; Ren, J.; Wang, L. Bi_2WO_6/Cu-0: A novel coupled system with enhanced photocatalytic activity by Fenton-like synergistic effect. *Catal. Commun.* **2011**, *12*, 834–838. [CrossRef]
26. Mao, J.; Quan, X.; Wang, J.; Gao, C.; Chen, S.; Yu, H.T.; Zhang, Y.B. Enhanced heterogeneous Fenton-like activity by Cu-doped $BiFeO_3$ perovskite for degradation of organic pollutants. *Front. Environ. Sci. Eng. Chin.* **2018**, *12*, 1–10. [CrossRef]
27. Zhang, L.L.; Nie, Y.L.; Hu, C.; Qu, J.H. Enhanced Fenton degradation of Rhodamine B over nanoscaled Cu-doped $LaTiO_3$ perovskite. *Appl. Catal. B* **2012**, *125*, 418–424. [CrossRef]
28. Xiao, P.; Li, H.L.; Wang, T.; Xu, X.L.; Li, J.L.; Zhu, J.J. Efficient Fenton-like La-Cu-O/SBA-15 catalyst for the degradation of organic dyes under ambient conditions. *RSC Adv.* **2014**, *4*, 12601–12604. [CrossRef]
29. Bonfim, D.P.F.; Santana, C.S.; Batista, M.S.; Fabiano, D.P. Catalytic Evaluation of CuO/[Si]MCM-41 in Fenton-like Reactions. *Chem. Eng. Technol.* **2019**, *42*, 882–888. [CrossRef]
30. Zuo, S.Y.; Xu, H.M.; Liao, W.; Yuan, X.J.; Sun, L.; Li, Q.; Zan, J.; Li, D.Y.; Xia, D.S. Molten-salt synthesis of g-C_3N_4-Cu_2O heterojunctions with highly enhanced photocatalytic performance. *Colloids Surf. A* **2018**, *546*, 307–315. [CrossRef]

31. Peng, B.Y.; Zhang, S.S.; Yang, S.Y.; Wang, H.J.; Yu, H.; Zhang, S.Q.; Peng, F. Synthesis and characterization of g-C_3N_4/Cu_2O composite catalyst with enhanced photocatalytic activity under visible light irradiation. *Mater. Res. Bull.* **2014**, *56*, 19–24. [CrossRef]
32. Liu, Q.; Guo, Y.R.; Chen, Z.H.; Zhang, Z.G.; Fang, X.M. Constructing a novel ternary Fe(III)/graphene/g-C_3N_4 composite photocatalyst with enhanced visible-light driven photocatalytic activity via interfacial charge transfer effect. *Appl. Catal. B* **2016**, *183*, 231–241. [CrossRef]
33. Pomilla, F.R.; Cortes, M.A.L.R.M.; Hamilton, J.W.J.; Molinari, R.; Barbieri, G.; Marci, G.; Palmisano, L.; Sharma, P.K.; Brown, A.; Byrne, J.A. An Investigation into the Stability of Graphitic C_3N_4 as a Photocatalyst for CO_2 Reduction. *J. Phys. Chem. C* **2018**, *122*, 28727–28738. [CrossRef]
34. Liu, J.; Liu, Y.; Liu, N.Y.; Han, Y.Z.; Zhang, X.; Huang, H.; Lifshitz, Y.; Lee, S.T.; Zhong, J.; Kang, Z.H. Metal-free efficient photocatalyst for stable visible water splitting via a two-electron pathway. *Science* **2015**, *347*, 970–974. [CrossRef] [PubMed]
35. Liu, Z.D.; Shen, Q.M.; Zhou, C.S.; Fang, L.J.; Yang, M.; Xia, T. Kinetic and Mechanistic Study on Catalytic Decomposition of Hydrogen Peroxide on Carbon-Nanodots/Graphitic Carbon Nitride Composite. *Catalysts* **2018**, *8*, 445. [CrossRef]
36. Fang, L.J.; Liu, Z.D.; Zhou, C.S.; Guo, Y.L.; Feng, Y.P.; Yang, M. Degradation Mechanism of Methylene Blue by H_2O_2 and Synthesized Carbon Nanodots/Graphitic Carbon Nitride/Fe(II) Composite. *J. Phys. Chem. C* **2019**, *123*, 26921–26931. [CrossRef]
37. Lyu, L.; Zhang, L.L.; He, G.Z.; He, H.; Hu, C. Selective H_2O_2 conversion to hydroxyl radicals in the electron-rich area of hydroxylated C-g-C_3N_4/ CuCo-Al_2O_3. *J. Mater. Chem. A* **2017**, *5*, 7153–7164. [CrossRef]
38. Zhu, J.J.; Xiao, P.; Li, H.L.; Carabineiro, S.A.C. Graphitic Carbon Nitride: Synthesis, Properties, and Applications in Catalysis. *ACS Appl. Mater. Interfaces* **2014**, *6*, 16449–16465. [CrossRef]
39. Jia, J.K.; Huang, W.X.; Feng, C.S.; Zhang, Z.; Zuojiao, K.C.; Liu, J.X.; Jiang, C.Y.; Wang, Y.P. Fabrication of g-C_3N_4/Ag_3PO_4-H_2O_2 heterojunction system with enhanced visible-light photocatalytic activity and mechanism insight. *J. Alloys Compd.* **2019**, *790*, 616–625. [CrossRef]
40. Xu, S.Q.; Zhu, H.X.; Cao, W.R.; Wen, Z.B.; Wang, J.N.; Francois-Xavier, C.P.; Wintgens, T. Cu-Al_2O_3-g-C_3N_4 and Cu-Al_2O_3-C-dots with dual-reaction centres for simultaneous enhancement of Fenton-like catalytic activity and selective H_2O_2 conversion to hydroxyl radicals. *Appl. Catal. B* **2018**, *234*, 223–233. [CrossRef]
41. Xu, D.; Zhang, L.L.; Liu, L.F. Fenton-like Catalytic Removal of Organic Pollutants in Water by Framework Cu in Cu-Al_2O_3. *Huanjing Kexue* **2017**, *38*, 1054–1060.
42. Ma, X.Z.; Zhang, J.T.; Wang, B.; Li, Q.G.; Chu, S. Hierarchical Cu_2O foam/g-C_3N_4 photocathode for photoelectrochemical hydrogen production. *Appl. Surf. Sci.* **2018**, *427*, 907–916. [CrossRef]
43. Dikdım, J.M.D.; Gong, Y.; Noumi, G.B.; Sieliechi, J.M.; Zhao, X.; Ma, N.; Yang, M.; Tchatchueng, J.B. Peroxymonosulfate improved photocatalytic degradation of atrazine by activated carbon/graphitic carbon nitride composite under visible light irradiation. *Chemosphere* **2019**, *217*, 833–842. [CrossRef]
44. Engelhart, W.; Dreher, W.; Eibl, O.; Schier, V. Deposition of alumina thin film by dual magnetron sputtering: Is it gamma-Al_2O_3? *Acta Mater.* **2011**, *59*, 7757–7767. [CrossRef]
45. Sareen, S.; Mutreja, V.; Singh, S.; Pal, B. Highly dispersed Au, Ag and Cu nanoparticles in mesoporous SBA-15 for highly selective catalytic reduction of nitroaromatics. *RSC Adv.* **2015**, *5*, 184–190. [CrossRef]
46. Wang, Y.; Li, J.; Sun, J.Y.; Wang, Y.B.; Zhao, X. Electrospun flexible self-standing Cu-Al_2O_3 fibrous membranes as Fenton catalysts for bisphenol A degradation. *J. Mater. Chem. A* **2017**, *5*, 19151–19158. [CrossRef]
47. Lyu, L.; Zhang, L.L.; Hu, C. Enhanced Fenton-like degradation of pharmaceuticals over framework copper species in copper-doped mesoporous silica microspheres. *Chem. Eng. J.* **2015**, *274*, 298–306. [CrossRef]
48. Sheng, Y.Y.; Sun, Y.; Xu, J.; Zhang, J.; Han, Y.F. Fenton-like degradation of rhodamine B over highly durable Cu-embedded alumina: Kinetics and mechanism. *AIChE J.* **2018**, *64*, 538–549. [CrossRef]
49. Merouani, S.; Hamdaoui, O.; Saoudi, F.; Chiha, M. Sonochemical degradation of Rhodamine B in aqueous phase: Effects of additives. *Chem. Eng. J.* **2010**, *158*, 550–557. [CrossRef]
50. Nieto, L.M.; Hodaifa, G.; Rodriguez, S.; Gimenez, J.A.; Ochando, J. Degradation of organic matter in olive-oil mill wastewater through homogeneous Fenton-like reaction. *Chem. Eng. J.* **2011**, *173*, 503–510. [CrossRef]
51. Hodaifa, G.; Ochando-Pulido, J.M.; Rodriguez-Vives, S.; Martinez-Ferez, A. Optimization of continuous reactor at pilot scale for olive-oil mill wastewater treatment by Fenton-like process. *Chem. Eng. J.* **2013**, *220*, 117–124. [CrossRef]

52. Ma, Q.B.; Hofmann, J.P.; Litke, A.; Hensen, E.J.M. Cu_2O photoelectrodes for solar water splitting: Tuning photoelectrochemical performance by controlled faceting. *Sol. Energy Mater. Sol. Cells* **2015**, *141*, 178–186. [CrossRef]
53. Aylmore, M.G.; Muir, D.M. Thermodynamic analysis of gold leaching by ammoniacal thiosulfate using Eh/pH and speciation diagrams. *Miner. Metall. Process* **2001**, *18*, 221–227. [CrossRef]
54. Lodha, S.; Jain, A.; Punjabi, P.B. A novel route for waste water treatment: Photocatalytic degradation of rhodamine B. *Arab. J. Chem.* **2011**, *4*, 383–387. [CrossRef]
55. Matheson, L.J.; Tratnyek, P.G. Reductive Dehalogenation of Chlorinated Methanes by Iron Metal. *Environ. Sci. Technol.* **2002**, *28*, 2045–2053. [CrossRef]
56. Bjorkbacka, A.; Yang, M.; Gasparrini, C.; Leygraf, C.; Jonsson, M. Kinetics and mechanisms of reactions between H_2O_2 and copper and copper oxides. *Dalton Trans.* **2015**, *44*, 16045–16051. [CrossRef]
57. Lousada, C.M.; Jonsson, M. Kinetics, Mechanism, and Activation Energy of H_2O_2 Decomposition on the Surface of ZrO_2. *J. Phys. Chem. C* **2010**, *114*, 11202–11208. [CrossRef]
58. Lousada, C.M.; Yang, M.; Nilsson, K.; Jonsson, M. Catalytic decomposition of hydrogen peroxide on transition metal and lanthanide oxides. *J. Mol. Catal. A Chem.* **2013**, *379*, 178–184. [CrossRef]
59. Kanel, S.R.; Greneche, J.M.; Choi, H. Arsenic(V) removal kom groundwater using nano scale zero-valent iron as a colloidal reactive barrier material. *Environ. Sci. Technol.* **2006**, *40*, 2045–2050. [CrossRef]
60. Cope, D.B.; Benson, C.H. Grey-Iron Foundry Slags As Reactive Media for Removing Trichloroethylene from Groundwater. *Environ. Sci. Technol.* **2009**, *43*, 169–175. [CrossRef]
61. Agrawal, A.; Tratnyek, P.G. Reduction of Nitro Aromatic Compounds by Zero-Valent Iron Metal. *Environ. Sci. Technol.* **1995**, *30*, 153–160. [CrossRef]
62. Sun, Y.K.; Li, J.X.; Huang, T.L.; Guan, X.H. The influences of iron characteristics, operating conditions and solution chemistry on contaminants removal by zero-valent iron: A review. *Water Res.* **2016**, *100*, 277–295. [CrossRef]
63. Li, H.; Liao, J.Y.; Zeng, T. A facile synthesis of CuO nanowires and nanorods, and their catalytic activity in the oxidative degradation of Rhodamine B with hydrogen peroxide. *Catal. Commun.* **2014**, *46*, 169–173. [CrossRef]
64. Zaman, S.; Zainelabdin, A.; Amin, G.; Nur, O.; Willander, M. Efficient catalytic effect of CuO nanostructures on the degradation of organic dyes. *J. Phys. Chem. Solids* **2012**, *73*, 1320–1325. [CrossRef]
65. Dragoi, M.; Samide, A.; Moanta, A. Discoloration of Waters Containing Azo Dye Congo Red by Fenton Oxidation Process Estimation of activation parameters. *Rev. Chim.* **2011**, *62*, 1195–1198.
66. Martin, R.; Navalon, S.; Delgado, J.J.; Calvino, J.J.; Alvaro, M.; Garcia, H. Influence of the Preparation Procedure on the Catalytic Activity of Gold Supported on Diamond Nanoparticles for Phenol Peroxidation. *Chem. Eur. J.* **2011**, *17*, 9494–9502. [CrossRef]
67. Yao, Z.P.; Chen, C.J.; Wang, J.K.; Xia, Q.X.; Li, C.X.; Jiang, Z.H. Iron Oxide Coating Fenton-like Catalysts: Preparation and Degradation of Phenol. *Chin. J. Inorg. Chem.* **2017**, *33*, 1797–1804.
68. Yang, M.; Zhang, X.; Grosjean, A.; Soroka, I.L.; Jonsson, M. Kinetics and Mechanism of the Reaction between H_2O_2 and Tungsten Powder in Water. *J. Phys. Chem. C* **2015**, *119*, 22560–22569. [CrossRef]
69. Ma, J.Q.; Yang, Q.F.; Wen, Y.Z.; Liu, W.P. Fe-g-C_3N_4/graphitized mesoporous carbon composite as an effective Fenton-like catalyst in a wide pH range. *Appl. Catal. B* **2017**, *201*, 232–240. [CrossRef]
70. Zhu, J.N.; Zhu, X.Q.; Cheng, F.F.; Li, P.; Wang, F.; Xiao, Y.W.; Xiong, W.W. Preparing copper doped carbon nitride from melamine templated crystalline copper chloride for Fenton-like catalysis. *Appl. Catal. B* **2019**, *256*, 117830. [CrossRef]
71. Ma, J.Q.; Jia, N.Z.F.; Shen, C.S.; Liu, W.P.; Wen, Y.Z. Stable cuprous active sites in Cu+-graphitic carbon nitride: Structure analysis and performance in Fenton-like reactions. *J. Hazard. Mater.* **2019**, *378*, 120782. [CrossRef]
72. Fan, J.W.; Jiang, X.; Min, H.Y.; Li, D.D.; Ran, X.Q.; Zou, L.Y.; Sun, Y.; Li, W.; Yang, J.P.; Teng, W.; et al. Facile preparation of Cu-Mn/CeO_2/SBA-15 catalysts using ceria as an auxiliary for advanced oxidation processes. *J. Mater. Chem. A* **2014**, *2*, 10654–10661. [CrossRef]
73. Sun, Y.; Tian, P.F.; Ding, D.D.; Yang, Z.X.; Wang, W.Z.; Xin, H.; Xu, J.; Han, Y.F. Revealing the active species of Cu-based catalysts for heterogeneous Fenton reaction. *Appl. Catal. B* **2019**, *258*, 117985. [CrossRef]
74. Zhang, H.; Zhao, L.X.; Geng, F.L.; Guo, L.H.; Wan, B.; Yang, Y. Carbon dots decorated graphitic carbon nitride as an efficient metal-free photocatalyst for phenol degradation. *Appl. Catal. B* **2016**, *180*, 656–662. [CrossRef]

75. Yu, D.Y.; Li, L.B.; Wu, M.; Crittenden, J.C. Enhanced photocatalytic ozonation of organic pollutants using an iron-based metal-organic framework. *Appl. Catal. B* **2019**, *251*, 66–75. [CrossRef]
76. He, Z.; Sun, C.; Yang, S.G.; Ding, Y.C.; He, H.; Wang, Z.L. Photocatalytic degradation of rhodamine B by Bi_2WO_6 with electron accepting agent under microwave irradiation: Mechanism and pathway. *J. Hazard. Mater.* **2009**, *162*, 1477–1486. [CrossRef]
77. Yu, K.; Yang, S.G.; He, H.; Sun, C.; Gu, C.G.; Ju, Y.M. Visible Light-Driven Photocatalytic Degradation of Rhodamine B over $NaBiO_3$: Pathways and Mechanism. *J. Phys. Chem. A* **2009**, *113*, 10024–10032. [CrossRef]
78. Martinez-de la Cruz, A.; Perez, U.M.G. Photocatalytic properties of $BiVO_4$ prepared by the co-precipitation method: Degradation of rhodamine B and possible reaction mechanisms under visible irradiation. *Mater. Res. Bull.* **2010**, *45*, 135–141. [CrossRef]
79. Guo, J.X.; Zhou, H.R.; Ting, S.; Luo, H.D.; Liang, J.; Yuan, S.D. Investigation of catalytic activity and mechanism for RhB degradation by $LaMnO_3$ perovskites prepared via the citric acid method. *New J. Chem.* **2019**, *43*, 18146–18157. [CrossRef]
80. Sun, M.; Li, D.Z.; Chen, Y.B.; Chen, W.; Li, W.J.; He, Y.H.; Fu, X.Z. Synthesis and Photocatalytic Activity of Calcium Antimony Oxide Hydroxide for the Degradation of Dyes in Water. *J. Phys. Chem. C* **2009**, *113*, 13825–13831. [CrossRef]
81. Sharma, G.; Dionysiou, D.D.; Sharma, S.; Kumar, A.; Al-Muhtaseb, A.H.; Naushad, M.; Stadler, F.J. Highly efficient Sr/Ce/activated carbon bimetallic nanocomposite for photoinduced degradation of rhodamine B. *Catal. Today* **2019**, *335*, 437–451. [CrossRef]
82. Natarajan, T.S.; Thomas, M.; Natarajan, K.; Bajaj, H.C.; Tayade, R.J. Study on UV-LED/TiO_2 process for degradation of Rhodamine B dye. *Chem. Eng. J.* **2011**, *169*, 126–134. [CrossRef]
83. Khandekar, D.C.; Bhattacharyya, A.R.; Bandyopadhyaya, R. Role of impregnated nano-photocatalyst ($SnxTi(1\text{-}x)O_2$) inside mesoporous silica (SBA-15) for degradation of organic pollutant (Rhodamine B) under UV light. *J. Environ. Chem. Eng.* **2019**, *7*, 103433. [CrossRef]
84. Wu, J.; Zhu, K.; Xu, H.; Yan, W. Electrochemical oxidation of rhodamine B by PbO_2/Sb-SnO_2/TiO_2 nanotube arrays electrode. *Chin. J. Catal.* **2019**, *40*, 917–927. [CrossRef]
85. Liu, Y.; Guo, H.G.; Zhang, Y.L.; Cheng, X.; Zhou, P.; Zhang, G.C.; Wang, J.Q.; Tang, P.; Ke, T.L.; Li, W. Heterogeneous activation of persulfate for Rhodamine B degradation with 3D flower sphere-like BiOI/Fe_3O_4 microspheres under visible light irradiation. *Sep. Purif. Technol.* **2018**, *192*, 88–98. [CrossRef]

Article

Effect of Zirconia on Hydrothermally Synthesized Co_3O_4/TiO_2 Catalyst for NO_x Reduction from Engine Emissions

Muhammad Habib Ur Rehman [1], Tayyaba Noor [2] and Naseem Iqbal [1,*]

[1] U.S-Pakistan Center for Advanced Studies in Energy (USPCAS-E), National University of Sciences and Technology (NUST), H-12, Islamabad 44000, Pakistan; engr.habib952@gmail.com

[2] School of Chemical & Materials Engineering (SCME), National University of Sciences & Technology (NUST), H-12, Islamabad 44000, Pakistan; tayyaba.noor@scme.nust.edu.pk

* Correspondence: naseem@uspcase.nust.edu.pk

Received: 13 December 2019; Accepted: 8 January 2020; Published: 9 February 2020

Abstract: Effect of zirconia on the 6 wt.% Co_3O_4/TiO_2 catalyst for NO_x reduction is investigated in this paper. Co_3O_4/TiO_2 catalyst was prepared by using hydrothermal method and then was promoted with zirconia by impregnation to get 8% wt. ZrO_2-Co_3O_4/TiO_2 catalyst. Catalysts were characterized by using XRD, SEM, and TGA. Catalysts real time activity was tested by coating them on stainless steel wire meshes, containing them in a mild steel shell and mounting them at the exhaust tailpipe of a 72 cm^3 motorcycle engine. Zirconia promoted catalyst showed higher conversion efficiency of NO_X than the simple Co_3O_4/TiO_2 catalyst due to small crystalline size, fouling inhibition and thermal stability.

Keywords: Co_3O_4/TiO_2 catalyst; ZrO_2-Co_3O_4/TiO_2 catalyst; NO_X reduction; CO and HC oxidation

---

1. Introduction

Pollutant gas emission from automotive vehicles is one of the biggest contributors to air pollution in most cities of the world [1–4]. Annually, 4.2 million people die due to the ambient air pollution throughout the world [5,6]. Since the introduction of the clean air act in 1970, a number of efforts have been made to reduce the engine exhaust pollution. Engine exhaust emissions contain three major pollutants—carbon monoxide (CO), nitrogen oxides (NOx), and hydrocarbons (HCs)—which need to be tackled [7]. Different methods, such as thermal reactors, diesel particulate filters (DPFs), selective catalytic reduction SCR, changes in engine design, oxygenated fuels, and catalytic converters had been adopted to truncate this concern to the minimum level [4,8]. Gasoline blended with 3% oxygen by weight reduces 30% of CO emissions but NO_X concentration increases which can swell ozone problems [9]. Conventionally catalytic converters employ Pt, Pd, and Rh as catalysts with innovations, such as the introduction of CeO_2 or CeO_2-ZrO_2 composite for managing the time lag from switching between lean to rich condition or vice versa [10]. These metals are emitted out due to high temperatures, mechanical friction, stresses, and chemical reactions to the roadside soils which cause disruption in plant growth [11]. Due to high solubility of PGEs with various compounds, the presence of these metals in the environment can cause many health threats, such as nausea, tumors, sensitization, pregnancy loss and other human health issues [12,13]. Moreover, due to thermal aging, Rh_2O_3 reacts with Al_2O_3 to form an inactive compound, Pt sintering occurs at 700 °C and at high temperature of 900 °C sintering of γ-AL_2O_3 occurs and it transforms to α-AL_2O_3 which has less surface area [14]. These metals are rarest of the elements present on the earth ranging from 5 to 15 ppm in ores mines and from 0.022 ppb for Ir to 0.52 ppb for Pd [15,16]. The automotive industry consumed 37% of platinum, 72% of palladium and 79% of rhodium in 2013 [15]. Due to the scarcity of these metals, increasing prices and health hazards, it is a requirement to find a replacement which is less rare, less expensive, and competitively active. Over time, many pure metals and metal oxides were tested as alternatives to these PGEs for redox reactions in catalytic converters i.e., Au, Ni, Cu, MnO_2, CoO, Co_3O_4 and CuO [15,17–19]. Co_3O_4 over different supports was investigated as a catalyst for oxidation of CO and HC in exhaust emission control system and demonstrated low temperature activity [20,21]. Wang et al. [22] experimented by preparing a cobalt oxide catalyst with three different supports, such as TiO_2, Al_2O_3 and SiO_2 by using the incipient wetness impregnation method. He expressed that the type of cobalt oxide and type of support used had a great effect on the activity and the surface area. He concluded that CoO_x/TiO_2 showed very significant results towards CO conversion. Hu et al. [23] performed the oxidation of CO on cobalt oxides nanobelts and nanocubes. He revealed that the shape Co_3O_4 had a great effect on the catalytic activity towards CO as nanobelts depicted more conversion efficiency than nanocubes. Jia et al. [24] prepared Co_3O_4-SiO_2 nanocomposite as catalyst with high surface area for oxidation of CO. He observed that the catalyst showed activity at a very low temperature of even −76 °C. He concluded that there was some adsorption of water molecules on the exterior of catalyst which caused abnormal behavior in the activity of catalyst. Xie et al. [25] reported that Co_3O_4 showed high activity for carbon monoxide conversion surprisingly at temperature of −77 °C and remained stable under the moist conditions. He observed that under the stoichiometric conditions, conversion efficiency of Co_3O_4 catalyst was 96% at temperature of 200 °C but the conversion efficiency decreased at temperature of 150 °C to 40% due the presence of moisture content.

On the other hand, due to high thermal stability, high surface area, and highly resistant towards poisoning TiO_2 was widely used for catalyst support as an alternate to alumina [26,27]. Titanium dioxide exists in three crystalline phas.

es that are anatase, brookite, and rutile however anatase and rutile play a role in applications of titanium dioxide. Anatase is less dense and less stable than rutile at ambient temperature and pressure conditions [28]. Zhu et al. [29] studied the effect of TiO_2 structure on catalytic properties of copper oxide supported by anatase or rutile for NO reduction by CO reaction. He concluded that copper oxide supported by rutile TiO_2 was more active than supported by anatase TiO_2. Presently, commercially available three way catalysis systems have ceria or ceria-zirconia along with alumina and precious metal to increase the conversion efficiency. The incorporation of ceria-zirconia mixture into automotive catalysts was to enhance the oxygen storage capacity (OSC) for the redox reactions by releasing/storing oxygen depending upon A/F ratio and other factors [30]. Extensive studies were carried on ceria, zirconia, and a mixture of both as OSC is very important for the increment of CO oxidation as well as the reduction of NOx at the same time. Transition metal oxides showed a great potential to have oxygen vacancies on the surface like TiO_2, ZrO_2 and V_2O_5 [31]. Yu et al. [32] studied the adsorption and oxidation of CO by O_2 at the surface of rutile TiO_2. Li et al. [33] calculated the oxygen vacancies on anatase and rutile TiO_2 and concluded that both phases have oxygen vacancies but in case of rutile more stable oxygen vacancy was found on the surface rather on subsurface. In the case of zirconia, oxygen vacancies are produced by doping but undoped zirconia also contains bulk concentration of oxygen vacancies [34,35]. In this work, we are going to present the effect of zirconia on hydrothermally produced Co_3O_4/TiO_2 catalyst for petrol engine emissions control system with Co_3O_4 being our oxidation catalyst and TiO_2 as reduction catalyst. Zirconia was impregnated on a catalyst and a new 8% wt. ZrO_2-Co_3O_4/TiO_2 was prepared. Both catalysts were coated on round wire meshes then placed in a mild steel clam shell and mounted on the exhaust of a motorcycle. The zirconia promoted catalyst showed more qualities of reductive catalyst as less NO_x were seen after the reaction inside the catalytic converter.

2. Results and Discussions

2.1. X-ray Diffraction (XRD)

All phases of the prepared catalyst were identified by using JCPDS powder diffraction file. In XRD analysis of Co_3O_4/TiO_2, the phase of titania identified was rutile and no peaks of other phases of titania were noticed. XRD pattern of Co_3O_4/TiO_2 catalyst is shown in Figure 1a. TiO_2 showed peaks at 2 theta values of 27.45°, 36.06°, 41.22°, 44.04°, 54.31°, 56.6°, 62.7°, 64.03°, 69.008° and 69.78°. Diffraction peaks of rutile titania are in complete match with JCPDS card no. 21-1276. Crystallite size of titania nanoparticles was calculated by Sherrer equation which was 65 nm on average. TiO_2 exhibited tetragonal crystal structure with the cell parameters as a = 4.593 Å, b = 4.593 Å and c = 2.959 Å. In XRD analysis of the catalyst the cubic crystal structure of cobalt oxide Co_3O_4 was identified with the cell parameters a = 8.084 Å, b = 8.084 Å and c = 8.084 Å. The pattern in Figure 1a shows strong peaks at the 2 theta values of 19°, 31.2°, 36.84°, 44.85°, 55.6° and 65.2° for Co_3O_4 matching with JCPDS card no. 42-1467. Crystallite size of cobalt oxide was determined by using Sherrer equation which came out to be 26 nm. No other peaks of $Co(OH)_2$ or any impurity were observed. XRD pattern of 8 wt.% ZrO_2-Co_3O_4/TiO_2 is shown in Figure 1b. Co_3O_4, TiO_2 and ZrO_2 were detected successfully. Peaks of monoclinic crystal structure of zirconia (ZrO_2) were observed at 24.04°, 24.44°, 28.17°, 31.46°, 34.15°, 34.38°, 49.26° and 50.55°. Crystallite size of zirconia was calculated as 32 nm on average. In 8 wt% ZrO_2-Co_3O_4/TiO_2 crystallite size of TiO_2 and Co_3O_4 calculated by Sherrer equation decreased to 58 nm and 22 nm respectively as compared to Co_3O_4/TiO_2.

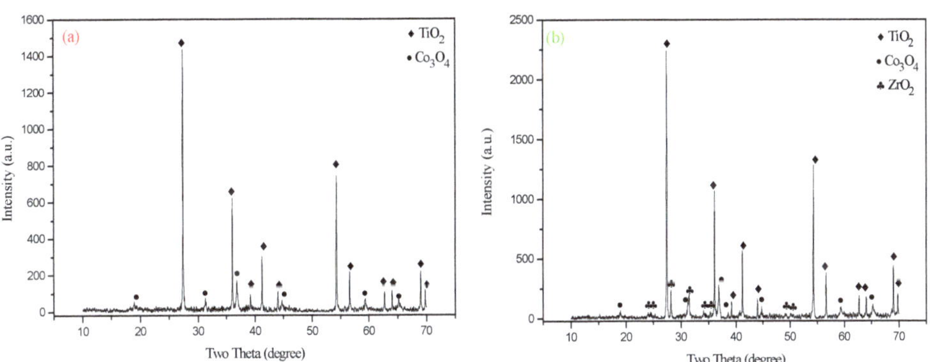

Figure 1. XRD patterns of catalysts (**a**) Co_3O_4/TiO_2 (**b**) 8% wt. ZrO_2-Co_3O_4/TiO_2.

2.2. Scanning Electron Microscopy (SEM)

SEM images of the samples are shown in Figure 2 with different scaling. It was observed that the particles of titania and both catalysts were spherical in shape and size, were in range of 50–170 nm. SEM images of titania nanoparticles and both catalysts are shown in Figure 2a–c. It was observed size and the shape of the particles were uniform without many variations. Titania nanoparticles image show that boundaries of particles are clear and well defined but after loading of Co_3O_4 clear structure starts diminishing. After addition of zirconia that smooth and well defined structure lessens very much. SEM images of a coated wire pieces before and after testing are shown in Figure 2d–g. Slight removal of the coated material was observed after testing and also there was some agglomeration of coated material after testing. This ascertains that coating technique needs to be upgraded. EDS (Energy-dispersive X-ray spectroscopy) patterns are shown in Figure 3. It was noticed in EDS analysis that there was no deposition of emission particles on coated wires with the catalysts after repetitive testing at varying conditions.

Figure 2. SEM images of (**a**) TiO_2 nanoparticles (**b**) Co_3O_4/TiO_2 (**c**) 8 wt.% ZrO_2-Co_3O_4/TiO_2 (**d**) Bef test 8 wt.% ZrO_2-Co_3O_4/TiO_2 coated wire (**e**) Aft test 8 wt.%. ZrO_2-Co_3O_4/TiO_2 coated wire (**f**) Bef test 8 wt.% ZrO_2-Co_3O_4/TiO_2 coated wire (**g**) Aft test 8 wt.% ZrO_2-Co_3O_4/TiO_2.

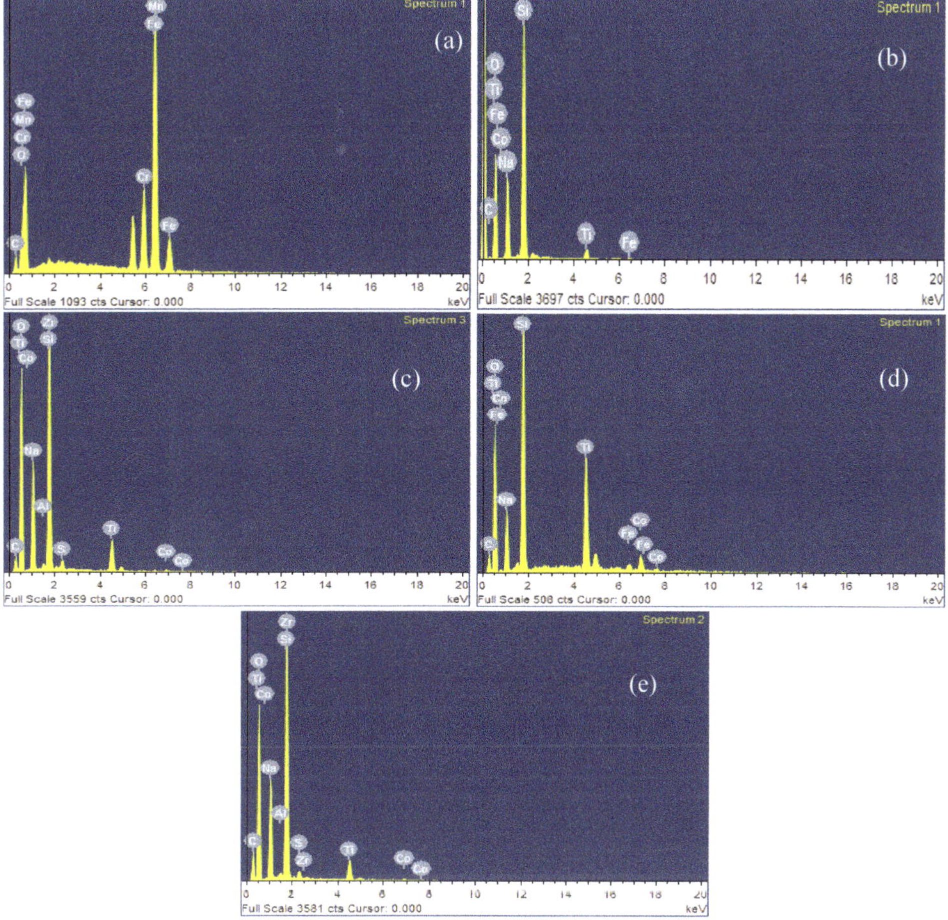

Figure 3. EDS patterns of (**a**) Uncoated wire (**b**) Co_3O_4/TiO_2 coated wire before test (**c**) 8 wt.% ZrO_2-Co_3O_4/TiO_2 coated wire before test (**d**) Co_3O_4/TiO_2 coated wire after test (**e**) 8 wt.% ZrO_2-Co_3O_4/TiO_2 coated wire after test.

2.3. Thermogravimetric Analysis (TGA)

Cyclic heating and cooling technique was carried out to find out the oxygen storage capacity (δ) of both catalysts as shown in Figure 4a,b. During real driving conditions when vehicle is accelerated, AFR fluctuates due to which NO_X emissions are increased. CeO_2 was used in conventional three way catalysts to overcome this fluctuating AFR issue as it has the ability to store/release oxygen during varying conditions. Zirconia with different mole fractions was added to give thermal stability and decrease sintering of CeO_2 [36]. In our case, we added 8 wt.% ZrO_2 in Co_3O_4/TiO_2 catalyst and investigated its effects on thermal stability and oxygen storage capacity of the catalyst. For this purpose, catalysts powders were put in the furnace of a TG analyzer turn by turn. Powders were heated from room temperature to 800 °C in the presence of air stream to provide atmospheric conditions with flow rate of 10 mL/min. Weight loss in the first heating cycle was calculated as 6.84% for Co_3O_4/TiO_2 catalyst and 7.22% for 8 wt.% ZrO_2-Co_3O_4/TiO_2 which corresponds to both loss of water and also oxygen molecules [37,38]. From 28 °C to 105 °C zirconia promoted catalyst lost 2.21% of weight while other catalyst experienced 0.51% of weight loss which indicates loss of moisture. After that, zirconia promoted catalyst exhibited more stability and lost 5.03% weight compared to 6.67% of simple catalyst. Then, they were cooled to 169 °C in the presence of an air stream with same flow during which zirconia promoted catalyst gathered 1.87% and simple catalyst gathered 3.37% of weight which corresponds to oxygen gaining of both catalysts [36]. The OSC value (δ) of zirconia promoted catalyst was calculated to be 1.4 μmol/g and for simple catalyst it was 1.6 μmol/g which are quite significant values as per literature [39]. The zirconia promoted catalyst displayed less oxygen storage capacity which is due to crystal structural defects; the introduction of Zr into a Co framework on a corresponding level would compensate the volume increase related to Co^{2+} reduction. This would decrease the influence of valence change; shorten the Co-O bond length in the ZrO_2-Co_3O_4/TiO_2 as a result it detains some of oxygen sites on cobalt oxide and titania [40]. There was 3.54% weight loss observed for simple catalyst in second heating cycle while zirconia promoted catalyst lost 2.2% weight as shown in Figure 4a,b which corresponds to the oxygen releasing capacity of the catalysts in atmospheric conditions. Results of TGA show that with addition of zirconia, which is highly stable molecule, catalyst gained thermal stability but oxygen gaining and releasing ability decreased which also affected the catalytic activity that is discussed in later part of the paper. A decrease in crystalline size was observed by addition of ZrO_2 on Co_3O_4/TiO_2, which may attribute to increase in thermal stability of ZrO_2-Co_3O_4/TiO_2. Hofmann et al. [31] studied oxygen vacancies on rutile TiO_2 (both active catalyst and support) and reported that it has significant ability to store and release oxygen. After addition of the zirconia in Co_3O_4/TiO_2 catalyst, it captures some of oxygen vacant sites of titania which results in a lower OSC value.

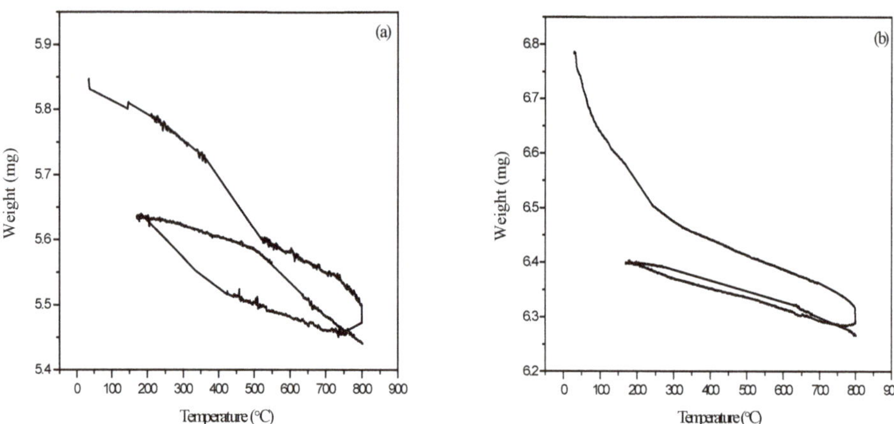

Figure 4. Plots for measurements of oxygen storage capacity (OSC) of (**a**) Co_3O_4/TiO_2 and (**b**) 8 wt.% ZrO_2-Co_3O_4/TiO_2.

2.4. Brunauer–Emmett–Teller (BET)

The surface area and pore volume for the Co_3O_4/TiO_2 catalyst were found to be 24.3 m^2/g and 0.14 cm^3/g, respectively, while the zirconia promoted catalyst surface area and pore volume were found to be 13.09 m^2/g and 0.071 cm^3/g, respectively. Reason behind decrement in surface area of zirconia promoted catalyst is that ZrO_2 which has higher density (5.68 gcm^{-3}) than titania, exhibits low surface area which is because of defects in crystal structural. This would decrease the influence of valence change; shorten the Co-O bond length in the ZrO_2-Co_3O_4/TiO_2, as a result it detains some of oxygen sites on cobalt oxide and titania may results in low surface area and low pore volume [41,42]. The N_2 adsoption-desorption isotherm of both catalysts is shown in Figure 5a,b.

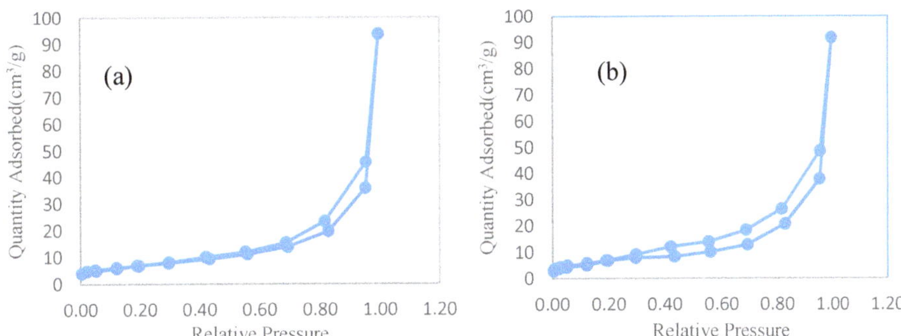

Figure 5. Nitrogen adsorption-desorption isotherm of (**a**) Co_3O_4/TiO_2 catalyst (**b**) 8 wt.% ZrO_2-Co_3O_4/TiO_2 catalyst.

2.5. Catalytic Activity

RPMs of the engine shaft increased manually from 1500 to 6000 at intervals of 1500. Temperature of the exhaust gases also increased giving values 121 °C, 257 °C, 284 °C, and 391 °C with respect to RPMs of 1500, 3000, 4500, and 6000. In the case of NO_X the maximum conversion efficiency was noticed at 6000 RPMs as shown in Figure 6 which was 71.4% for simple catalyst but zirconia promoted catalyst showed higher efficiency of 75% for NO_X reduction shown in Figure 7. The highest efficiency of zirconia promoted catalyst is due to absence of moisture content at high temperature, the poisoning of catalyst was diminished when temperature reach above 300 °C thus it provides more sites for reaction. Furthermore, NO_X amount in the gas stream was maximum in its range. While lambda value at this point was 1.323 which showed that the air to fuel ratio (AFR) inside the combustion chamber of motorcycle engine was 19.5:1 and the air fuel mixture was lean. Shah et al. [40] coated wire meshes with zirconia and reported that it reduced HCs in exhaust stream by 35%. In the case of CO, maximum conversion efficiency was noticed at 1500 RPMs which was 78.15% for simple catalyst shown in Figure 8 and slightly higher 78.65% for zirconia promoted catalyst as shown in Figure 9. CO oxidation is strongly dependent on morphology of catalysts and oxidation state of transition metals in catalyst. The oxidation state of Co in the prepared Co_3O_4/TiO_2 and ZrO_2-Co_3O_4/TiO_2 was expected to be +2 in both cases with tetrahedral and octahedral geometries respectively. Despite similar Co coordination environments, the catalytic activity and selectivity was considerably improved by the Zr modification of the Co_3O_4/TiO_2. This was endorsed to the change in oxygen donor ability and Co–O bond strength of the ≡TiO–Zr–O sites of Co–Zr/TiO_2 compared with the ≡TiO– ligands in Co/TiO_2. The tuning of the support TiO_2 oxygen donation ability by use of an anchoring site (e.g., ≡TiO-Zr-O^-) can be used to alter both rate and selectivity of conversion efficiency i.e., (NO_X, CO, HC). Possibly Co^{2+} active sites prefer to associate with ≡TiO–Zr–O^- sites as compare to ≡TiO– [42,43] as shown in below Figure 10. Furthermore, zirconia promoted catalyst efficiency is attributed by the presence of zirconia that effect on redox properties through dispersion of active phases which oxidize CO more efficiently. Moreover, at this point CO amount in exhaust gas was maximum in its range which decreases drastically with temperature increment from 1500 to 6000 RPMs. The inlet lambda value was noticed to be same 1.323 thus the AFR value of 19.5:1. Higher AFR indicating that air fuel mixture is lean thus oxidation reactions are more favorable. The outlet lambda value after the catalyst was recorded as 6.02 which mean that catalyst is working properly.

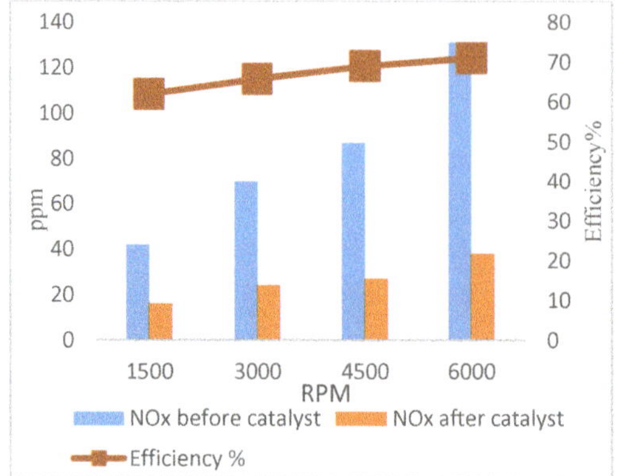

Figure 6. Conversion of NO_X with Co_3O_4/TiO_2.

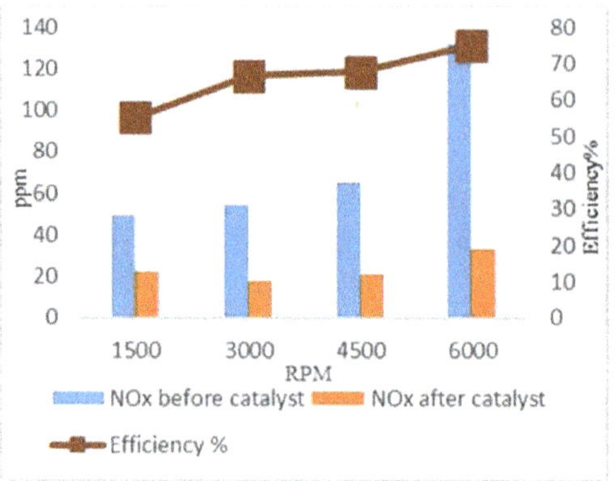

Figure 7. Conversion of NO_X with 8 wt.% ZrO_2-Co_3O_4/TiO_2.

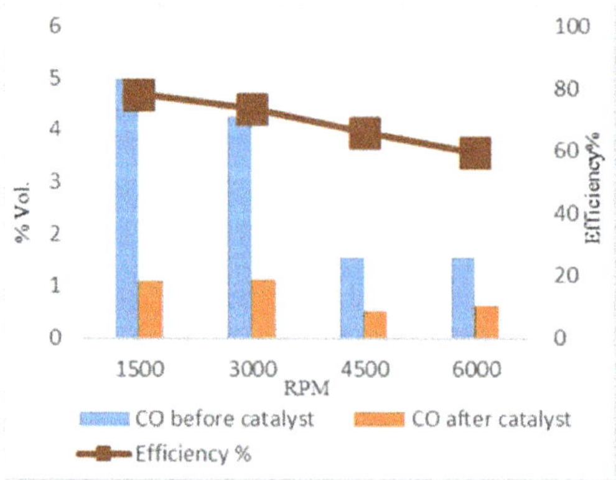

Figure 8. Conversion of CO with Co_3O_4/TiO_2.

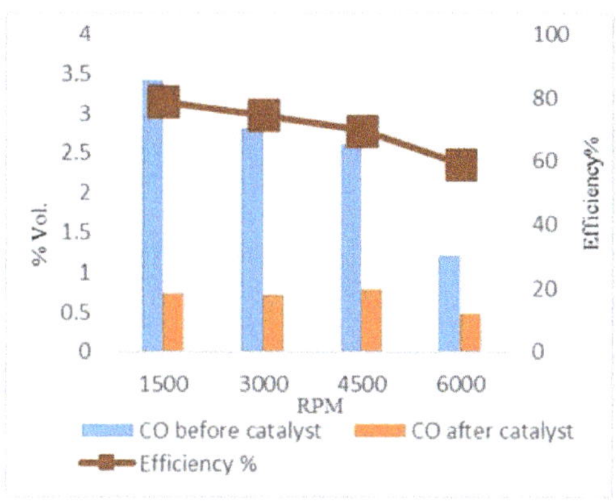

Figure 9. Conversion of CO with 8 wt.% ZrO_2-Co_3O_4/TiO_2.

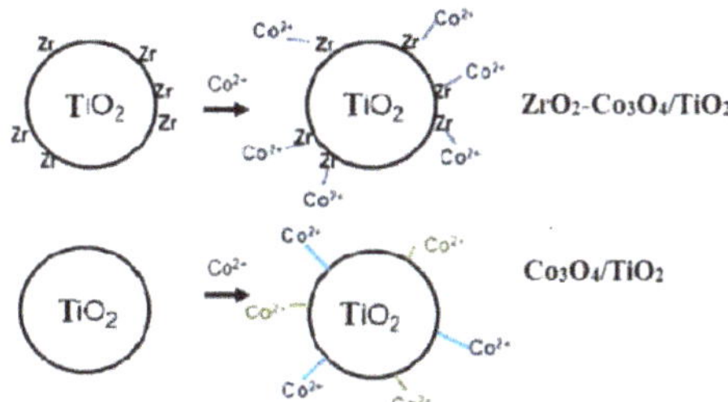

Figure 10. Graphical diagram of 8 wt.% ZrO_2-Co_3O_4/TiO_2 and Co_3O_4/TiO_2.

Furthermore, the type of basic sites together with whole basicity and whole acidity of the ZrO_2-based catalysts also effect the HC, CO, and NO_x conversion on active sites. The Lewis basic sites of ZrO_2 consist of coordinately unsaturated O^{-2} species and the Lewis acidic sites is Zr^{4+} species [42–46]. Overall reactions on the surface of catalysts were given below in Figure 11:

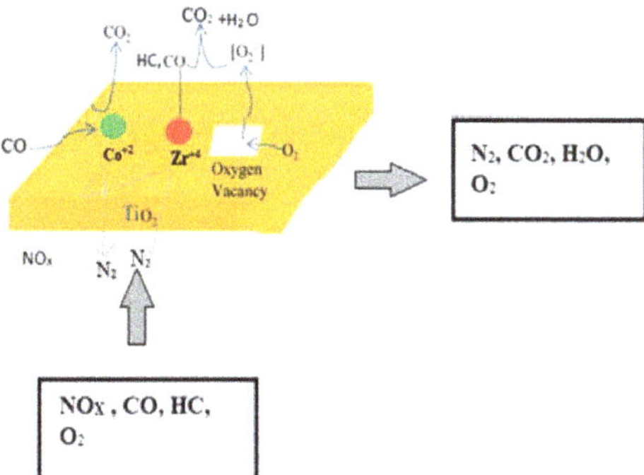

Figure 11. Conversions of NO_x, CO, HC to N_2, CO_2, and H_2O on 8 wt.% ZrO_2-Co_3O_4/TiO_2.

Amin et al. [39] developed a copper-based catalytic converter and reported that it converted HCs by 33% and CO by 66% at full load. Patel et al. [47] made a copper-based catalytic converter and reported conversion efficiencies as 50–62% for CO, 20–27% for NO_X and 35–40% HCs. Makwana et al. fabricated a catalytic converter by coating nickel on steel wire meshes and reported efficiencies as 40% and 35% for HCs, and CO, respectively. Hydrocarbons conversion efficiency was noticed as

82.5% for simple catalyst at 1500 RPMs as shown in Figure 12 but in case of promoted catalyst HC conversion decreased and gave highest value of 81% shown in Figure 13. The decrease in efficiency is due to presence of moisture content that absorb very easily on the Lewis acid site of the catalyst at a lower temperature i.e., 121 °C, As a result it lowers the vacancies of oxygen for oxidation on zirconia promoted Co_3O_4. Furthermore, the zirconia promoted catalyst displayed less oxygen storage capacity which is due to zirconia captures some of oxygen sites on cobalt oxide and titania. Enhanced catalytic performance of ZrO_2 promoted catalyst may be attributed to small crystalline size (calculated by the sSherrer equation). As crystalline small decrease, the no of catalytic active sites increases accordingly. Therefore, it leads to the better conversion efficiency of the reactants towards product formation. Also due to increase in thermal stability of ZrO_2 promoted catalyst at higher temperatures, catalyst becomes more resistance towards sintering process which could lead to better catalytic performance. ZrO_2 is also reported to prevent deactivation of catalyst in catalytic converters, which is caused by fouling process due to incomplete combustion of fuel. The catalytic activity results are summarized in Tables 1–3.

Table 1. Real time conversion results of NO_X with both catalysts.

RPM	**NO_X (ppm) for Co_3O_4/TiO_2**		-	**NO_X (ppm) for 8 wt.% ZrO_2- Co_3O_4/TiO_2**		-
-	**Before**	**After**	**Efficiency (%)**	**Before**	**After**	**Efficiency (%)**
1500	42	16	61.9	49	22	55.1
3000	70	24	65.7	54	18	66.6
4500	87	27	69.8	65	19	70.7
6000	131	38	71	132	33	75

Table 2. Real time conversion results of CO with both catalysts.

RPM	**CO (% vol.) for Co_3O_4/TiO_2**		-	**CO (% vol.) for 8 wt.% ZrO_2- Co_3O_4/TiO_2**		-
-	**Before**	**After**	**Efficiency (%)**	**Before**	**After**	**Efficiency (%)**
1500	4.99	1.09	78.1	3.42	0.73	78.6
3000	4.24	1.13	73.3	2.82	0.72	74.8
4500	1.55	0.53	65.8	2.47	0.79	68
6000	1.51	0.63	58.2	1.2	0.49	59.1

Table 3. Real time conversion results of HCs with both catalysts.

RPM	**HCs (ppm) for Co_3O_4/TiO_2**		-	**HCs (ppm) for 8 wt.% ZrO_2- Co_3O_4/TiO_2**		-
-	**Before**	**After**	**Efficiency (%)**	**Before**	**After**	**Efficiency (%)**
1500	414	72	82.6	418	78	81.3
3000	198	45	77.2	237	56	76.3
4500	124	38	69.3	128	42	67.1
6000	39	15	61.5	40	14	65

With increasing RPMs, the temperature also increased and the ppm value of hydrocarbons production inside combustion chamber decreased. Thus, confirming that amount of CO, NOx and HCs producing inside combustion chamber heavily depends on the temperature inside combustion chamber.

After addition of zirconia in catalyst, there was a negative effect on oxidation reactions as it suppressed the oxidation reaction of HCs. HC conversion efficiency was less in ZrO_2 promoted catalyst compared to the other one. This is because zirconia captures some of the active sites of Co_3O_4 which is our oxidation catalyst. Moreover, mostly transition metals oxides have the tendency to work as reduction catalysts [48]. TiO_2 was used as our reduction catalyst for NO_X, so addition of ZrO_2 itself being a metal oxide which develops the reduction property at higher energies, contributed towards NOx reduction. It should be noted that about (50%–80%) of HCs are emitted during the cold start period which is about 200 s from the starting of an engine [49].

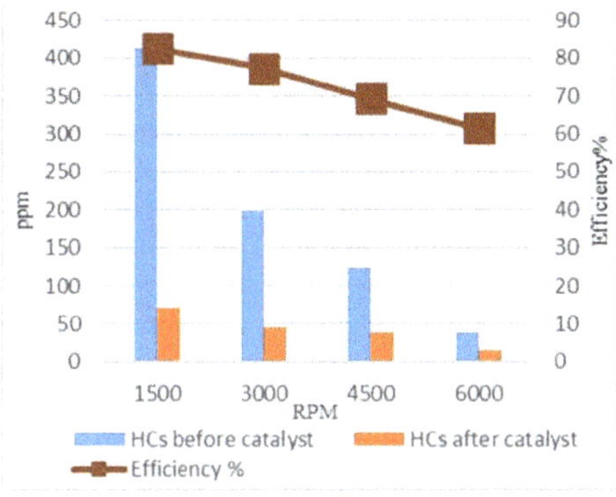

Figure 12. HC Conversions with Co_3O_4/TiO_2.

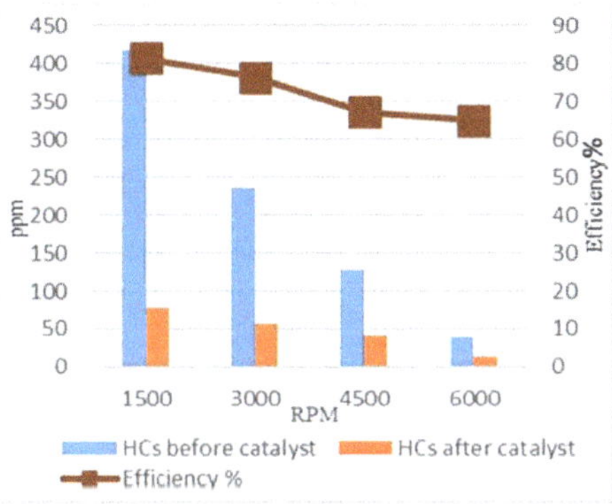

Figure 13. HC conversions with 8 wt.% ZrO_2- Co_3O_4/TiO_2.

Therefore, a good catalytic converter should convert maximum unburnt hydrocarbons during this span of time, maintained its high metal dispersion and high catalytic activity. In our case, both the catalysts showed maximum conversion efficiency for HCs in the beginning at the lowest RPMs. Uncoated wires mesh showed zero conversion of gases.

3. Experimental Section

3.1. Catalyst Preparation

For synthesis of TiO_2 nanoparticles, 60 g of titania powder (BDH) was added 400 mL of distilled water. The mixture was stirred for 24 h at 800 rpm and then it was allowed to settle down for 12 h. Then settled material was dried in oven for 12 h at 100 °C. The dried material was continuously crushed by using a pestle and mortar and then it was allowed to calcine at 500 °C for 6 h in a furnace. Co_3O_4/TiO_2 Catalyst was prepared by using hydrothermal synthesis method. For that purpose, 2.75 g of cobalt nitrate hexahydrate $Co(NO_3)_2.6H_2O$ (Panreac, Barcelona, Spain) was added to 30 mL of distilled water while stirring at room temperature to give 6 wt.% loading of Co_3O_4 over TiO_2 support. A mixture of potassium hydroxide KOH and water was added dropwise as precipitating agent for the formation of $Co(OH)_2$. The color of the mixture changed from pink to purple after addition of KOH. The pH value of the solution was continuously monitored while adding KOH so that it should reach 8. Because as per literature [36] pH value from 8–9 gives a uniform size of Co_3O_4 while increasing pH to 11 or 12 give an irregular shape. Moreover, the condensation of $Co(OH)_2$ occurs at higher pH value due to which there are substantial chances of agglomeration of nanoparticles. After that, a 30% mass fraction of H_2O_2 (DAEJUNG, Busan, South Korea) was added dropwise in the solution as oxidant to convert $Co(OH)_2$ into spinel Co_3O_4 [36]. Color of the mixture started turning into brown from purple after addition of hydrogen peroxide. 4 g of titania nanoparticles were stirred in 40mL of water for one hour and then poured into a stainless steel autoclave along with the suspension already prepared. The autoclave was sealed and was put in a furnace at 180 °C for 10 h. Distilled water and ethanol were respectively used to wash the obtained material from autoclave and then it was dried in an oven at 100 °C for 8 h. 8% wt. zirconia (UNI-CHEM, Belgrade, Serbia) which is 0.15 g for 1 g of Co_3O_4/TiO_2 catalyst, was dissolved in required amount of distilled water and was dropped on the already prepared Co_3O_4/TiO_2 catalyst so that it completely soaked the powder. Then it was kept in oven at 100 °C for 5 h and calcined at 400 °C for 6 h to have 8% wt. ZrO_2-Co_3O_4/TiO_2.

3.2. Characterization

Powder X-ray diffraction (XRD) patterns were recorded by using Bruker D8 (Karlsruhe, Germany) Advance using Cu-kα radiation operating at 40 kV and 30 mA with 0.02° step size. Samples were scanned ranging the 2θ values from 10° to 70°. Tescan Vega3 scanning electron microscope (Brno, Czech Republic) was used to analyze the morphology of prepared catalysts and coated wires operating at 20 kV. Schimadzu DTG-60H (Kyoto, Japan) with an alumina pan was used to analyze the thermal stability and the oxygen storage capacity (OSC) of the zirconia promoted catalyst with temperature ranging from 28 °C to 800 °C with heating and cooling rate of 10 °C/min. Brunauer–Emmett–Teller (BET) surface areas of both catalysts were studied using N_2 adsorption and desorption isotherms with the help of Quantachrome, NOVA 2200e at −195.8 °C. Samples were degassed in vacuum at 300 °C for 3 h.

3.3. Fabrication and Testing of Catalytic Converter

3.3.1. Preparation of Wire Meshes

A SS-304 wire mesh sheet of 4 × 4 feet size was cut into 48 circular pieces of 6 cm diameter each as shown in Figure 14 by using a manual table cutter. Before coating the catalyst on the wire meshes, they were pretreated with 10% HCL. For this purpose wire meshes were kept dipping in 10% HCL for half an hour. After that distilled water was used to wash them and kept in an oven for drying for an hour at 100 °C. In this way, impurities were removed from the stainless steel structure. Cell density of wire meshes was 64 cells per square inch.

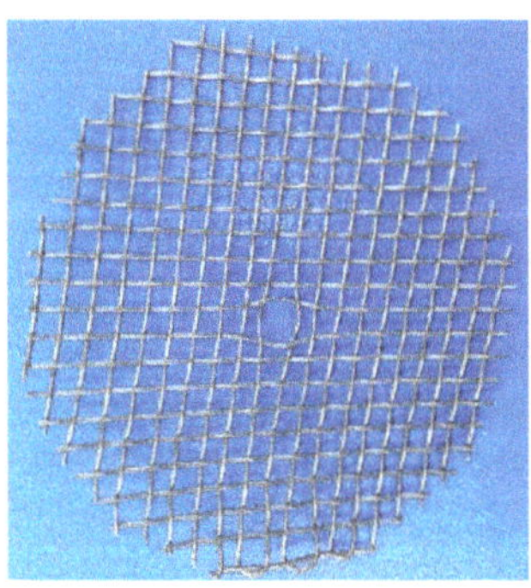

Figure 14. Wire mesh piece.

3.3.2. Catalyst Slurry

Catalyst slurry was prepared for coating it onto the wire mesh substrate. For this purpose, 3 g sodium metabisulfite (BDH) and 270 g of sodium silicate solution (Sigma-Aldrich, St. Louis, MI, USA) were added together for each catalyst separately while stirring shown in Figure 15. 30 g of each catalyst was added to this mixture and stirred for 12 h then it was coated on wire meshes.

Figure 15. Catalyst slurry preparation.

3.3.3. Catalyst Coating

30 g of each catalyst was deposited on the pretreated wire meshes by using the dip coating method. After immersing, a blower was used to remove the extra material from the wires of structure. This immersion and blowing was repeated three times to achieve uniformity. The coated wire meshes were calcined in a furnace for 5 h at 400 °C to remove the impurities and then cooled at the room temperature.

3.3.4. Wire Meshes Arrangement

24 circular wire meshes (for each catalyst) after being coated were arranged on a threaded bar with 4 mm diameter with 1mm thick washers between them to keep them away from intermingling. Both sides of the bar were closed by using hexagonal nut. These catalysts coated wire mesh structures were respectively placed in mild steel clam shell as shown in Figure 16 for activity testing with an inner diameter of 6.2 cm and sealed with Teflon lining to prevent any leakage of gases.

Figure 16. Coated wire meshes arrangement

3.4. Activity Test

The fabricated catalytic converters were mounted at the exhaust pipe of a 70 cc petrol engine as shown in Figure 17 with 72 cm^3 displacement, 47 mm bore, and 41.4 mm stroke length was used. Test readings were taken by using Crypton's gas analyzer (Birmingham, UK) for CO and HC conversions but for NO_x readings E instruments' E4500-2 was used. Catalytic activity readings were taken at 1500, 3000, 4500, and 6000 revolutions per minute which were manually set. Engine shaft angular speed was measured by a DT-2234B photo type digital tachometer.

Figure 17. Catalytic converter mounted on engine exhaust for testing

4. Conclusions

Two alternative catalysts, Co_3O_4/TiO_2 and 8% wt. ZrO_2 -Co_3O_4/TiO_2 along with a wire mesh-based substrate, were successfully developed and found very active for CO, HC, and NO_X conversions. Zirconia promoted catalyst showed more promising towards NO_X conversion. The cobalt supported by the titania Co_3O_4/TiO_2 catalyst shows a performance towards conversion of carbon monoxide, nitrogen oxides and unburnt hydrocarbons to a value of 78.1%, 61.9%, and 82.6% efficiency at 1500 RPM. Whereas, the conversion efficiency of zirconia promoted ZrO_2 -Co_3O_4/TiO_2 catalyst is 81.3%, 78.6%, and 55.1% towards HCs, CO, and NOx respectively at 1500 RPM value. Due to small crystalline size, thermal stability, and fouling inhibition, the ZrO_2 promoted Co_3O_4/TiO_2 catalyst showed better conversion efficiency towards CO and NOx. The slightly lower efficiency of zirconia promoted catalyst towards HCs is due to the non-availability of the vacancies of oxygen for oxidation on Co_3O_4. Both the catalysts showed selectivity towards CO, NO_X and HC and have comparable performance with respect to be the activity of a conventional catalyst.

Author Contributions: Conceptualization, M.H.U.R., T.N. and N.I.; data curation, methodology, and investigation, M.H.U.R.; supervision T.N., and N.I.; original draft preparation, M.H.U.R., T.N. and N.I.; writing—reviews and editing, all authors. All authors have read and agreed to the published version of the manuscript.

Funding: This research received no external funding.

Acknowledgments: U.S-Pakistan Center for Advanced Studies in Energy (USPCAS-E) at National University of Sciences and Technology (NUST) is highly acknowledged by authors for providing lab facilities.

Conflicts of Interest: The authors declare no conflict of interest.

References

1. Xia, T.; Nitschke, M.; Zhang, Y.; Shah, P.; Crabb, S.; Hansen, A. Traffic-related air pollution and health co-benefits of alternative transport in Adelaide, South Australia. *Environ. Int.* **2015**, *74*, 281–290. [CrossRef] [PubMed]
2. Cheng, Y.; Fan, Y.; Pei, Y.; Qiao, M. Graphene-supported metal/metal oxide nanohybrids: Synthesis and applications in heterogeneous catalysis. *Catal. Sci. Tech.* **2015**, *5*, 3903–3916. [CrossRef]
3. Lelieveld, J.; Evans, J.S.; Fnais, M.; Giannadaki, D.; Pozzer, A. The contribution of outdoor air pollution sources to premature mortality on a global scale. *Nature* **2015**, *525*, 367. [CrossRef] [PubMed]
4. Seigneur, C. *Air Pollution: Concepts, Theory, and Applications*; Cambridge University Press: Cambridge, UK, 2019.
5. World Health Organization. *World Health Statistics 2016: Monitoring Health for the SDGs Sustainable Development Goals*; World Health Organization: Washington, DC, USA, 2016.

6. Franco, V.; Kousoulidou, M.; Muntean, M.; Ntziachristos, L.; Hausberger, S.; Dilara, P. Road vehicle emission factors development: A review. *Atmos. Environ.* **2013**, *70*, 84–97. [CrossRef]
7. Lee, S.E. ARCHEOLOGY: Temple Gateways in South India: The Architecture and Iconography of the Cidambaram Gopuras. James, C.; Harle. *Am. Anthropol.* **1964**, *66*, 1445–1446. [CrossRef]
8. Cooper, C.D.; Alley, F.C. *Air Pollution Control: A Design Approach*, 4th ed.; Waveland Pr Inc.: Long Grove, IL, USA, 2010.
9. Schifter, I.; Gonzalez, U.; Diaz, L.; Sánchez-Reyna, G.; Mejía-Centeno, I.; González-Macías, C. Comparison of performance and emissions for gasoline-oxygenated blends up to 20 percent oxygen and implications for combustion on a spark-ignited engine. *Fuel* **2017**, *208*, 673–681. [CrossRef]
10. Gallini, N.T. The economics of patents: Lessons from recent US patent reform. *J. Econ. Perspect.* **2002**, *16*, 131–154. [CrossRef]
11. Zereini, F.; Wiseman, C.L. *Platinum Metals in the Environment*; Springer: Berlin/Heidelberg, Germany, 2015.
12. Wang, Y.; Li, X. Health risk of platinum group elements from automobile catalysts. *Procedia Eng.* **2012**, *45*, 1004–1009. [CrossRef]
13. LaMer, V.K.; Dinegar, R.H. Theory, production and mechanism of formation of monodispersed hydrosols. *J. Am. Chem. Soc.* **1950**, *72*, 4847–4854. [CrossRef]
14. Xu, Q.; Kharas, K.C.; Croley, B.J.; Datye, A.K. The contribution of alumina phase transformations to the sintering of Pd automotive catalysts. *Top. Catal.* **2012**, *55*, 78–83. [CrossRef]
15. Escrig-Olmedo, E.; Rivera-Lirio, J.M.; Muñoz-Torres, M.J.; Fernández-Izquierdo, M.Á. Integrating multiple ESG investors' preferences into sustainable investment: A fuzzy multicriteria methodological approach. *J. Clean. Prod.* **2017**, *162*, 1334–1345. [CrossRef]
16. Mudd, G.M.; Jowitt, S.M.; Werner, T.T. Global platinum group element resources, reserves and mining—A critical assessment. *Sci. Total Environ.* **2018**, *622*, 614–625. [CrossRef] [PubMed]
17. Scurrell, M.S. Thoughts on the use of gold-based catalysts in environmental protection catalysis. *Gold Bull.* **2017**, *50*, 77–84. [CrossRef]
18. Zhao, F.; Harrington, D.L.; Lai, M.-C. *Automotive Gasoline Direct-Injection Engines*; Society of Automotive Engineers: Warrendale, PA, USA, 2002; Volume 400.
19. Guerrero, L.M.; Mendoza, J.F.; Ong, K.T.V.; Olegario-Sanchez, E.M.; Ferrer, E.L. Copper-Exchanged Philippine Natural Zeolite as Potential Alternative to Noble Metal Catalysts in Three-Way Catalytic Converters. *Arab. J. Sci. Eng.* **2019**, *44*, 5581–5588. [CrossRef]
20. Li, L.; Yao, Y.; Tang, Z.; Ji, W.; Dai, Y.; Shen, X. Size Effect of Co3O4 Nanoparticles as Catalysts for CO Oxidation. *J. Nanosci. Nanotechnol.* **2016**, *16*, 7573–7578. [CrossRef]
21. Ma, L.; Seo, C.Y.; Chen, X.; Sun, K.; Schwank, J.W. Indium-doped Co3O4 nanorods for catalytic oxidation of CO and C3H6 towards diesel exhaust. *Appl. Catal. B Environ.* **2018**, *222*, 44–58. [CrossRef]
22. Wang, C.-B.; Tang, C.-W.; Tsai, H.-C.; Chien, S.-H. Characterization and catalytic oxidation of carbon monoxide over supported cobalt catalysts. *Catal. Lett.* **2006**, *107*, 223–230. [CrossRef]
23. Teng, Y.; Kusano, Y.; Azuma, M.; Haruta, M.; Shimakawa, Y. Morphology effects of Co 3 O 4 nanocrystals catalyzing CO oxidation in a dry reactant gas stream. *Catal. Sci. Tech.* **2011**, *1*, 920–922. [CrossRef]
24. Jia, C.-J.; Schwickardi, M.; Weidenthaler, C.; Schmidt, W.; Korhonen, S.; Weckhuysen, B.M.; Schüth, F. Co3O4–SiO2 nanocomposite: A very active catalyst for CO oxidation with unusual catalytic behavior. *J. Am. Chem. Soc.* **2011**, *133*, 11279–11288. [CrossRef]
25. Xie, X.; Li, Y.; Liu, Z.-Q.; Haruta, M.; Shen, W. Low-temperature oxidation of CO catalysed by Co 3 O 4 nanorods. *Nature* **2009**, *458*, 746. [CrossRef]
26. Yao, X.; Zhao, R.; Chen, L.; Du, J.; Tao, C.; Yang, F.; Dong, L. Selective catalytic reduction of NOx by NH3 over CeO2 supported on TiO2: Comparison of anatase, brookite, and rutile. *Appl. Catal. B Environ.* **2017**, *208*, 82–93. [CrossRef]
27. Xu, W.; Yu, Y.; Zhang, C.; He, H. Selective catalytic reduction of NO by NH3 over a Ce/TiO2 catalyst. *Catal. Commun.* **2008**, *9*, 1453–1457. [CrossRef]
28. Rui, Z.; Wu, S.; Peng, C.; Ji, H. Comparison of TiO2 Degussa P25 with anatase and rutile crystalline phases for methane combustion. *Chem. Eng. J.* **2014**, *243*, 254–264. [CrossRef]
29. Zhu, H.; Dong, L.; Chen, Y. Effect of titania structure on the properties of its supported copper oxide catalysts. *J. Colloid Interface Sci.* **2011**, *357*, 497–503. [CrossRef] [PubMed]

30. Fornasiero, P.; Dimonte, R.; Rao, G.R.; Kaspar, J.; Meriani, S.; Trovarelli, A.; Graziani, M. Rh-loaded CeO2-ZrO2 solid-solutions as highly efficient oxygen exchangers: Dependence of the reduction behavior and the oxygen storage capacity on the structural-properties. *J. Catal.* **1995**, *151*, 168–177. [CrossRef]
31. Ganduglia-Pirovano, M.V.; Hofmann, A.; Sauer, J. Oxygen vacancies in transition metal and rare earth oxides: Current state of understanding and remaining challenges. *Surf. Sci. Rep.* **2007**, *62*, 219–270. [CrossRef]
32. Yu, Y.-Y.; Gong, X.-Q. CO oxidation at rutile TiO2 (110): Role of oxygen vacancies and titanium interstitials. *ACS Catal.* **2015**, *5*, 2042–2050. [CrossRef]
33. Li, H.; Guo, Y.; Robertson, J. Calculation of TiO2 surface and subsurface oxygen vacancy by the screened exchange functional. *J. Phys. Chem. C* **2015**, *119*, 18160–18166. [CrossRef]
34. Malakooti, R.; Mahmoudi, H.; Hosseinabadi, R.; Petrov, S.; Migliori, A. Facile synthesis of pure non-monoclinic zirconia nanoparticles and their catalytic activity investigations for Knoevenagel condensation. *RSC Adv.* **2013**, *44*, 22353–22359. [CrossRef]
35. Ahmad, W.; Noor, T.; Zeeshan, M. Effect of synthesis route on catalytic properties andperformance of Co3O4/TiO2 for carbon monoxide and hydrocarbon oxidation under real engine operating conditions. *Catal. Commun.* **2017**, *89*, 19–24. [CrossRef]
36. Yang, Y.; Huang, K.; Liu, R.; Wang, L.; Zeng, W.; Zhang, P. Shape-controlled synthesis of nanocubic Co3O4 by hydrothermal oxidation method. *Trans. Nonferrous Met. Soc. China* **2007**, *17*, 1082–1086. [CrossRef]
37. Matsumoto, S.I. Recent advances in automobile exhaust catalysts. *Catal. Today* **2004**, *3–4*, 183–190. [CrossRef]
38. Vera, D.; Crossan, M. Improvisation and innovative performance in teams. *Organ. Sci.* **2005**, *3*, 203–224. [CrossRef]
39. Li, W.; Nie, X.; Jiang, X.; Zhang, A.; Ding, F.; Liu, M.; Liu, Z.; Guo, X.; Song, C. ZrO2 support imparts superior activity and stability of Co catalysts for CO2 methanation. *Appl. Catal. B Environ.* **2018**, *220*, 397–408. [CrossRef]
40. Li, P.; Chen, X.; Li, Y.; Schwank, J.W. A review on oxygen storage capacity of CeO2-based materials Influence factors, measurement techniques, and applications in reactions related to catalytic automotive emissions control. *Catal. Today* **2019**, *327*, 90–115. [CrossRef]
41. Priya, N.S.; Somayaji, C.; Kanagaraj, S. Optimization of ceria-zirconia solid solution based on OSC measurement by cyclic heating process. *Procedia Eng.* **2013**, *64*, 1235–1241. [CrossRef]
42. Jigish, J.G.; Pravin, P.R.; Krunal, P.S. Exhaust Analysis of C.I Engine by Using Zirconium Dioxide Coated Wire Mesh Catalytic Converter. *Int. J. Sci. Res. Dev.* **2014**, *1*, 763–766.
43. Zhao, Y.; Sohn, H.; Hu, B.; Niklas, J.; Poluektov, O.G.; Tian, J.; Delferro, M.; Hock, A.S. Zirconium Modification Promotes Catalytic Activity of a Single-Site Cobalt Heterogeneous Catalyst for Propane Dehydrogenation. *ACS Omega* **2018**, *3*, 11117–11127. [CrossRef]
44. Zhang, H.; Ruan, S.; Feng, C.; Xu, B.; Chen, W.; Dong, W. Photoelectric Properties of TiO2-ZrO2 Thin Films Prepared by Sol-Gel Method. *J. Nanosci. Nanotechnol.* **2011**, *11*, 10003–10006. [CrossRef]
45. Viinikainen, T.; Rönkkönen, H.; Bradshaw, H.; Stephenson, H.; Airaksinen, S.; Reinikainen, M.; Simell, P.; Krause, O. Acidic and basic surface sites of zirconia-based biomass gasification gas clean-up catalysts. *Appl. Catal. A Gen.* **2009**, *362*, 169–177. [CrossRef]
46. Yu, J.J.; Cheng, J.; Ma, C.Y.; Wang, H.L.; Li, L.D.; Hao, Z.P.; Xu, Z.P. NOx decomposition, storage and reduction over novel mixed oxide catalysts derived from hydrotalcite-like compounds. *J. Colloid Interface Sci.* **2009**, *333*, 423–430. [CrossRef] [PubMed]
47. Amin, C.M.; Chavda, K.; Gadhia, U. Exhaust analysis of four stroke single cylinder diesel engine using copper based catalytic converter. *Int. J. Sci. Res. Dev.* **2013**, *1*, 493–497.
48. Reiter, M.S.; Kockelman, K.M. The problem of cold starts: A closer look at mobile source emissions levels. *Transp. Res. Part D Transp. Environ.* **2016**, *43*, 123–132. [CrossRef]
49. Paresh, D.P.; Pradip, M.P.; Nimesh, A.P.; Dhanajay, D.V. Experimental Investigation on Exhaust Emission with & Without MOLY Coated Piston Ring in C. I. Engine. *Int. J. Sci. Res. Dev.* **2014**, *1*, 570–573.

Article

Heterogeneous Nanomagnetic Catalyst from Cupriferous Mineral Processing Gangue for the Production of Biodiesel

Wighens I. Ngoie [1,2], Pamela J. Welz [1], Daniel Ikhu-Omoregbe [2] and Oluwaseun O. Oyekola [2,*]

1 Institute of Biomedical and Microbial Biotechnology, Cape Peninsula University of Technology, Cape Town 7535, South Africa; wghngoie@gmail.com (W.I.N.); welzp@cput.ac.za (P.J.W.)
2 Department of Chemical Engineering, Cape Peninsula University of Technology, Cape Town 7535, South Africa; IkhuOmoregbeD@cput.ac.za
* Correspondence: oyekolas@cput.ac.za; Tel.: +27-(0)-21-959-6799

Received: 14 October 2019; Accepted: 26 November 2019; Published: 10 December 2019

Abstract: The commercialisation of biodiesel as an alternative energy source is challenged by high production costs. The cost of feedstock, catalyst and separation of the dissolved catalyst (homogeneous catalyst) from the product are the major contributors to the total manufacturing cost of biodiesel. This study investigated the potential of a heterogeneous catalyst produced from mineral processing waste for biodiesel production. Tailings from the concentration of cupriferous minerals served as the starting material for synthesis of the catalyst. The nanomagnetic catalysts were prepared using co-precipitation (CMCO) and sol-gel (CMSG) methods, combined with zero-valent iron nanoparticles (ZVINPs) to form a hydride catalyst (CMSG/ZVINPs). Catalyst properties were assessed using SEM, TEM, BET and EDX. The catalyst activity was enhanced by a large number of basic sites that were afforded by the presence of calcite and magnesite. Good surface areas and particle sizes of 58.9 m^2/g and 15.4 nm, and 52.6 m^2/g and 16.9 nm were observed for the catalysts that were prepared using the CMSG and CMCO methods, respectively. 173 emu/g mass magnetisation was obtained for CMSG/ZVINPs, which was sufficient for the catalyst to be regenerated and reused for biodiesel production by exploiting the magnetic properties. The maximum yield obtained with this catalyst was 88% and an average of 27% decrease in biodiesel yield was observed after four reaction cycles. The physicochemical properties of the biodiesel produced complied with the ASTM standard specification. The results showed that mineral processing tailings are a viable starting material for catalyst preparation in biodiesel production.

Keywords: biodiesel; heterogeneous nano-magnetic catalyst; edible oil wastewater sludge

1. Introduction

The main purpose of ore beneficiation in the mineral processing industries is to increase the metal content. Pyrometallurgy or hydrometallurgy processes are employed depending on their nature (sulfide, oxides, or mixed), or on the initial metal purity in the ore [1]. Hydrometallurgy is the most suitable process for low-grade ores [2]. The essential concentration operation after the ore has been reduced into an appropriate size, for valuable mineral liberation from the unwanted material (gangue), is froth flotation. This involves the chemical separation of gangue from the mineral (fraction containing the valuable metal of interest) [1,2]. Gangue can be acidic (mainly containing silica and quartz), as in the case of sulfide ores or alkaline (mainly dolomite and calcite), as found in oxide and mixed ores [2]. Dolomite is mainly comprised of calcium and magnesium, and it is the dominant material in the tailings effluent from froth flotation process [2,3].

Catalysts that are used for biodiesel production are classified vis-à-vis their mode of action and reaction phases, into homogeneous, heterogeneous, and enzymatic catalysts [4]. The selection of a catalyst for any process depends on a number of parameters, including, cost, availability, activity, selectivity, and stability [5,6]. Homogeneous-catalysed transesterification, which is the conventional method that is used to produce biodiesel from vegetable oils, utilises potassium hydroxide (KOH) or sodium hydroxide (NaOH) as homogenous base catalysts to produce high yields of biodiesel under moderate conditions. This approach is impaired by soap formation in the presence of water, especially when low-cost feedstocks are utilised. This results in separation difficulties during biodiesel purification [4,7]. Magnesium oxide (MgO) and calcium oxide (CaO) have been shown to have high catalytic activities, good robustness, and favourable resistance to acid [8]. It is envisaged that the use of heterogeneous magnetic catalysts produced from MgO and CaO for biodiesel production will enable more efficient mass transfer than conventional heterogeneous catalysts due to the high surface area and ferromagnetic properties of the nanoparticles. This might reduce catalyst loading and improve the biodiesel yield, ultimately reducing biodiesel production costs [9]. Amani et al. [8] identified MgO as a promising heterogeneous catalyst for the transesterification of soybean and vegetable oils into biodiesel, with a maximum biodiesel yield of 79% with 10 wt.% catalyst loading. It has been shown that these oxides do not require magnetic cations to become ferromagnetic, and a novel type of magnetism, which is tentatively called interface magnetism, has been recognised. Under certain conditions, magnetism arises in the absence of transition metal elements, and this phenomenon confers potential ferromagnetic properties to MgO and CaO [8,10].

The use of nanomagnetic catalysts is a promising alternative in the transesterification process, because the magnetic separation generally avoids the loss of catalyst and increases its reusability in comparison to filtration or centrifugation separation [11,12]. Nanomagnetic catalysts have been widely applied in the fields of photocatalysis [13], biocatalysis [14], and phase-transfer catalysis [15], in addition to a large specific surface area and high catalytic activity. Ying and Chen [16] stabilised the cells of lipase-producing *Bacillus subtilis* on the net of a hydrophobic carrier with Fe_3O_4 magnetic particles. for the transesterification of waste cooking oils with methanol, while Xie and Ma [17] immobilised lipase on Fe_3O_4 nanoparticles for biodiesel production. In these instances, the magnetic particles did not possess catalytic properties; they merely acted as immobilisation matrices for enhancing catalyst recovery. Ali et al. [18] synthesised a nanocatalyst of CaO supported by Fe_3O_4 magnetic particles using the chemical precipitation method. A maximum biodiesel yield of 69.7% was obtained under optimum conditions (65 °C, 300 min., 20 methanol/oil molar ratio, and 10 wt.% of CaO/Fe_3O_4 catalyst loading) using palm seed oil as feedstock.

The synthesis of oxide nanoparticles can be grouped into two main streams, based on the liquid-solid and gas-solid nature of the transformations [7]. Liquid-solid transformations are the most broadly used methods. Catalyst preparation aims at ensuring suitable activity, selectivity, and stability. These are related to the physical and chemical properties of the starting material, as well as the parameters that are inherent in the preparation method [19]. In this study, the CMSG and CMCO methods were investigated for the catalyst synthesis from mineral processing wastes, due to their low-cost when compared to other techniques.

The co-precipitation method (CMCO): Co-precipitation involves dissolving a salt precursor (mainly chloride or nitrate) in water or other solvents to precipitate the oxo-hydroxide [O(OH)] with the help of a base. The control of size and chemical homogeneity in the case of mixed metal oxides are difficult to achieve. This may be overcome by the use of surfactants and high-gravity reactive precipitation methods [8,20].

The sol-gel method (CMSG): The method prepares metal oxides via hydrolysis of precursors, usually alkoxides in alcohol solution, resulting in the corresponding [O(OH)] group. The condensation of molecules by dehydration leads to the formation of a network of the metal hydroxide. Hydroxyl-species undergo polymerisation by condensation and form a dense porous gel. Appropriate drying and calcination lead to ultrafine porous oxides [7,15]. Irrespective of the preparation method that was used to obtain ultrafine nano-oxides, crystallisation does not follow a traditional nucleation and growth mechanism [21]. The crystallisation mechanisms may be a function of the chemical nature of the oxide and the process temperature [22].

In the conventional biodiesel production process, oil is derived from oilseed crops, which is associated with the food versus fuel dilemma [23,24]. In this study, the oil and alcohol that were employed for the transesterification reaction for biodiesel production were extracted from edible oil wastewater sludge. The heterogeneous nanomagnetic catalyst, which was used in mediating the reaction, was prepared from waste without magnetite support, in the presence of ZVINPs to enhance the catalyst mass transfer abilities during this process [23].

2. Results and Discussion

2.1. Gangue Characterisation

EDX analysis was conducted to assess the elemental composition and purity of the gangue material and the prepared catalysts. The results that were obtained from the EDX analysis of the gangue material showed predominance of magnesium (4 *cps/ev*) and calcium (4.2 *cps/ev*) (Figure 1). The predominance of magnesium or other related transitional metals in heterogeneous catalysts has been widely shown to have a positive impact on their mass transfer capabilities in the transesterification process, subsequently enhancing the biodiesel production at lower reaction rates [22].

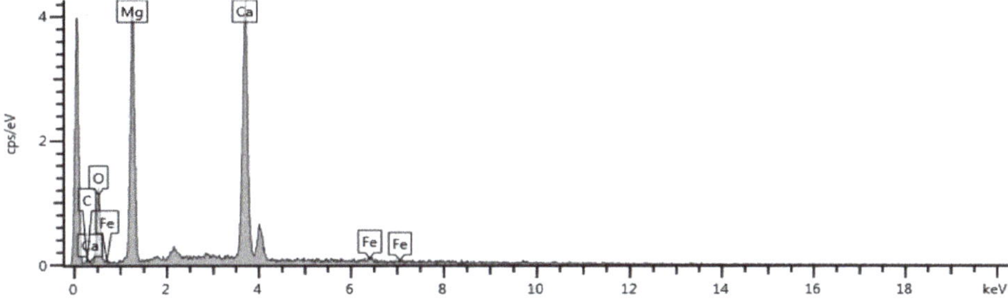

Figure 1. EDX results of the raw material (gangue) used to prepare the catalyst.

On the Linus Pauling scale, magnesium and calcium are characterised by electronegativity values of 1.32 and 1, respectively [25,26]. This is an important factor that can enable the catalyst to exhibit higher basicity and improved catalytic activity for better biodiesel yield during the transesterification reaction. The peak maxima were obtained at 1.3 and 3.6 keV for magnesium and calcium, respectively (Figure 1).

2.2. Brunauer-Emmett-Teller (BET) Surface Area Analysis and Barrett-Joyner-Halenda (BJH) Pore Size and Volume Analysis

The surface area of the catalysts was determined using BET analyses, while the pore size and specific pore volume were determined using the adsorption and desorption techniques (BJH analyses). Results showed that an increase in nitric acid concentration increased the surface area of all catalysts (Table 1).

Table 1. Comparison of surface areas and pore sizes of the prepared catalysts with previous studies.

Sample	Surface Area (m^2/g)	Average Pore Size (Å)	References
$CMCO_2$ *	52.6 ± 0.37	135.6 ± 0.13	This study
$CMSG_3$ *	58.7 ± 0.55	169.3 ± 0.21	This study
$CMSG_3/ZVINPs_2$ *	53.5 ± 1.52	148.5 ± 0.09	This study
CaO/Fe_3O_4	59.1	8.5	[26]
KF/Ca-Mg-Al hydrotalcite	108.4	3.7	[14]

* Only the best results are represented in this table. Results are expressed as averages ± standard deviation (n = 3); With CMCO = Catalyst prepared using the co-precipitation method, CMSG = Catalyst prepared using the sol-gel method, ZVINPs = Zerovalent iron nanoparticles.

The results also indicated (Table 2) that $CMSG_3$ had the highest surface area of 58.7 m^2/g. The adsorption average pore width (4 V/A by BET) was 169.3 Å for $CMCO_2$ with a BJH adsorption average pore width (4 V/A) of 146.7 Å and a BJH desorption average pore width (4V/A) of 148.5 Å, compared to 155.1 Å for $CMSG_3$ with BJH Adsorption average pore width (4V/A) of 135.6 Å and BJH desorption average pore width (4 V/A) of 153.9 Å.

Table 2. Mass magnetisation of catalysts prepared.

Catalysts	Average Mass Magnetisation (emu/g)
$CMCO_1$	153 ± 0.71
$CMCO_2$	168 ± 1.45
$CMCO_3$	159 ± 0.97
$CMSG_1$	76 ± 1.94
$CMSG_2$	91 ± 1.06
$CMSG_3$	84 ± 0.56
$CMSG_3/ZVINPs_1$	134 ± 2.52
$CMSG_3/ZVINPs_2$	173 ± 0.36
$CMSG_3/ZVINPs_3$	162 ± 1.27

The effect of synthesis methods on the surface area of CMCOs and CMSGs was significant, with $CMSG_3$ maintaining a higher surface area when compared to the $CMCO_2$ sample. Cross-sectional area measurements showed that all of the tested samples were mesoporous. The results that were obtained in the current study agree with previous reports. Report by Xie and Fan [27] investigated the production of biodiesel using tetra alkyl ammonium hydroxides that were immobilised on mesoporous silica as a solid catalyst. Using the co-precipitation method, the catalyst samples prepared were characterised by maximum surface area of 61 m^2/g. Similarly, Rashtizadeh et al. [28] and Teo et al. [29] achieved a surface area of 59 m^2/g when nanocomposite and mesoporous materials containing calcium oxide were prepared.

2.3. Microscopic Observations

SEM micrographs at 20 nm scale (Figure 2A–F) show the three-dimensional (3-D) morphology of the CMCO and CMSG catalysts, while Figure 3A–F are TEM micrographs of the same catalysts at 200 nm scale and 5.00 KX magnification. The catalysts prepared using the sol-gel method had a more elongated shape and uniform distribution, while those that were prepared using the co-precipitation method exhibited strong agglomeration and a higher degree of variation in size. For the CMSG catalysts, the micrographs were homogeneous microstructures (Figure 3). This can be attributed to the fact that, in the sol-gel process, the open structure of the gelatinous state formed allows for free crystal agglomeration, resulting in a more spherical and uniform structure [25]. Taufiq-Yap and Lee reported similar observations [20]. They showed that sol-gel catalysts were characterised by better homogeneity and phase purity with sintering at low temperatures (≤200 °C), as opposed to the co-precipitated catalyst often described as lower phase purity catalysts [25].

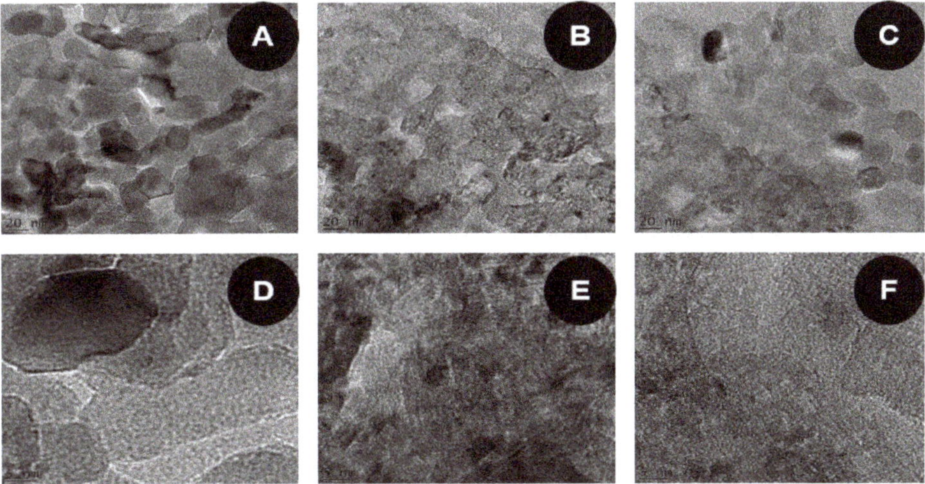

Figure 2. SEM results of the prepared catalyst (**A**: $CMCO_1$/**B**: $CMCO_2$/**C**: $CMCO_3$/**D**: $CMSG_1$/**E**: $CMSG_2$/**F**: $CMSG_3$) [EHT = 5.00 keV, WD = 4.7 mm, Mag = 5.00 K X, 20 nm scale].

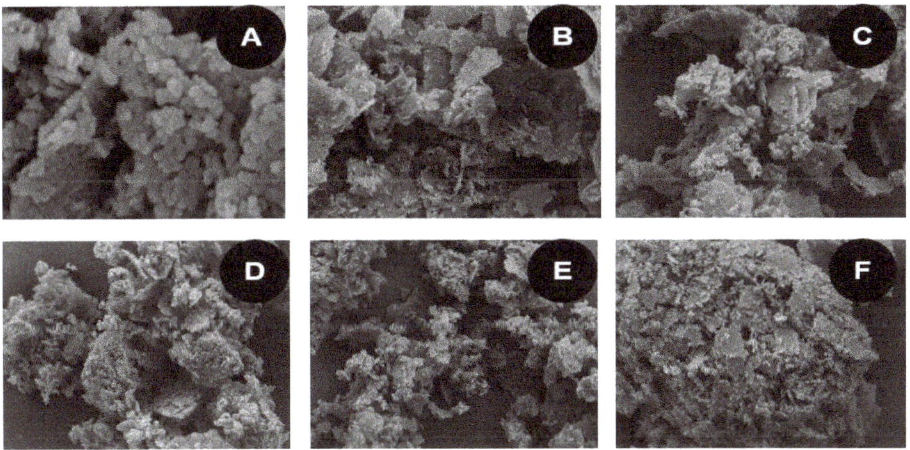

Figure 3. TEM results of the prepared catalyst (**A**: $CMCO_1$/**B**: $CMCO_2$/**C**: $CMCO_3$/**D**: $CMSG_1$/**E**: $CMSG_2$/**F**: $CMSG_3$) [EHT = 5.00 keV, WD = 4.7 mm, Mag = 5.00 K X, 200 nm and 1 μm scale].

Large pores separated the CaO and MgO aggregates in the CMCOs samples (Figure 3). An approximate quantification of $CMSG_3$ samples indicated that the size of aggregates, as well as the width of pores between those aggregates, were in the range of 16 to 17 nm in comparison to 15 to 17 nm for the CMSG samples. For the catalysts that were prepared by the sol-gel technique (Figure 3F), there was a clear indication of localisation of metal nanoparticles in very small pores of the support structure. Farooq et al. [9] and Rahstizadeh et al. [28] reported a similar observation. Additionally, there was a uniform material contrast, justifying the clear visibility of calcium and iron as compared to magnesium based on a single peak that was represented in the spectrum (Figure 3).

The pH was increased from $CMCO_1$ to $CMCO_3$ by decreasing the HNO_3 concentration from 1.5 M to 0.5 M. Alhassan et al. [30] suggested that pH decreases before reaction completion might adversely affect the precipitation stage of the metal oxide catalyst, so the pH was carefully monitored. The pH adjustment was carried out in an attempt to enhance co-precipitation, which is favoured in a more alkaline environment. The optimum concentration was 1M HNO_3 for the $CMCO_2$ catalyst. It was unclear why the microscopic characteristics and surface area of $CMCO_3$ were inferior to those of $CMCO_2$. The increase in ethylene glycol concentration positively contributed to the good quality of the gel that was obtained using the CMSG method (Figure 3F). This agrees with the reports by Zebarjad et al. [31]. In their study, high purity MgO sol-gel particles were synthesised with sizes that ranged from 30 nm to several μm. The optimal CMSG catalyst exhibited more desirable structural properties than its CMCO counterpart, being more agglomerated and interconnected (Figure 3D–F).

2.4. Magnetic Susceptibility and Mass Magnetisation Calculations

In order to quantify the sample's magnetic strength, the magnetic susceptibility of each catalyst sample was converted into mass magnetisation using Gouw's principle [8,10] (Table 2). The CMCO sample exhibited higher magnetic than the CMSG catalyst. This might be attributed to (i) the presence of iron originating from the magnetite in the sample and/or (ii) the size of the catalysts, which were smaller than the super paramagnetic critical size of 27 nm, as described by Zebarjad et al. [31]. There was no major hysteresis in the magnetisation for either sample, which suggests that the particles were paramagnetic.

The CMSG samples were coupled to ZVINPs to boost the saturation magnetisation of the CMSG catalyst while conserving the optimum surface area for mass transfer. Maximum and minimum mass magnetisations were obtained for $CMSG_3/ZVINPs_2$ (173 emu/g) and $CMSG_3$ (84 emu/g), respectively, indicating that the magnetic separation and regeneration of $CMSG_3/ZVINPs_2$ would be most favourable.

In this study, paramagnetism was confirmed by demonstrating that the catalyst particles in a bottle were attracted to a neodymium magnet outside the bottle. Paramagnetism (i.e., responsiveness to an applied magnetic field without permanent magnetisation) was an essential property, because it allowed for the $CMSG_3/ZVINPs_2$ to be separated from the mixture of biodiesel and glycerol under an external magnetic field, which was coupled with electrostatic separation to improve the sedimentation of glycerol.

2.5. Evaluation of Catalytic Performance

Biodiesel obtained from the transesterification process mediated by the $CMSG_3/ZVINPs_2$ catalyst was analysed to assess its quality. As previously described [23], the biodiesel was complied with selected standard specifications (Table 3).

Table 3. Comparison of the synthesised biodiesel (obtained using $CMSG_3/ZVINPs_2$) quality and the commercial diesel [23].

Properties	Unit	Measurement Standards	Commercial Diesel (50ppm Sulphur)	Biodiesel [23]
Viscosity at 40 °C	m^2/s	ASTM D445	$3.0 * 10^{-6} \pm 0.87$	$3.7 * 10^{-6} \pm 0.71$
Density at 15 °C	kg/m^3	ASTM D941	830.0 ± 0.78	832.62 ± 0.69
HHV	MJ/kg	ASTM D2015	48.12 ± 1.59	45.75 ± 1.21
Flash point	°C	ASTM D93	50.4 ± 0.73	61.3 ± 0.64

Results are expressed as averages ± standard deviation ($n = 2$).

Selected catalysts were used to produce biodiesel, and regenerated over four reaction cycles (Figure 4A–D). All of the catalysts were able to mediate biodiesel production (83–88%) up until the fourth reaction cycle. In the case of the catalysts that were not coupled with ZVINPs, the biodiesel yield that was obtained using $CMCO_2$ (Figure 4A) decreased from a maximum of 88% biodiesel yield to a minimum of 64%. In contrast, $CMSG_3$ exhibited the worst performance (Figure 4B), losing 39% of its activity after four transesterification cycles. This indicated that the co-precipitation method yielded a more robust catalyst than the sol-gel method in this instance. The hydride catalysts (CMSG/ZVINPs) showed an enhancement in performance (Figure 4C,D) over the CMSG catalyst. This was attributed to the enhanced mass magnetisation (Table 2) that was conferred by the ZVINPs. Factors, such as loss of catalyst during the separation [32], catalyst surface saturation [11,19], and catalyst leaching, are likely to have contributed to the reduction in the biodiesel yield [9] when using the regenerated catalysts.

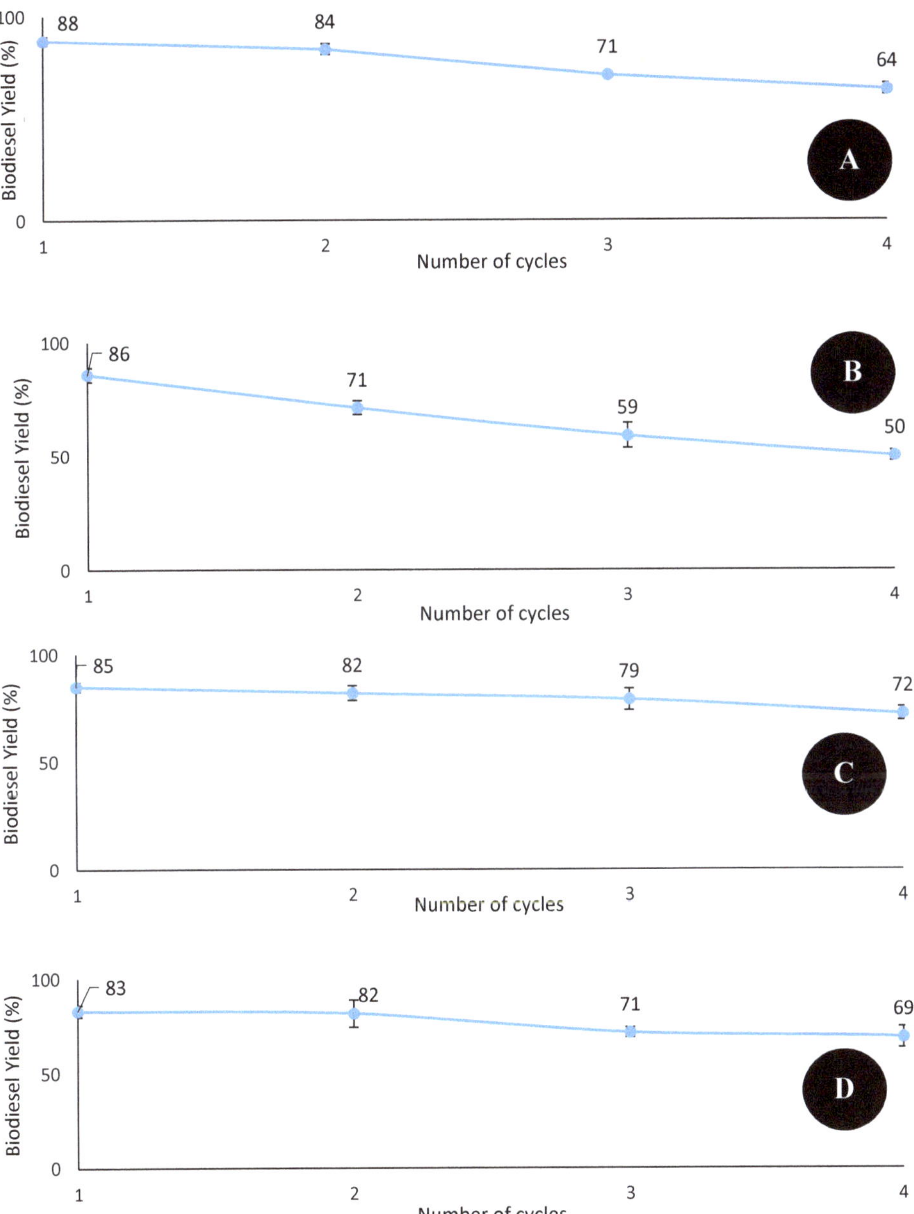

Figure 4. (**A**) Recycled $CMCO_2$; (**B**) $CMSG_3$; (**C**) $CMSG_3/ZVINPs_2$; and, (**D**) $CMSG_3/ZVINPs_3$. Effect on the biodiesel yield (Error bars represent the standard deviation from the mean (n = 3)).

Vishal et al. [33] investigated the use of *n*-hexane to improve the catalytic activity of the catalyst that was used to convert jatropha oil into biodiesel. A reduction in biodiesel yield from 98 to 89% after five reaction cycles was attributed to the loss of catalyst material during the separation process and the oversaturation of the catalyst's pores. Devarapaga et al. [34] described a decrease in biodiesel yield during three reaction cycles when using a novel β-tricalcium phosphate catalyst; the authors attributed the loss in performance to a high FFA concentration in the feedstock that neutralised the functional basic sites of the catalyst. In this instance, the catalyst efficiency was affected by the ability of that particular catalyst to influence the reaction rate, the biodiesel yield through the transesterification process, as well as the extent to which the catalyst could be regenerated while conserving its catalytic activity for biodiesel production. While the sol-gel method was easier to perform and less time consuming, and it resulted in a catalyst with good specific surface areas, the biodiesel yield was negatively affected. This could be due to the lower stability of CMSGs during the reaction. The catalyst sample could have disaggregated when stirring was applied into the vessel [34].

3. Materials and Methods

3.1. Materials

Sodium hydroxide pellets (98%), anhydrous hexane (95%), sodium chloride (anhydrous ≥99%), anhydrous ethylene glycol ($C_2H_6O_2$) (99%), ferric sulphate, sodium borohydride, and nitric acid (70%) used in this study were obtained from Sigma Aldrich (Sigma-Aldrich, St Louis, MO, USA), distilled water. Gangue material was obtained from a mineral processing plant in the Rustenburg region (South Africa) from cupriferous minerals that were generated from the concentration (floatation process) of sulfide ores.

3.2. Catalyst Preparation

The gangue sample that originated from cupriferous mineral processing was obtained from the outlet of bulk oil flotation cells and these samples were characterised by EDX to identify the phases and unit cell dimensions in the samples. This was then subjected to the different catalyst synthesis methods, as described in Sections 3.2.1–3.2.3.

3.2.1. Sol-gel Method

The catalyst was prepared following the method described by Chen et al. [25] with the following modifications: 10 g of gangue was dissolved in warm (60 °C) 1 mol/L nitric acid (HNO_3) into a 500 cm^3 beaker. The slurry was placed into an evaporative basin until crystals were formed. Nine g of the crystals were dissolved in 100 mL of distilled water, added to 15 mL of $C_2H_6O_2$ to obtain catalyst $CMSG_1$, to 20 mL of $C_2H_6O_2$ to obtain catalyst $CMSG_2$, and to 25 mL of $C_2H_6O_2$ to synthesise catalyst $CMSG_3$. The mixing time for all the catalysts was kept constant at 2 h. During mixing, drops of 2 M NaOH was added until a clear white gel was obtained. The gel was then kept static for 2 h for the reaction to complete, and then washed with distilled water. After filtration using Whatman's filter paper PTFE (pore size 0.5 µm), the washed, pH-adjusted gel (pH = 10) was dried by heating at 80 °C for 2 h and then placed in a desiccator for 1 h. The dry gel was milled to form a white powder, which was decarbonised by gradually heating to 800 °C in an oven.

3.2.2. Co-precipitation Method

Three CMCO catalysts were prepared using the co-precipitation method that was adapted from Lu et al. [35] based on different hematite-double nitrate of calcium and magnesium ratios $[(Fe_3O_4)/(Ca,Mg)(NO_3)_2]$ samples were accordingly labelled as $CMCO_1$ (1:2); $CMCO_2$ (1:3); and, $CMCO_3$ (1:4). The Fe_3O_4 value was varied to investigate the enhancement of the magnetic susceptibility of the catalyst for separation purposes

3.2.3. Catalyst Mixed with Zero-Valent Iron Nanoparticles (ZVINPs)

ZVINPs were prepared according to the stoichiometry shown in Equation (1) using the sulphate method [27]; 0.5 M sodium borohydride was added to an equal volume of 0.28 M ferrous sulphate at 0.15 L/min.

$$2Fe^{2+}{}_{(aq)} + BH_4{}^{-}{}_{(aq)} + 3H_2O_{(l)} \rightarrow 2Fe_{(s)} + H_2BO_3{}^{-}{}_{(aq)} + 4H^{+}{}_{(aq)} + 2H_{2\,(g)} \quad (1)$$

The $CMSG_3$ catalyst was then mixed with ZVINPs to form a hybrid catalyst. Three catalysts were obtained depending on the mixing ratio: $CMSG_3/ZVINPs_1$ (2:1), $CMSG_3/ZVINPs_2$ (3:1), and $CMSG_3/ZVINPs_3$ (4:1). CMSG catalysts were amalgamated with ZVINPs to improve their magnetic susceptibility for separation purposes. In theory, magnetic separation can reduce the production costs by eliminating the need for more expensive separation methods, such as centrifugation. CMCO catalysts were not amalgamated with ZVINPs, because the co-precipitation method was significantly longer.

3.3. Biodiesel Production and Separation

Biodiesel was produced through a transesterification reaction using oil extracted from edible oil wastewater sludge and bioethanol (originating from the edible oil wastewater sludge residues after oil extraction). The process was separately mediated using different catalysts ($CMCO_2$, $CMSG_3$, $CMSG_3/ZVINPs_2$, and $CMSG_3/ZVINPs_3$), as previously described [23,24]. The triglycerides extracted from waste that was mainly contained monounsaturated fats. The catalysts were CaO and MgO-based. The use of these two compounds are known to result in higher biodiesel yield, at faster reaction rates, lower temperatures, and low catalyst loading [8,21,23,24]. The transesterification process was carried out in a 500 mL flask using a 1:6 oil/ethanol molar ratio and catalyst dosages were selected at 3, 5, and 8 wt.%. The temperature was set at 75 °C for 2 h. All of the runs were performed in triplicate. After the reaction was completed, the collected catalyst was washed with ethanol and then dried at 80 °C for 2 h before being used.

The electrostatic method was used to separate the biodiesel from the glycerol using LG electronics Neon transformer (Seoul, South Korea). The transformer electrodes were immersed in a flask containing 1 L biodiesel/glycerol mixture. The transformer was then set at 8 kV and adjusted for 25 mA. This separation process was completed after 20 min. This technique was adapted from Mayvan et al. [36].

The yield of biodiesel obtained was calculated using Equations (2) and (3) relating the weight of oil in the raw material and the weight of biodiesel obtained after glycerol separation. These findings were compared with the values that were obtained using a GC-FID 6980A that was manufactured by Agilent (Santa Clara, CA, USA). The column used was a HP88 (60 m × 150 μm, 0.250 μm) manufactured by Agilent Technologies (Santa Clara, CA, USA).

$$\text{Yield}\ (\%) = \frac{Weight_{\ Biodiesel}}{Weight_{\ Oil}} \times 100 \quad (2)$$

$$\text{Conversion}\ (\%) = \frac{Weight_{\ Glycerol} - Weight_{\ Biodiesel}}{Weight_{\ Biodiesel}} \times 100 \quad (3)$$

3.4. Recovery of Catalyst from Biodiesel

A magnetic field that was provided by a lightweight (75 g) neodymium magnet assembly with a 2 cm air gap and magnetic field of approximately 0.22 T (Tesla) was used to recover the catalyst. This magnet assembly was immersed into the flasks containing 500 mL biodiesel at ambient temperature for 10 min, while stirring at 50 rpm. The collected catalyst was then washed with ethanol and then dried in an oven at 80 °C for 2 h [23,24].

3.5. Catalyst Characterisation and Performance

The morphology and particle sizes of the raw materials, as well as the catalysts, were assessed by EDX (STADI P, STOE model, Darmstadt, Germany). An Auriga (Carl Zeiss model, Jena, Germany). High-resolution scanning electron microscope (HRSEM) was used with a resolution of 1.5 nm to provide details from a focused electron beam across the surface of the sample. These images were magnified using an Auriga TF 20 high-resolution transmission electron microscope (Carl Zeiss model, Jena, Germany) with a resolution of 0.25 nm for 45 min. to 2 h for sample mapping and imaging. Furthermore, an Aztec series (Hitachi High-Technologies, Tokyo, Japan) TM 4000 energy dispersive X-ray analyser with a 30 mm^2 detector) was employed to provide the elemental identification and quantitative composition of the prepared catalyst. Catalyst surface areas were studied using a VF-Sorb 2400CE BET (Brunauer-Emmett-Teller) surface area analyser (Beijing, China). The magnetic susceptibilities of the catalysts were calculated based on Gouw's principle [32]. These results were confirmed by the PS PRO Gauss meter (Corby, United Kingdom). The performance of these catalysts was evaluated based on their abilities to improve the yield of biodiesel that is produced from edible oil wastewater sludge at different catalyst dosages (3, 5, and 8 wt.%) and ethanol-oil ratios (3:1, 6:1, and 9:1), respectively. The particle surface area was confirmed using BET analysis.

4. Conclusions

The catalysts that were obtained from cupriferous mineral processing wastes were successfully prepared and characterised using TEM, SEM, EDX, and BET. The mass magnetisation analysis showed that the nanocatalyst samples were magnetic. The quality of the CMSG catalysts was not affected by the increase in ethylene glycol concentrations. Further, as for the catalysts that were prepared using the co-precipitation method CMCO, the magnetite ratio chosen in this study did not considerably increase their respective mass magnetisation. The reusability of the catalyst preparation over four cycles of transesterification reaction has the potential to reduce the cost of biodiesel production. The maintenance of 50–72% catalytic activities, as demonstrated by the biodiesel yields, after four cycles suggest that the catalysts can be used for more cycles. Properties of the biodiesel that were produced from the edible oil wastewater sludge using these catalysts were within the biodiesel standard specifications.

Author Contributions: Conceptualisation and validation, W.I.N., O.O.O., P.J.W.; methodology and software, W.I.N.; Formal analysis, W.I.N., P.J.W., O.O.O., D.I.-O.; Writing—Original draft preparation, W.I.N., P.J.W. and O.O.O.; writing—review and editing, W.I.N., P.J.W. and O.O.O.; supervision, P.J.W., O.O.O. and D.I.O.; project administration W.I.N., P.J.W., O.O.O. and D.I.-O.

Funding: The authors wish to extend their sincere appreciation to the Water Research Commission of South Africa (WRC-Project K5/2404) and the Council for Scientific and Industrial Research (CSIR/HCD-IBS programme) for funding this project. The content does not necessarily reflect the views and policies of the funding organisations.

Acknowledgments: The authors would also like to acknowledge the Department of Chemical Engineering for additional facility to conduct the research at the Cape Peninsula University of Technology (CPUT); Agrifood Station at the Department of Food Technology (CPUT) for wastewater analysis; The Department of Physics at the University of Western Cape (UWC) for the catalyst characterisation.

Conflicts of Interest: The authors declare no conflict of interest. The content does not necessarily reflect the views and policies of the funding organisations.

References

1. Cabri, L.J.; Rudashevsky, N.S.; Rudashevsky, V.N. Current approaches for the process mineralogy of platinum-group element ores and tailings. In Proceedings of the Ninth International Congress for Applied Mineralogy (ICAM 2008), Kerala State, India, 18–21 February 2008; Volume 8, pp. 9–17. [CrossRef]
2. Arens, V.Z.; Chernyak, S.A. Hydrometallurgy in the mining industry. *Metallurgist* **2008**, *52*, 3–10. [CrossRef]
3. Cairncross, B. History of the Okiep Copper District, Namaqualand, Northern Cape Province, South Africa. *Mineral. Rec.* **2004**, *35*, 289–317. [CrossRef]
4. Demirbas, A. Biodiesel Fuels from Vegetable Oils via Catalytic and Non-Catalytic Supercritical Alcohol Transesterifications and Other Methods: A Survey. *Energy Convers. Manag.* **2003**, *44*, 2093–2109. [CrossRef]

5. Chen, S.Y.; Mochizuki, T.; Abe, Y.; Toba, M.; Yoshimura, Y. Ti-incorporated SBA-15 mesoporous silica as an efficient and robust Lewis solid acid catalyst for the production of high-quality biodiesel fuels. *Appl. Catal. B Environ.* **2014**, *148*, 344–356. [CrossRef]
6. Hu, S.; Guan, H.; Wang, Y.; Han, H. Nano-magnetic catalyst KF/CaO–Fe_3O_4 for biodiesel production. *Appl. Energy* **2011**, *88*, 2688. [CrossRef]
7. Montero, J.M.; Brown, R.; Gai, P.L.; Lee, A.F.; Wilsonc, K. In situ studies of structure-reactivity relations in biodiesel synthesis over nanocrystalline MgO. *Chem. Eng. J.* **2010**, *161*, 332–336. [CrossRef]
8. Amani, H.; Ahmad, Z.; Hameed, B.H. Synthesis of fatty acid methyl esters via the methanolysis of palm oil over Ca3.5xZr0.5yAlxO_3 mixed oxide catalyst. *Renew. Energy* **2014**, *66*, 680–685. [CrossRef]
9. Farooq, M.; Ramli, A.; Subbarao, D. Biodiesel production from waste cooking oil using bifunctional heterogeneous solid catalysts. *J. Clean. Prod.* **2013**, *59*, 131–140. [CrossRef]
10. Elfimov, I.S.; Yunoki, S.; Sawatzky, G.A. Possible path to a new class of ferromagnetic and half-metallic ferromagnetic materials. *Phys. Rev. Lett.* **2002**, *89*, 216403. [CrossRef]
11. Mardhiah, H.H.; Ong, H.C.; Masjuki, H.H.; Lim, S.; Lee, H.V. A review on latest developments and prospects of heterogeneous catalyst in biodiesel production from non-edible oils. *Renew. Sustain. Energy. Rev.* **2017**, *67*, 1225–1236. [CrossRef]
12. Konaka, A.; Tago, T.; Yoshikawa, T.; Shitara, H.; Nakasaka, Y.; Masuda, T. Conversion of biodiesel-derived crude glycerol into useful chemicals over a zirconia–iron oxide catalyst. *Ind. Eng. Chem. Res.* **2013**, *52*, 15509–15515. [CrossRef]
13. Bobade, V.V.; Kulkarni, K.S.; Kulkarni, A.D. Application of Heterogeneous Catalyst for the Production of Biodiesel. *Int. J. Adv. Eng. Technol.* **2011**, *2*, 184–185. [CrossRef]
14. Gao, X.; Yu, K.M.K.; Tam, K.Y.; Tsang, S.C. Colloidal stable silica encapsulated Nanomagnetic composite as a novel bio-catalyst carrier. *Chem. Commun.* **2003**, *24*, 2998. [CrossRef] [PubMed]
15. Wen, L.B.; Wang, Y.; Lu, D.L.; Hu, S.Y.; Han, H.Y. Preparation of KF/CaO nanocatalyst and its application in biodiesel production from Chinese tallow seed oil. *Fuels* **2010**, *89*, 2267. [CrossRef]
16. Ying, M.; Chen, G. Study on the production of biodiesel by magnetic cell biocatalyst based on lipase-producing Bacillus subtilis. *Appl. Biochem. Biotechnol.* **2007**, *137*, 793–803. [CrossRef]
17. Xie, W.; Ma, N. Immobilized lipase on Fe_3O_4 nanoparticles as biocatalyst for biodiesel production. *Energy Fuels* **2009**, *23*, 1347. [CrossRef]
18. Ali, M.A.; Al-Hydary, I.A.; Al-Hattab, T.A. Nano-Magnetic Catalyst CaO-Fe_3O_4 for Biodiesel Production from Date Palm Seed Oil. *Bull. Chem. React. Eng. Catal.* **2017**, *12*, 460–468. [CrossRef]
19. Di Serio, M.; Tesser, R.; Pengmei, L.; Santacesaria, E. Heterogeneous catalysts for Biodiesel production. *Energy Fuels* **2008**, *207*, 17–22. [CrossRef]
20. Taufiq-Yap, Y.H.; Teo, S.H.; Rashid, U.; Islam, A.; Hussien, M.Z.; Lee, K.T. Transesterification of Jatropha curcas crude oil to biodiesel on calcium lanthanum mixed oxide catalyst: Effect of stoichiometric composition. *Energy Convers. Manag.* **2014**, *88*, 1290–1296. [CrossRef]
21. Kouzu, M.; Kajita, A.; Fujimori, A. Catalytic activity of calcined scallop shell for rapeseed oil transesterification to produce biodiesel. *Fuels* **2016**, *182*, 220–226. [CrossRef]
22. Marinkovic, D.M.; Avramovic', J.M.; Stankovic, M.V.; Stamenkovic, O.S.; Jovanovića, D.M.; Veljković, V.B. Synthesis and characterization of spherically-shaped CaO/γ-Al_2O_3 catalyst and its application in biodiesel production. *Energy Convers. Manag.* **2017**, *144*, 399–413. [CrossRef]
23. Ngoie, I.W.; Welz, P.J.; Oyekola, O.O.; Ikhu-Omoregbe, D. Valorisation of edible oil wastewater sludge: Bioethanol and biodiesel production. *Waste Biomass Valorization* **2019**. [CrossRef]
24. Ngoie, W.I.; Welz, P.J.; Oyekola, O.O.; Ikhu-Omoregbe, D.I. Qualitative Assessment of Biodiesel Produced from Primary Edible Oil Wastewater Sludge. *Waste Biomass Valorization* **2019**. [CrossRef]
25. Di Serio, M.; Cozzolino, M.; Giordano, M.; Tesser, R.; Patrono, P.; Santacesaria, E. From homogeneous to heterogeneous catalysts in biodiesel production. *Ind. Eng. Chem. Res.* **2007**, *46*, 6379–6384. [CrossRef]
26. Chen, H.; Zhang, P.; Duan, Y.; Zhao, C. Reactivity enhancement of calcium based sorbents by doped with metal oxides through the sol-gel process. *Appl. Energy* **2016**, *162*, 390–400. [CrossRef]
27. Coulson, J.M.; Richardson, J.F. *Coulson's and Richardson's Chemical Engineering Handbook*; Chhabra, R.P., Gurappa, B., Eds.; Oxford: Butterworth-Heinemann, UK, 2002.
28. Rashtizadeh, E.; Farzaneh, F.; Talebpour, Z. Synthesis and characterization of $Sr_3Al_2O_6$ nanocomposite as catalyst for biodiesel production. *Bioresour. Technol.* **2014**, *154*, 32–37. [CrossRef] [PubMed]

29. Hu, S.; Guan, Y.; Wang, Y.; Han, H. Nano-magnetic catalyst KF/CaO-Fe_2O_3 for biodiesel production. *Appl. Energy* **2011**, *88*, 2685–2690. [CrossRef]
30. Xie, W.; Fan, M. Biodiesel production by transesterification using tetraalkylammonium hydroxides immobilized onto SBA-15 as a solid catalyst. *Chem. Eng. J.* **2014**, *239*, 60–67. [CrossRef]
31. Teo, S.H.; Rashid, U.; Taufiq-Yap, Y.H. Biodiesel production from crude Jatropha Curcas oil using calcium based mixed oxide catalysts. *Fuels* **2014**, *136*, 244–252. [CrossRef]
32. Alhassan, F.H.; Rashid, U.; Taufiq-Yap, Y.H. Synthesis of waste cooking oil-based biodiesel via effectual recyclable bi-functional Fe_2O_3MnOSO42−/ZrO_2 nanoparticle solid catalyst. *Fuels* **2014**, *142*, 38–45. [CrossRef]
33. Lu, A.H.; Salabas, E.L.; Schiith, F. Magnetic nanoparticles: Synthesis, protection, functionalization, and application. *Cheminternationale* **2007**, *46*, 1222. [CrossRef] [PubMed]
34. Vishal, T. Biodiesel—An Alternative Method for Energy Crisis: A Review. *J. Biol. Chem. Chron.* **2016**, *2*, 14–26. [CrossRef]
35. Devarapaga, M.; Chavan, S.; Singh, V.; Singh, B.; Sharma, Y.C. An economically viable synthesis of biodiesel from a crude Millettia pinnata oil of Jharkhand, India as feedstock and crab shell derived catalyst. *Bioresour. Technol.* **2016**, *214*, 210–217. [CrossRef]
36. Mayvan, A.A.; Ghobadian, B.; Najafi, G. Electrostatic coagulation for separation of crude glycerin from biodiesel. *Adv. Environ. Biol.* **2014**, *8*, 321–324.

Article

Low-Temperature Selective Catalytic Reduction of NO with NH_3 over Natural Iron Ore Catalyst

Naveed Husnain, Enlu Wang * and Shagufta Fareed

Institute of Thermal Energy Engineering, School of Mechanical Engineering, Shanghai Jiao Tong University, Shanghai 200240, China; naveed69@sjtu.edu.cn (N.H.); shagufta91@sjtu.edu.cn (S.F.)

* Correspondence: elwang@sjtu.edu.cn; Tel.: +86-21-3420-5683

Received: 28 October 2019; Accepted: 12 November 2019; Published: 14 November 2019

Abstract: The selective catalytic reduction of NO with NH_3 at low temperatures has been investigated with natural iron ore catalysts. Four iron ore raw materials from different locations were taken and processed to be used as catalysts. The methods of X-ray diffraction (XRD), X-ray fluorescence (XRF), Brunauer-Emmett-Teller (BET), X-ray photoelectron spectroscopy (XPS), hydrogen temperature-programmed reduction (H_2-TPR), ammonia temperature-programmed desorption (NH_3-TPD), scanning electron microscopy (SEM) and Fourier-transform infrared spectroscopy (FTIR) were used to characterize the materials. The results showed that the sample A (comprised mainly of α-Fe_2O_3 and γ-Fe_2O_3), calcined at 250 °C, achieved excellent selective catalytic reduction (SCR) activity (above 80% at 170–350 °C) and N_2 selectivity (above 90% up to 250 °C) at low temperatures. Suitable calcination temperature, large surface area, high concentration of surface-adsorbed oxygen, good reducibility, lots of acid sites and adsorption of the reactants were responsible for the excellent SCR performance of the iron ore. However, the addition of H_2O and SO_2 in the feed gas showed some adverse effects on the SCR activity. The FT-IR analysis indicated the formation of sulfate salts on the surface of the catalyst during the SCR reaction in the presence of SO_2, which could cause pore plugging and result in the suppression of the catalytic activity.

Keywords: natural iron ore; NO; low-temperature; selective catalytic reduction

---

1. Introduction

Current global major environmental problems, such as smog, rain, fine particle pollution, and ozone depletion are credited to nitrogen oxides (NO_x, x = 1, 2) [1]. Around the world, NO_x legislation has become more and more stringent due to the increased awareness in society about protecting the environment [2]. Therefore, in the modern era, a lot of focus is given to the removal of NO_x by the researchers. To meet these stringent regulations of NO_x abatement for point sources (coal-fired power plants), selective catalytic reduction (SCR) of NO_x with ammonia (NH_3) has become the most effective and extensively used NO_x reduction technology [3]. Vanadium-based catalysts (V_2O_5/TiO_2 or V_2O_5-WO_3/TiO_2) are the most common industrial catalysts for SCR of NO_x with NH_3, for which the most efficient reaction temperature range is 300~400 °C [4]. Since the operating temperature range is high, that is why the catalyst must be placed upstream of the desulfurizing section and the particulate removal device, to avoid reheating the flue gas; but it can cause deactivation due to the high concentration of sulfur dioxide and dust [5]. Other issues include high activity for the oxidation of SO_2 to SO_3 [6], the formation of N_2O at high temperatures [7] and the toxicity of vanadium to the environment [8]. That is why recently, environmental catalysis has put great efforts into the development of V-free NH_3-SCR catalysts.

Some other transition metal oxide-based catalysts (Cr, Mn, Fe, Ni, Zr, Co, Cu, La, etc.) have shown good NH_3-SCR performance [9]. For example, Ma et al. prepared TiO_2-supported iron oxide catalysts. It was found that $Fe_2(SO_4)_3/TiO_2$ catalysts have shown excellent NO_x conversions in the NH_3-SCR reaction at low temperatures [10]. Chen et al. synthesized a series of Fe–Mn oxide-based catalysts by three different synthesis methods: the citric acid method, the coprecipitation method and the solid reaction method [11]. The results exhibited that the catalytic activity of the catalyst prepared by the citric acid method ($Fe_{(0.4)}$-MnO_x(CA-500)) was higher (around 98.8% NO_x conversion at 120 °C) than other catalysts. Sankar et al. applied two preparation techniques (impregnation and deposition–precipitation) to synthesize the Mn–Fe/TiO_2 catalysts, and concluded that the deposition–precipitation technique-based catalyst (25 wt % $Mn_{0.75}Fe_{0.25}Ti$) showed superior catalytic activity (97% at 200 °C) [12]. Likewise, Wu et al. prepared a $FeMnTiO_x$-mixed oxide catalyst by cetyltrimethylammonium bromide (CTAB)-assisted coprecipitation method [13,14]. It was found that the NO_x conversion was around 80~100% at 100~350 °C. Zhang et al. prepared catalysts by loading Mn–FeO_x and MnO_2–Fe_2O_3–CeO_2–Ce_2O_3 onto carbon nanotubes (CNTs) support, respectively [15,16]. The NO_x conversion reached almost 100% from 140~180 °C. Although some catalysts could provide high NO_x conversion at low temperatures, yet there exist some difficulties in their sophisticated preparation methods, as well as the fact that the preparation cost of these catalysts is very high.

For many centuries, the oxides of iron have been serving humankind. The first use of yellow and red ochres was in prehistoric paintings in the caves. After that, the role of iron oxides in our lives has seen an unbelievable expansion. Their primary applications include the steel industry (as a precursor of iron and steel), the ecosphere (their activity as absorbents), and chemical reactions (their ability to catalyze various chemical reactions). Around 80% of the world's iron ore production consists of sedimentary iron ore, and the significant proportion of these sedimentary iron ore deposits in the world is made up of iron oxides [17].

Previously, the NO conversion of manganese ore has been investigated in some studies. Park et al. used natural manganese ore for NO reduction and studied its catalytic activity [18]. Similarly, Zhu et al. also studied the performance of natural manganese ore catalysts and showed that the NH_3-SCR activity of the manganese-based catalyst reached 80% at 120 °C [19]. Gui et al. studied the performance of iron and iron ore particles as catalysts in a magnetically fluidized bed [20]. The results showed that the catalytic activity of the iron ore was not affected much by the magnetic field. However, some of these studies were unable to clarify why the temperature of NO conversion was reduced by those catalysts.

Inspired by the idea of finding a catalyst with low preparation cost, high NO conversion and nominal environmental pollution, natural iron ore as a De–NO_x catalyst is investigated in this study. The methods of XRD, XRF, BET, XPS, H_2-TPR, NH_3-TPD, SEM and FT-IR were used to characterize the materials. The results showed that the sample A which calcined at 250 °C achieved excellent SCR activity (above 80% at 170–350 °C) and N_2 selectivity (above 90% up to 250 °C) at low temperatures. Suitable calcination temperature, large surface area, high concentration of surface-adsorbed oxygen, good reducibility, lots of acid sites and adsorption of the reactants, were responsible for the excellent SCR performance of the iron ore. However, the addition of H_2O and SO_2 in the feed gas showed some adverse effects on the SCR activity. The FT-IR analysis indicated the formation of sulfate salts on the surface of the catalyst during the SCR reaction in the presence of SO_2, which could cause pore plugging and result in a reduction of the catalytic activity.

2. Results and Discussion

2.1. NH_3-SCR Activity and Selectivity

For different samples calcined at 250 °C, the low-temperature SCR removal activity, and N_2 selectivity at different reaction temperatures, are shown in Figure 1A and Table 1, respectively. It can be seen that sample A exhibited the best activity and selectivity among other samples in a wide temperature range with NO conversion above 90% from 200–290 °C and the N_2 selectivity above 90% up to 250 °C.

It can be concluded that the catalytic activity and selectivity of all the samples were different from each other, which could result from the difference in their material structure, surface properties and composition of the iron ore. The catalytic activity and selectivity of the catalysts at low temperatures followed the particular sequence: Sample D < Sample B < Sample C < Sample A. In addition, it can be seen that the NO conversion of sample A is higher than the Mn–Ce–γ–Al_2O_3 catalyst [21] within 150–300 °C, even though the gas hourly space velocity (GHSV) of our experiments is higher than that of the literature.

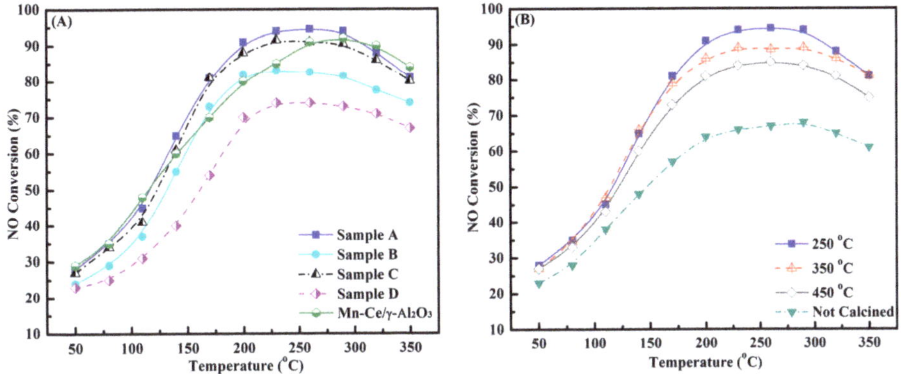

Figure 1. (**A**) NO conversion of different iron ore samples calcined at 250 °C and the Mn–Ce–γ–Al_2O_3 catalyst reported in literature [21]; (**B**) Influence of calcination temperature on the NO conversion of sample A. Our reaction conditions: 500 ppm NO, 500 ppm NH_3, 3 vol % O_2, N_2 balance, the flow rate of 145 mL·min^{-1}, and gas hourly space velocity (GHSV) = 20,000 h^{-1}. Reaction conditions of literature [21]: 700 ppm NO, 700 ppm NH_3, 3 vol % O_2, N_2 as balance, and GHSV = 10,000 h^{-1}.

Table 1. N_2 selectivity of the calcined samples (at 250 °C) at different reaction temperatures.

Samples	50 °C	150 °C	250 °C	350 °C
Sample A	96.66%	95.43%	91.98%	85.45%
Sample B	96.66%	94.50%	83.45%	70.35%
Sample C	96.66%	94.52%	86.35%	80.84%
Sample D	96.66%	94.13%	78.78%	59.47%

Calcination of a catalyst can help to obtain a large specific surface area and better dispersion of the active ingredients of the catalyst [2]. To find out the suitable calcination temperature for iron ore catalysts, sample A was thermally treated to different temperatures (250 °C, 350 °C, 450 °C) and the NO conversions are shown in Figure 1B. The results showed that the NO conversions of the thermally-treated samples were better than the sample without thermal treatment. For example, at 200 °C reaction temperature, the NO conversion of sample A that calcined at 250 °C was 91%, whereas the catalytic activity of sample A without calcination was just 64%. Additionally, the sample A calcined at 250 °C showed better NO conversion than the other samples calcined at 350 °C and 450 °C, respectively, indicating the possible severe morphology changes in the iron ore sample after high calcination temperatures of 350 and 450 °C (as shown in SEM results). Higher calcination temperatures cause a decrease of the surface area and the mobility of lattice oxygen, which may be the reason for the lower catalytic activity of these samples [19,21].

2.2. NO, and NH_3 Oxidation Activity

It was reported that with the increase in the NO_2/NO molar ratio in the feeding gas fast SCR reaction: $NO + NO_2 + 2NH_3 \rightarrow 2N_2 + 3H_2O$ could be promoted, which can significantly boost the low-temperature SCR activity of the catalyst [16,22]. That is why separate NO oxidation activities of the different samples were tested and the results are demonstrated in Figure 2A.

It can be seen that with the rise in temperature the NO oxidation activities increased, first up to the 300 °C reaction temperature, and then decreased. Some evident differences could be seen between the NO oxidation activity and SCR activity of the samples which could be ascribed to the role of NH_3 during SCR reaction; in a real SCR reaction, the presence of NH_3 would increase the NO to NO_2 conversion as compared to the pure NO oxidation. NO conversion to NO_2 was proven to be a slow step for an SCR reaction, and the NO_2 generated was rapidly consumed in the existence of ammonia [22]. Sample A comprising the larger weight fraction of γ-Fe_2O_3 exhibited the highest NO conversion to NO_2 as well as the NH_3–SCR activity. Figure 2B shows that the NH_3 oxidation increased with the rise in temperature. This increase in the NH_3 oxidation at high temperatures is considered as a side reaction which aids in the formation of NO in the NH_3–SCR system and also causes the deactivation of the catalysts at high temperatures [13].

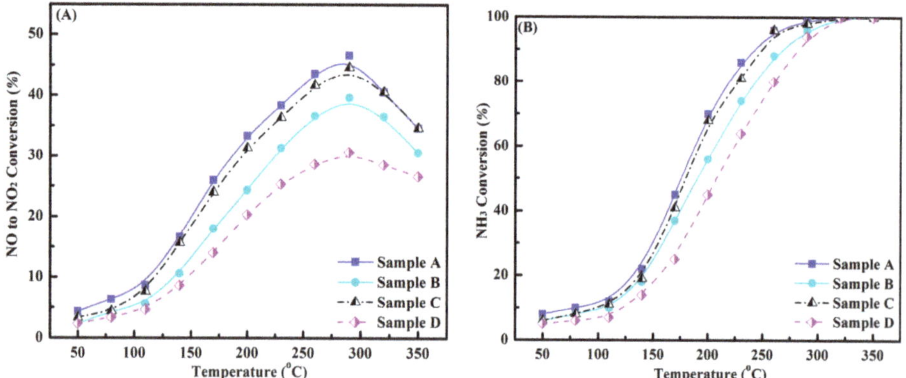

Figure 2. (**A**) NO oxidation activity; (**B**) NH_3 oxidation activity. Reaction conditions: 500 ppm NO or 500 ppm NH_3, 3 vol % O_2, N_2 balance, flow rate of 145 mL·min^{-1}, and GHSV = 20,000 h^{-1}.

2.3. Characterization

XRD patterns of the different iron ore samples without calcination are shown in Figure 3. It can be seen that the samples consist of three minerals of iron: γ-Fe_2O_3 (maghemite, JCPDS 00-39-1346), α-Fe_2O_3 (hematite, JCPDS 01-089-8104) and α-FeOOH (goethite, JCPDS 00-017-0536).

The diffraction spectra of sample A indicated that three main phases of ferric oxide (α-Fe_2O_3, α-FeOOH and γ-Fe_2O_3,) were present in it, and through quantitative analysis by the Rietveld method it was found that the ratio of these three phases in sample A was 59.4:11.5:29.1, respectively. Similarly, in sample C the ratio was 67.7:24.1:8.2, respectively. Diffraction spectra of sample B showed that α-Fe_2O_3 and α-FeOOH were the two phases present in the sample and existed in 27.4:72.6, respectively. Whereas, in sample D the ratio was 7.7:92.3, respectively.

The XRD spectra in Figure 3 depicts that the sample A comprised mainly of α-Fe_2O_3 and γ-Fe_2O_3 and the weight fraction of γ-Fe_2O_3 was found highest among other samples. With a change in the temperature range, the catalytic influence of these two components varies. The optimum temperature of the NO conversion of γ-Fe_2O_3 is relatively low as compared to α-Fe_2O_3 [23]. It is known that the lattice of γ-Fe_2O_3 has point defects created by its cation vacancies [24]. These point lattice defects could be useful in providing more active sites for the catalytic reaction to take place, as well as contribute to the surface electron transfer in the SCR reaction. So the existence of γ-Fe_2O_3 could be crucial for the low-temperature SCR activity of the iron ore.

Partly magnified XRD spectra of the 'not calcined' as well as 'calcined' sample A at different temperatures is shown in Figure 4. Sample A without calcination exhibited sharp crystalline peaks of α-Fe_2O_3 and γ-Fe_2O_3, indicating good crystallinity (Figure 4A). After calcining the sample to 250 °C, both phases of the Fe_2O_3 were still intact, but the intensities of the peaks were much lower, indicating a poor crystallinity (amorphous structure formation verified by SEM) which was beneficial for obtaining high SCR activity at low temperatures. However, with a further rise in calcination temperature above 250 °C, the γ-Fe_2O_3 phase (peak at 63.99°) in the sample has been converted into the α-Fe_2O_3 phase (peak at 64.17°). Cao et al. also concluded that after increasing the calcination temperature above 250 °C, the γ-Fe_2O_3 particles started to engulf their surrounding crystals and the different sizes of the α-Fe_2O_3 particles formed [25], which resulted in the decrease of surface area. According to the Scherrer formula (Equation (5)), the half-width of the diffraction peak and the crystallite size of the material are inversely proportional to each other. So, the average crystallite sizes of the sample that was calcined at 350 °C and 450 °C, respectively are larger than the sample calcined at 250 °C, which could be related to the sintering (frittage) phenomenon at higher temperatures and the formation of α-Fe_2O_3. That is why the samples calcined at a higher calcination temperature (350 °C, 450 °C) exhibited lower catalytic activity and selectivity than the sample calcined at 250 °C (Figure 1, Table 1).

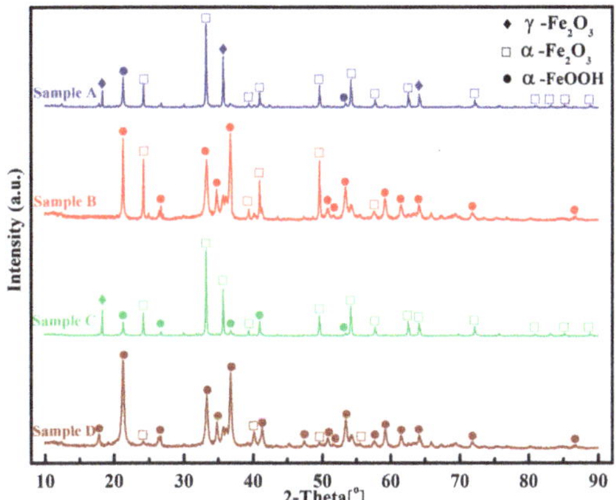

Figure 3. XRD spectra of different iron ore samples.

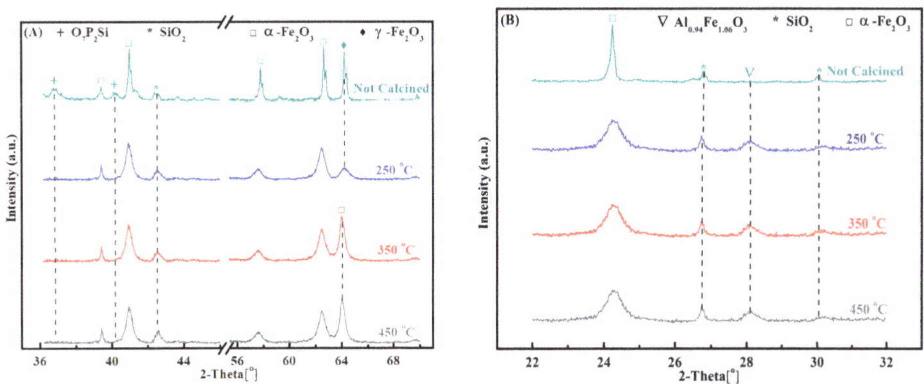

Figure 4. Partly magnified XRD spectra of sample A: (**A**) transformation of γ-Fe_2O_3 to α-Fe_2O_3 above 250 °C; (**B**) formation of $Al_{0.94}Fe_{1.06}O_3$ in the calcined samples.

In addition, it can be seen in Figure 4A (not calcined sample), apart from α-Fe_2O_3 peaks, two small peaks appear at 36.8° and 40.25° which correspond to the O_7P_2Si crystals phase (JCPDS 00-022-1274), and one peak at 42.46° represents the presence of SiO_2 (JCPDS 00-046-1045). In calcined samples, the peaks of O_7P_2Si are not present. However, the peaks of $Al_{0.94}Fe_{1.06}O_3$ (JCPDS 00-011-0562) at 28.36°, and SiO_2 at 26.7° are detected (Figure 4B). These results exhibited that the O_7P_2Si was decomposed to SiO_2, and after calcination $Al_{0.94}Fe_{1.06}O_3$ was formed. This formation of $Al_{0.94}Fe_{1.06}O_3$ may have aided in a better NO reduction of sample A.

The elemental compositions of the iron ore samples show that Fe, O, Al and Si are the major elements present in these samples (Table 2). Natural iron ore was composed mainly of ferric oxide (XRD). Ferric oxide, particularly maghemite (γ-Fe_2O_3) particles, exhibited good SCR activity at low temperatures [26], and the interactions of some major elements like Al and Si with ferric oxide could be beneficial for the low-temperature SCR activity [27–29]. XRD results indicated the formation of $Al_{0.94}Fe_{1.06}O_3$ and SiO_2 (from the decomposition of O_7P_2Si) after calcining the catalyst, which may have assisted in a better NO reduction of sample A.

Table 2. Iron ore samples with their major elements and their mass percentages.

Elements	Sample A (%)	Sample B (%)	Sample C (%)	Sample D (%)
Fe	60.961	68.099	67.853	64.32
O	31.564	26.630	25.801	28.95
Al	1.706	2.043	2.243	2.13
Si	3.386	1.080	1.881	2.95
P	0.029	0.066	0.082	0.087

BET surface areas and average crystallite size of the catalysts are shown in Tables 3 and 4. It can be seen from Table 3 that BET surface areas of the calcined samples are much higher than the non-calcined sample, and a decline in the surface area of sample A was observed with a rise in calcination temperature (350 °C and 450 °C). In addition, it can be seen from Table 3 that the surface area of the catalysts and the average crystallite size are inversely proportional to each other. Table 4 is comprised of BET surface areas and the average crystallite size of the different samples of the iron ore calcined at 250 °C. Sample A exhibited the highest surface area among other samples. Larger surface area can help to provide more active sites on the surface of the catalyst by which flue gas adsorption could be promoted. That is why the catalytic activity and selectivity of the sample A calcined at 250 °C was the best among other samples (Figure 1, Table 1).

Table 3. Sample A BET surface area at different calcination temperatures.

Calcination Temperature (°C)	BET Surface Area ($m^2 \cdot g^{-1}$)	Average Crystallite Size (nm)
Not calcined	21.63	43.24
250	42.52	19.62
350	36.76	33.98
450	28.14	38.24

Table 4. Four ore samples and their BET surface area (calcined at 250 °C).

Samples	BET Surface Area ($m^2 \cdot g^{-1}$)	Average Crystallite Size (nm)
Sample A	42.52	19.63
Sample B	26.81	39.64
Sample C	38.87	29.74
Sample D	22.84	42.04

XPS analysis was performed to identify the oxidation state of the surface elements of the iron ore samples. For samples A and D, the binding energy values (in eV) and valence state ratios of O are shown in Table 5. The XPS spectrum of the samples in the Fe2p and O1s regions are shown in Figure 5.

Figure 5A shows the Fe2p spectra of the samples A and D. For sample A, the characteristic peaks of Fe $2p_{3/2}$, Fe $2p_{3/2}$ satellite, Fe$2p_{1/2}$ and Fe $2p_{1/2}$ satellite were located at 710.8, 718.7, 724.7 and 732.5

eV, respectively [30]. Whereas for sample D, the characteristics peaks have shown a little higher BE values (Fe $2p_{3/2}$ (710.9), Fe $2p_{3/2}$ satellite (718.9), Fe$2p_{1/2}$ (724.8) and Fe $2p_{1/2}$ satellite (732.9) (in eV)). The Fe$2p_{3/2}$ envelopes of both the samples showed multiple splitting spectra and are in good agreement with the Gupta and Sen (GS) multiples [30]. The fitted $2p_{3/2}$ envelopes in the samples correspond well to the presence of iron oxide in the Fe^{3+} state [31]. The result is in good agreement with the XRD data (Figures 3 and 4). The XPS spectrum of the Samples A and D in the O1s region is shown in Figure 5B. After deconvolution of the peaks, the O1s spectra peaks of both of the samples can be separated into three peaks. According to the literature [32], the peak at the lower BE around 529.9 eV corresponds to the lattice oxygen (denoted as Oβ); the second one at a bit higher BE value around 531.4 eV corresponds to the surface adsorbed oxygen (denoted as Oα); the third peak around 532.6 corresponds to the peak originates from chemisorbed water (indicated as Oγ). It is reported that surface chemisorbed oxygen (Oα) is highly active in the oxidation reaction because of its higher mobility than the lattice oxygen (Oβ) [33], and the high SCR activity could be correlated with the high relative concentration ratio of Oα/Oα + Oβ + Oγ on the surface of the catalyst. As shown in Table 5, the concentration ratio of Oα/(Oα + Oβ + Oγ) of sample A was 44%; whereas, the concentration ratio of sample D was 39%. Therefore, the higher NO conversion of sample A could be due to higher values of surface chemisorbed oxygen (Oα).

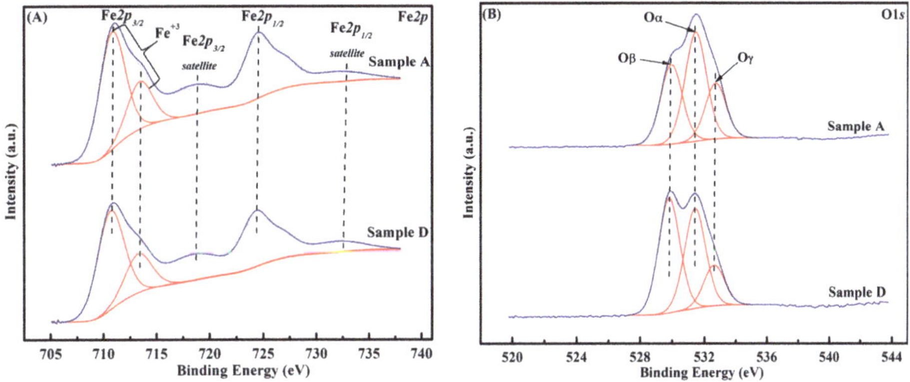

Figure 5. XPS spectrum of (**A**) Fe2p; (**B**) O1s of sample A and sample D, calcined at 250 °C.

Table 5. Comparison of samples A and D: binding energies and valance state ratio of O.

Samples	Oβ		Oα		Oγ		AO_α/A(Oα + Oβ + Oγ) (%)
	BE (eV)	Area	BE (eV)	Area	BE (eV)	Area	
Sample A	529.9	7841	531.4	10784	532.6	5483.9	44
Sample D	529.8	7016	531.4	6117	532.6	2469.5	39

To examine the redox properties of samples A and D, temperature-programmed reduction (H_2-TPR) experiments were conducted. Generally, the hydrogen consumption peaks located at 250–500 °C are attributed to surface lattice oxygen, and peaks at 500–700 °C are attributed to bulk lattice oxygen [34]. Figure 6A shows the H_2-TPR profiles of samples A and D. It can be seen that two well-separated reduction peaks were present in the H_2-TPR profiles of the catalysts. Surface lattice oxygen was credited with the first stage of the reduction process from hematite to magnetite (α-Fe_2O_3 → Fe_3O_4) in the low-temperature peak, and the bulk lattice oxygen was responsible for the reduction of magnetite to metallic iron (Fe_3O_4 → Fe) in the high-temperature peak. In point of fact, with a sample heating rate of 5.5 °C per minute and higher, two heavily overlapped peaks could be observed in the high-temperature region (as shown in Figure 6A). That is why the reduction of magnetite was believed to happen in a two-step magnetite reduction sequence, i.e., $2Fe_3O_4$ → 6FeO → 6Fe [35]. Sample D was characterized by four different reduction peaks at 297 °C, 368 °C, 528 °C and 638 °C, corresponding to the reduction of Fe_2O_3 to Fe_3O_4 (297 °C, 368 °C), Fe_3O_4 to FeO (528 °C), and FeO to Fe (638 °C) [36,37]. However, the results exhibited that the corresponding temperatures to the reduction peaks have shifted to lower values for the sample A, which indicated higher mobility of the oxygen species in the sample A. Since the reducibility can be indicated by the reduction peak temperature, the lower reduction peak temperature indicated the stronger reducibility. Therefore, the catalytic activity of sample A was better than that of sample D.

One of the main processes of the NH_3–SCR reaction is the adsorption and activation of ammonia on the acid sites that are present on a catalyst's surface. To examine this, NH_3–TPD experiments were conducted. It can be seen from Figure 6B that the samples have three NH_3 desorption peaks from 50~500 °C. For sample A, two desorption peaks can be observed at 135 °C and 192 °C, which can be attributed to NH^{4+} bound to weak Brønsted acid sites [38]. Whereas, the third peak is centered at 267 °C (strong peak), respectively. The peak above 200 °C can be ascribed to NH^{4+} bound to strong Brønsted acid sites, and the coordinated NH_3 bound to Lewis acid sites [39,40]. So NH_3–TPD results showed the presence of both Lewis and weak Brønsted acid sites in the catalysts for ammonia adsorption. It is interesting to note that the peak area of Lewis acid sites and weak Brønsted acid sites are obviously larger than those of sample D. The peak area implies the amount of ammonia adsorption in the sample [19], and this indicated that sample D has less number of acid sites than the sample A. Therefore, the catalytic activity and selectivity of the sample A was higher than the sample D.

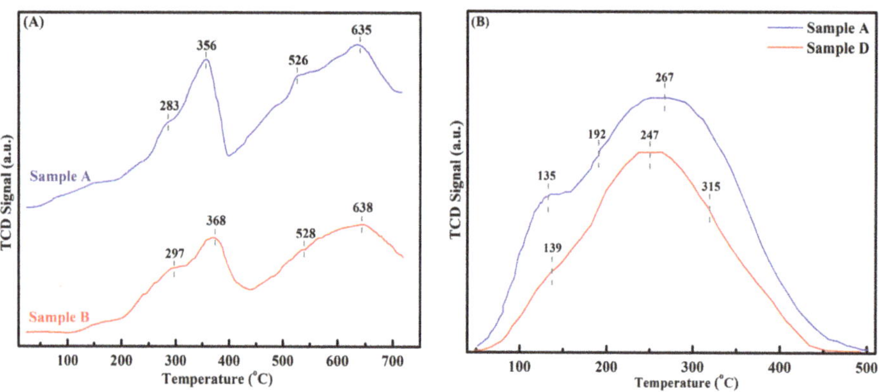

Figure 6. (**A**) H_2–TPR profiles; (**B**) NH_3–TPD profiles of samples A and D, calcined at 250 °C.

A scanning electron microscope was used to observe the influence of calcination on the micromorphology of the iron ore sample A (Figure 7). It can be seen from Figure 7A,B that calcination of the sample at 250 °C resulted in the formation of an amorphous structure of the iron ore, which is also shown by the XRD results (Figure 4A). An amorphous structure usually has a higher surface area than a crystallized structure [21].

Therefore, the BET surface area of the sample A calcined at 250 °C is much higher than that of the non-calcined sample (Table 3). However, with a rise in calcination temperature the aggregated particles appeared due to the sintering phenomenon which becomes a worse further increase in calcination temperature (Figure 7C,D). This sintering phenomenon caused a decrease in the surface area (Table 3) and consequently the number of active sites for the SCR reaction. So, the calcination temperature of 250 °C was found suitable for the iron ore sample to achieve optimum NH_3–SCR performance at low temperatures.

Figure 7. SEM images of sample A: (**A**) Not Calcined; calcined at (**B**) 250 °C; (**C**) 350 °C; and (**D**) 450 °C.

2.4. Influence of H_2O and SO_2 on SCR Activity

The influence of SO_2 and H_2O on the SCR activity was investigated at 200 °C (Figure 8). To test SO_2 tolerance, 150 ppm SO_2 was added to the simulated flue gas after 50 min of stable reaction. It can be seen that the NO conversion showed a negligible decrease and the NO conversion still remained above 85% after 100 min. However, the NO conversion decreased from 85% to 65% in 200 min, and after that, the NO conversion became almost stable. When the SO_2 feed was stopped after 400 min, the NO conversion was restored to some extent. Compared with the previous reports [41,42], our experiments exhibited that the iron ore sample had better tolerance to SO_2 at 200 °C in one hour of added SO_2. However, after this time, substantial deactivation occurred due to the formation of sulfates on the surface of the catalyst, which resulted in a sharp decrease in NO conversion. The coexistence of H_2O and SO_2 led to more catalyst deactivation to the addition of a single gas (SO_2), because H_2O can compete with the gaseous NH_3 for the active sites [43]. When 5 vol % H_2O and 150 ppm SO_2 were added to the simulated flue gas after 50 min of stable reaction, the synergy effects of the H_2O and SO_2 could be seen on the NO conversion of the iron ore catalyst.

To authenticate the formation of ammonium sulfates the FT-IR analysis of the fresh and the deactivated sample was carried out, and the results are shown in Figure 9. By comparison, some new bands at 1117, 1400 and 3250 cm^{-1} were found in the spectra of the sample poisoned by the SO_2 gas. The new bands at 1117 and 1400 cm^{-1} are credited to the adsorption peak of SO_4^{2-} [44,45], whereas, the band at 3250 cm^{-1} is ascribed to the corresponding N–H stretching vibration of NH^{4+} ions [42]. These results showed that sulfate salts such as $(NH_4)_2SO_4$ and NH_4HSO_4 were formed during the SCR reaction on the iron ore sample, which could cause pore plugging and result in a reduction of the catalytic activity.

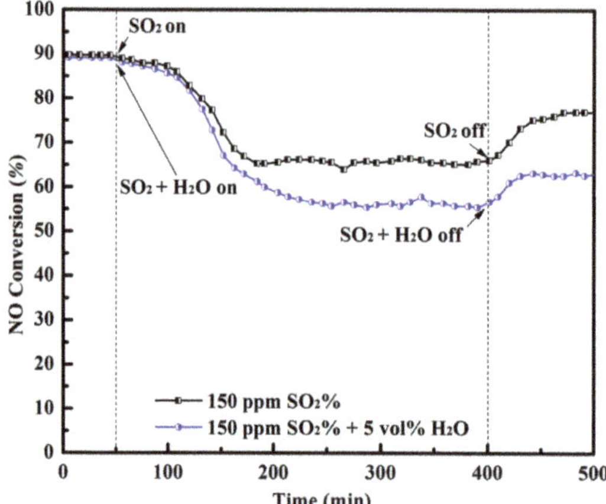

Figure 8. Influence of H_2O and SO_2 on SCR activity at 200 °C. Reaction conditions: 500 ppm NO, 500 ppm NH_3, 3 vol % O_2, 150 ppm SO_2, 5 vol % H_2O (when used), flow rate of 145 mL·min^{-1}, N_2 balance, and GHSV = 20,000 h^{-1}.

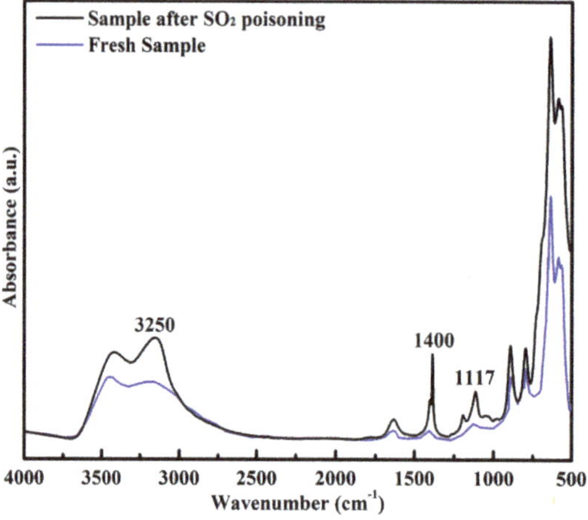

Figure 9. FT-IR spectra of fresh sample and the sample after SO_2 poisoning.

3. Materials and Methods

3.1. Materials and Reagents

In this study, four iron ore raw materials from different locations were taken and processed to be used as catalysts. Among them, two were mined in Australia and the other two in Brazil. To prepare the catalysts, the raw materials were processed as follows: At first, the iron ore raw materials were dried at 105 °C for 4 h, and then ZM-200 Ultra Centrifugal Mill (Retsch, Germany) was used to ground the ore to make 100-mesh powder. Then, these ground samples were again dried at 105 °C for 5 h. Finally, these ground samples were calcined at 250 °C, 350 °C and 450 °C, respectively, for 6 h to obtain the catalysts.

3.2. Experimental Setup and Governing Equations

The experimental setup consisted of a simulated flue gas system, an electrically heated test rig, a fixed-bed quartz reactor and a flue gas analyzer system. The reactor was made of quartz glass (6 mm i.d × 500 mm length). A K-type thermocouple of 2 mm diameter with an accuracy of 2.5 °C was inserted into the reactor to obtain the data of the flue gas temperature from the reactor inlet and outlet. The catalyst powder (500 mg) was loaded inside the reactor and was heated to the desired temperature by an electrically heated test rig. The reaction gas consisted of 500 ppm NO, 500 ppm NH_3, 3 vol % O_2, 150 ppm SO_2 (when used), 5 vol % H_2O (when used), with balance N_2 at a total flow rate of 145 mL·min^{-1} and gas hourly space velocity (GHSV) of 20,000 h^{-1}. Water vapors were generated by passing N_2 through a heated gas-wash bottle containing deionized water. Mass flow controllers (MFCs) (CS200A, CS200D, Sevenstar, Beijing, China) were used to control the flow of simulated flue gas. A flue gas analyzer (Testo 350, Lenzkirch, Germany) was used to constantly monitor the concentrations of NO, NO_2, O_2 and SO_2. To test N_2 selectivity, the outlet gas compositions were detected by GC-14C with Porapak Q column (Shimadzu, Kyoto, Japan) and FT-IR spectrometer (Vertex70v, Billerica, USA) with a scanning range of 4000–400 cm^{-1}, 0.15 cm^{-1} resolution and an average of 32 scans for each spectrum. By using the concentration of gases at steady state, the NO conversion and N_2 selectivity were calculated according to the following equations:

$$\text{NO conversion } (\%) = \frac{[\text{NO}]_{in} - [\text{NO}]_{out}}{[\text{NO}]_{in}} \times 100\% \quad (1)$$

$$\text{N}_2 \text{ selectivity } (\%) = \frac{[\text{NO}]_{in} + [\text{NH}_3]_{in} - [\text{NO}]_{out} - [\text{NH}_3]_{out} - 2[\text{N}_2\text{O}]_{out}}{[\text{NO}]_{in} + [\text{NH}_3]_{in} - [\text{NO}]_{out} - [\text{NH}_3]_{out}} \times 100\% \quad (2)$$

where $[NO]_{in}$ represents the NO concentration at the inlet of the reactor (ppm), and $[NO]_{out}$ represents the NO concentration at the outlet of the reactor (ppm).

Both NO and NH_3 oxidation tests were conducted in the same reactor with a 500 mg sample. For NO oxidation, the feed gas consisted of 500 ppm NO, 3 vol % O_2, and balance N_2. For NH_3 oxidation, the feed gas consisted of 500 ppm NH_3, 3 vol % O_2, and balance N_2. The total flow rate was 145 mL·min^{-1}. The NO to NO_2 conversion percentage was calculated using the following equation:

$$\text{NO to NO}_2 \text{ conversion } (\%) = \frac{[\text{NO}_2]_{out}}{[\text{NO}]_{in}} \times 100\% \quad (3)$$

The NH_3 conversion percentage was calculated using the following equation:

$$\text{NH}_3 \text{ conversion } (\%) = \frac{[\text{NH}_3]_{in} - [\text{NH}_3]_{out}}{[\text{NH}_3]_{in}} \times 100\% \quad (4)$$

where $[NH_3]_{in}$ and $[NH_3]_{out}$ were the concentration of NH_3 in the inlet and outlet flue gases, respectively (ppm).

The Scherrer formula is shown as follows.

$$D = 0.94 \times \lambda / (\beta \times Cos\theta) \quad (5)$$

where D represents the average diameter of crystallite, λ denotes the wavelength of the incident X-ray, β is the half-width of the diffraction peak, and θ is the diffraction angle.

3.3. Characterization Used

X-ray powder diffraction (XRD) measurements were carried out to determine the crystalline structures of the catalysts with CuKα radiation on a D8 Advance X-ray diffractometer (Bruker, Billerica,

USA). The scan rate of diffraction pattern was 1° min^{-1}, with a resolution of 0.02° and the diffraction pattern was taken in a 2θ range of 10~90°.

An XRF-1800 sequential X-ray fluorescence spectrometer (Shimadzu, Japan) was used to carry the elemental analysis of the ore samples.

An Autosorb-IQ3 (Quantachrome; Anton Paar, Austria) analyzer was used to determine the Brunauer-Emmett-Teller (BET) surface properties of the catalysts. To examine the surface characteristics of the catalysts, the samples were undergone by N_2 adsorption at 77 K, and after that, the samples were degassed under vacuum for 12 h at 180 °C.

X-ray photoelectron spectroscopy (XPS) analysis was performed on an AXIS Ultra DLD (Shimadzu Kratos, Kyoto, Japan) X-ray photoelectron spectrometer with a spherical mirror and concentric hemispherical detector operating at constant pass energy (PE = 46.95 eV). All binding energies (BE) were referenced to the C1s line at 284.6 eV.

A Scanning electron microscope (SEM) S-3400 (Hitachi, Tokyo, Japan) was used to study the micromorphologies of the catalysts.

An AutoChem II 2920 (Micrometrics, Norcross, USA) instrument was used to conduct temperature-programmed reduction (H_2-TPR) experiments. Each catalyst (100 mg) was placed in a quartz U-tube reactor to conduct the experiment. The samples were pretreated in He at 100 °C for 1 h before reduction and then cooled to a temperature of 50 °C. Then the samples were heated from 50~700 °C with a heating rate of 5.5 °C·min^{-1} and were simultaneously introduced to a mixture of gases (90% Ar and 10% H_2) with a flow rate of 0.03 L·min^{-1}. A thermal conductivity detector (TCD) was used to determine the content of H_2 in the effluent gas.

An AutoChem II 2920 (Micrometrics, USA) instrument was used to conduct temperature-programmed desorption (NH_3–TPD) experiments. Each catalyst (100 mg) was placed in a quartz U-tube reactor to conduct the experiment. The samples were pretreated at 100 °C in He gas atmosphere with a flow rate of 0.03 L·min^{-1} for 2 h. After that, the samples were cooled to a temperature of 50 °C, and at this temperature, the samples were fed with a mixture of gases (10% NH_3 and 90% He) until saturation. Then, He gas was used for cleaning the samples. After this, the samples were heated to 550 °C with a heating rate of 10 °C·min^{-1}.

A Fourier transform infrared spectrometer Nicolet 6700 (Thermo Fisher, Waltham, USA) with spectral range 4000–400 cm^{-1} was used to identify the sulfur-based species present on the surface of the deactivated catalyst.

4. Conclusions

In this work, iron ore raw materials from different locations were taken and processed to be used as catalysts for the NH_3–SCR of NO at low temperatures. The existence of the γ-Fe_2O_3 (maghemite) phase along with a suitable calcination temperature was found crucial for the NH_3–SCR performance of the iron ore at low temperatures. It was found that sample A (composed mainly of α-Fe_2O_3 and γ-Fe_2O_3) calcined at 250 °C exhibited poor crystallinity, large surface area, high concentration of surface-adsorbed oxygen, good reducibility, lots of acid sites, as well as adsorption of the reactants, which brought about excellent SCR activity (above 80% at 170–350 °C) and N_2 selectivity (above 90% up to 250 °C) at low temperatures. However, the addition of H_2O and SO_2 in the feed gas showed some adverse effects on the SCR activity.

FT-IR analysis confirmed the presence of sulfate salts ($(NH_4)_2SO_4$ and NH_4HSO_4) during the SCR reaction in the presence of SO_2, which could cause pore plugging and result in the suppression of the catalytic activity.

Author Contributions: Conceptualization, E.W.; methodology, N.H., E.W.; characterization, N.H., S.F; investigation, N.H., E.W.; writing—original draft preparation, N.H., E.W.; writing—review and editing, E.W., N.H.; visualization, S.F., N.H.; supervision, E.W; funding acquisition, E.W.

Funding: This research was funded by the National Natural Science Foundation of China (grant no. 50676057).

Acknowledgments: The authors would like to thank National Natural Science Foundation of China for the research funding. In addition, the authors are highly thankful to Muhammad Tuoqeer Anwar (SJTU), Nabila Mehwish (SJTU), and Muhammad Ashraf (SJTU) for providing the technical support for this study.

Conflicts of Interest: The authors state no conflict of interest.

References

1. Skalska, K.; Miller, J.S.; Ledakowicz, S. Trends in NO_x abatement: A review. *Sci. Total Environ.* **2010**, *408*, 3976–3989. [CrossRef] [PubMed]
2. Husnain, N.; Wang, E.; Li, K.; Anwar, M.T.; Mehmood, A.; Gul, M.; Li, D.; Jinda, M. Iron oxide-based catalysts for low-temperature selective catalytic reduction of NO_x with NH_3. *Rev. Chem. Eng.* **2019**, *35*, 239–264. [CrossRef]
3. Damma, D.; Ettereddy, P.R.; Reddy, B.M.; Simirniotis, P.G. A review of low temperature NH_3-SCR for removal of NO_x. *Catalyst* **2019**, *9*, 349. [CrossRef]
4. Phil, H.H.; Reddy, M.P.; Kumar, P.A.; Ju, L.K.; Hyo, J.S. SO_2 resistant antimony promoted V_2O_5/TiO_2 catalyst for NH_3-SCR of NO_x at low temperatures. *Appl. Catal. B Environ.* **2008**, *78*, 301–308. [CrossRef]
5. Rivas, F.C.; Iznaga, I.R.; Berlier, G.; Ferro, D.T.; Rosabal, B.C.; Petranovskii, V. Fe speciation in iron modified natural zeolites as sustainable environmental catalysts. *Catalysts* **2019**, *9*, 866. [CrossRef]
6. Dunn, J.P.; Koppula, P.R.; Stenger, H.G.; Wachs, I.E. Oxidation of sulfur dioxide to sulfur trioxide over supported vanadia catalysts. *Appl. Catal. B Environ.* **1998**, *19*, 103–117. [CrossRef]
7. Bauerle, G.L.; Wu, S.C.; Nobe, K. Catalytic reduction of nitric oxide with ammonia on vanadium oxide and iron-chromium oxide. *Ind. Eng. Chem. Pro. Res. Dev.* **1975**, *14*, 268–273. [CrossRef]
8. Balle, P.; Geiger, B.; Kureti, S. Selective catalytic reduction of NO_x by NH_3 on Fe/HBEA zeolite catalysts in the oxygen-rich exhaust. *Appl. Catal. B Environ.* **2009**, *85*, 109–119. [CrossRef]
9. Liu, C.; Shi, J.W.; Gao, C.; Niu, C. Manganese oxide-based catalysts for low-temperature selective catalytic reduction of NO_x with NH_3: A review. *Appl. Catal. A Gen.* **2016**, *522*, 54–69. [CrossRef]
10. Ma, L.; Li, J.; Ke, R.; Fu, L. Catalytic performance, characterization, and mechanism study of $(Fe_2SO_4)_3/TiO_2$ catalyst for selective catalytic reduction of NO_x by ammonia. *J. Phys. Chem C.* **2011**, *115*, 7603–7612. [CrossRef]
11. Zhihang, C.; Wang, F.; Li, H.; Wang, L.; Li, X. Low-temperature selective catalytic reduction of NO_x with NH_3 over Fe-Mn mixed-oxide catalysts containing $Fe_3Mn_3O_8$ phase. *Ind. Eng. Chem. Res.* **2011**, *51*, 202–212.
12. Putluru, S.S.R.; Schill, L.; Jensen, A.D.; Siret, B.; Tabaries, F.; Fehrmann, R. Mn/TiO_2 and Mn-Fe/TiO_2 catalysts synthesized by deposition precipitation-promising for selective catalytic reduction of NO with NH_3 at low temperatures. *Appl. Catal. B Environ.* **2015**, *165*, 628–635. [CrossRef]
13. Wu, S.; Yao, X.; Zhang, L.; Cao, Y.; Zou, W.; Li, L.; Ma, K.; Tang, C.; Gao, F.; Dong, L. Improved low-temperature NH_3-SCR performance of $FeMnTiO_x$ mixed oxide with CTAB-assisted synthesis. *Chem. Commun.* **2015**, *51*, 3470–3473. [CrossRef] [PubMed]
14. Wu, S.; Zhang, L.; Wang, X.; Zou, W.; Cao, Y.; Sun, J.; Tang, C.; Gao, F.; Deng, Y.; Dong, L. Synthesis, characterization and catalytic performance of $FeMnTiO_x$ mixed oxides catalyst prepared by a CTAB-assisted process for mid-low temperature NH_3-SCR. *Appl. Catal. A Gen.* **2015**, *505*, 235–242. [CrossRef]
15. Zhang, Y.; Zheng, Y.; Wang, X.; Lu, X. Preparation of Mn-FeO_x/CNTs catalysts by redox co-precipitation and application in low-temperature NO reduction with NH_3. *Catal. Commun.* **2015**, *62*, 57–61. [CrossRef]
16. Zhang, Y.; Zheng, Y.; Zou, H.; Zhang, X. One-step synthesis of ternary MnO_2-Fe_2O_3-CeO_2-Ce_2O_3/CNT catalysts for use in low-temperature NO reduction with NH_3. *Catal. Commun.* **2015**, *71*, 46–50. [CrossRef]
17. Salama, W.; El, A.M.; Gaupp, R. Spectroscopic characterization of iron ores formed in different geological environments using FTIR, XPS, Mössbauer spectroscopy and thermoanalyses. *Spectrochim. Acta Part A Mol. Biomol. Spectrosc.* **2015**, *136*, 1816–1826. [CrossRef]
18. Park, T.S.; Jeong, S.K.; Hong, S.H.; Hong, S.C. Selective catalytic reduction of nitrogen oxides with NH_3 over natural manganese ore at low-temperature. *Ind. Eng. Chem. Res.* **2001**, *40*, 4491–4495. [CrossRef]
19. Zhu, B.; Yin, S.; Sun, Y.; Zhu, Z.; Li, J. Natural manganese ore catalyst for low-temperature selective catalytic reduction of NO with NH_3 in the coke-oven flue gas. *Environ. Sci. Pollut. Res.* **2017**, *24*, 24584–24592. [CrossRef]
20. Gui, K.; Liang, H.; Wang, F.; Yao, G. Low-temperature selective catalytic reduction of NO on an iron ore catalyst in a magnetically fluidized bed. *Chem. Eng. Tech.* **2015**, *38*, 1537–1542. [CrossRef]

21. Cao, F.; Su, S.; Wang, P.; Hu, S.; Sun, L.; Zhang, A. The activity and mechanism study of Fe–Mn–Ce/γ-Al_2O_3 catalyst for low-temperature selective catalytic reduction of NO with NH_3. *Fuel* **2015**, *139*, 232–239. [CrossRef]
22. Liu, F.; He, H.; Zhang, C. Selective catalytic reduction of NO_x with NH_3 over iron titanate catalyst: Catalytic performance and characterization. *Appl. Catal. B Environ.* **2010**, *96*, 408–420. [CrossRef]
23. Liu, C.; Yang, S.; Ma, L.; Peng, Y.; Hamidreza, A.; Chang, H.; Li, J. Comparison on the performance of α-Fe_2O_3 and γ-Fe_2O_3 for selective catalytic reduction of nitrogen oxides with ammonia. *Catal. Lett.* **2013**, *143*, 697–704. [CrossRef]
24. Wang, D.; Yang, Q.; Li, X.; Peng, Y.; Li, B.; Si, W.; Lu, C.; Gan, L. Preparation of γ-Fe_2O_3 catalysts and their de-NO_x performance: Effects of precipitation conditions. *Chem. Eng. Technol.* **2018**, *41*, 1019–1026. [CrossRef]
25. Cao, D.; Li, H.; Pan, L.; Li, J.; Wang, X.; Jing, P.; Cheng, X.; Wang, W.; Wang, J.; Liu, Q. High saturation magnetization of γ-Fe_2O_3 nano-particles by a facile one-step synthesis approach. *Sci. Rep.* **2016**, *6*, 1–9. [CrossRef]
26. Wang, X.; Gui, K. Fe_2O_3 particles as superior catalysts for low temperature selective catalytic reduction of NO with NH_3. *J. Environ. Sci.* **2013**, *25*, 2469–2475. [CrossRef]
27. Gao, F.; Kollar, M.; Kukkadapu, R.K.; Washton, N.M.; Wang, Y.; Szanyi, J.; Peden, C.H.F. Fe/SSZ-13 as an NH_3-SCR catalyst: A reaction kinetics and FTIR/Mössbauer spectroscopic study. *Appl. Catal. B Environ.* **2015**, *164*, 407–419. [CrossRef]
28. Gao, F.; Zheng, Y.; Kukkadapu, R.K.; Wang, Y.; Walter, E.D.; Schwenzer, B.; Szanyi, J.; Peden, C.H.F. Iron loading effects in Fe/SSZ-13 NH_3-SCR catalysts: Nature of the Fe ions and structure-function relationships. *ACS Catal.* **2016**, *6*, 2939–2954. [CrossRef]
29. Wang, A.; Wang, Y.; Walter, E.D.; Washton, N.M.; Guo, Y.; Lu, G.; Peden, C.H.F.; Gao, F. NH_3-SCR on Cu, Fe and Cu + Fe exchanged beta and SSZ-13 catalysts: Hydrothermal aging and propylene poisoning effects. *Catal. Today* **2019**, *320*, 91–99. [CrossRef]
30. Omran, M.; Fabritius, T.; Elmahdy, A.M.; Abdel-Khalek, N.A.; Aref, M.E.; Elmanawi, A.H. XPS and FTIR spectroscopic study on microwave treated highphosphorus iron ore. *Appl. Surf. Sci.* **2015**, *345*, 127–140. [CrossRef]
31. Grosvenor, A.P.; Kobe, B.A.; Biesinger, M.C.; Mclntyre, N.S. Investegation of multiplet splitting of Fe 2p XPS spectra and bonding in iron compounds. *Surf. Interface Anal.* **2004**, *36*, 1564–1574. [CrossRef]
32. Xu, L.; Yang, Q.; Hu, L.; Wang, D.; Peng, Y.; Shao, Z.; Lu, C.; Li, J. Insights over titanium modified $FeMgO_x$ catalysts for selective catalytic reduction of NO_x with NH_3: Influence of precursors and crystalline structures. *Catalysts* **2019**, *9*, 560. [CrossRef]
33. Wu, Z.; Jin, R.; Liu, Y.; Wang, H. Ceria modified MnO_x/TiO_2 as a superior catalyst for NO reduction with NH_3 at low temperature. *Catal. Commun.* **2008**, *9*, 2217–2220. [CrossRef]
34. Pineau, A.; Kanari, N.; Gaballah, I. Kinetics of reduction of iron oxides by H_2. *Thermochim. Acta* **2006**, *447*, 89–100. [CrossRef]
35. Jozwiak, W.K.; Kaczmarek, E.; Maniecki, T.P.; Ignaczak, W.; Maniukiewicz, W. Reduction behavior of iron oxides in hydrogen and carbon monoxide atmospheres. *Appl. Catal. A Gen.* **2007**, *326*, 17–27. [CrossRef]
36. Yang, S.; Guo, Y.; Yan, N.; Wu, D.; He, H.; Qu, Z.; Yang, C.; Zhou, Q.; Jia, J. Nanosized cation-deficient Fe-Ti spinel: A novel magnetic sorbent for elemental mercury capture from flue gas. *Appl. Mater. Interface* **2011**, *3*, 209–217. [CrossRef] [PubMed]
37. Hou, B.; Zhang, H.; Li, H.; Zhu, Q. Study on the kinetics of iron oxide reduction by hydrogen. *Chin. J. Chem. Eng.* **2012**, *20*, 10–17. [CrossRef]
38. Ning, R.; Chen, L.; Li, E.; Liu, X.; Zhu, T. Applicability of V_2O_5-WO_3/TiO_2 catalysts for the SCR denitrification of alumina calcining flue gas. *Catalysts* **2019**, *9*, 220. [CrossRef]
39. Cheng, L.S.; Yang, R.T.; Chen, N. Iron oxide and chromia supported on titania-pillared clay for selective catalytic reduction of nitric oxide with ammonia. *J. Catal.* **1996**, *164*, 70–78. [CrossRef]
40. Long, R.Q.; Yang, R.T. Selective catalytic oxidation of ammonia to nitrogen over Fe_2O_3–TiO_2 prepared with a sol-gel method. *J. Catal.* **2002**, *207*, 158–165. [CrossRef]
41. Li, Q.; Yang, H.; Ma, Z.; Zhang, X. Selective catalytic reduction of NO with NH_3 over CuOX-carbonaceous materials. *Catal. Commun.* **2012**, *17*, 8–12. [CrossRef]
42. Chen, W.; Hu, F.; Qin, L.; Han, J.; Zhao, B.; Tu, Y.; Yu, F. Mechanism and Performance of the SCR of NO with NH_3 over Sulfated Sintered Ore Catalyst. *Catalysts* **2019**, *9*, 90. [CrossRef]

43. Yang, S.; Wang, C.; Chen, J.; Peng, Y.; Ma, L.; Chang, H. A novel magnetic Fe–Ti–V spinel catalyst for the selective catalytic reduction of NO with NH_3 in a broad temperature range. *Catal. Sci. Technol.* **2012**, *2*, 915–917. [CrossRef]
44. Won, P.; Hee, K.; Chang, S. SO_2 durability enhancement of ball-milled V/TiO_2 catalyst. *Ind. Eng. Chem. Res.* **2010**, *16*, 283–287.
45. Pietrogiacomi, D.; Magliano, A.; Ciambelli, P.; Sannino, D.; Campa, M.C.; Indovina, V. The effect of sulphation on the catalytic activity of CoO_x/ZrO_2 for NO reduction with NH_3 in the presence of O_2. *Appl. Catal. B Environ.* **2009**, *89*, 33–40. [CrossRef]

Article

Fe Speciation in Iron Modified Natural Zeolites as Sustainable Environmental Catalysts

Fernando Chávez Rivas [1], Inocente Rodríguez-Iznaga [2,*], Gloria Berlier [3,*], Daria Tito Ferro [4], Beatriz Concepción-Rosabal [2] and Vitalii Petranovskii [5]

1 Instituto Politécnico Nacional, ESFM, Departamento de Física, UPALM, Zacatenco, Ciudad de México 07738, Mexico; fchavez@esfm.ipn.mx

2 Instituto de Ciencia y Tecnología de Materiales (IMRE)-Universidad de La Habana, Zapata y G s/n, La Habana, C.P. 10400, Cuba; beatriz@imre.uh.cu

3 Dipartimento di Chimica and NIS Centre, Università di Torino, Via P. Giuria 7, 10125 Torino, Italy

4 Centro Nacional de Electromagnetismo Aplicado (CNEA), Universidad de Oriente, Ave. Las Américas s/n, Santiago de Cuba, C.P. 90400, Cuba; dariat@uo.edu.cu

5 Departamento de Nanocatálisis, Centro de Nanociencias y Nanotecnología (CNyN), Universidad Nacional Autónoma de México, Carretera Tijuana-Ensenada, Km 107. Ensenada, C.P. 22860 BC, Mexico; vitalii@cnyn.unam.mx

* Correspondence: inocente@imre.uh.cu (I.R.-I.); gloria.berlier@unito.it (G.B.)

Received: 24 September 2019; Accepted: 16 October 2019; Published: 19 October 2019

Abstract: Natural purified mordenite from Palmarito de Cauto (ZP) deposit, Cuba, was subjected to a hydrothermal ion exchange process in acid medium with Fe^{2+} or Fe^{3+} salts (Fe^{2+}ZP and Fe^{3+}ZP). The set of samples was characterized regarding their textural properties, morphology, and crystallinity, and tested in the NO reduction with CO/C_3H_6. Infrared spectroscopy coupled with NO as a probe molecule was used to give a qualitative description of the Fe species' nature and distribution. The exchange process caused an increase in the iron loading of the samples and a redistribution, resulting in more dispersed Fe^{2+} and Fe^{3+} species. When contacted with the NO probe, Fe^{2+}ZP showed the highest intensity of nitrosyl bands, assigned to NO adducts on isolated/highly dispersed Fe^{2+}/Fe^{3+} extra-framework sites and Fe_xO_y clusters. This sample is also characterized by the highest NO sorption capacity and activity in NO reduction. Fe^{3+}ZP showed a higher intensity of nitrosonium (NO^+) species, without a correlation to NO storage and conversion, pointing to the reactivity of small Fe_xO_y aggregates in providing oxygen atoms for the NO to NO^+ reaction. The same sites are proposed to be responsible for the higher production of CO_2 observed on this sample, and thus to be detrimental to the activity in NO SCR.

Keywords: natural zeolite; mordenite; Iron exchange; FTIR-NO; HRTEM; NO reduction

1. Introduction

Transition metal ion-exchanged zeolites are being considered for practical applications in diesel and stationary combustion sources' emission control due to good stability at high temperatures and other qualities. Among other metals, iron has shown interesting catalytic properties and it is particularly attractive due to its low cost and non-toxic nature [1]. The activity of Fe-zeolites in the selective catalytic reduction (SCR) of NO_x has been studied in two main processes, i.e., with ammonia (NH_3-SCR) [2–8] and hydrocarbons (HC-SCR) as reducing agents. In the latter case, a variety of compounds, such as C_3H_6 [9], iso and n-C_8H_{18} [10], C_3H_8 [11], iso-C_4H_{10} [12], and C_2H_5OH [13], have been evaluated. Even though NH_3-SCR is recognized among the most efficient deNO_x processes, HC-SRC has the main advantage of using a gas mixture very similar to that found in exhausts [14,15], avoiding the use and related safety/environmental concerns of NH_3 sources. Natural zeolites, such as mordenite, have also drawn attention as sustainable materials due to their ion exchange properties, thermal stability, availability, and low price. Therefore, efficient and low-cost catalysts could be developed from natural mordenite modified with transition metal ions, such as iron.

Several types of iron species, such as isolated or binuclear ions, and iron oxide clusters stabilized in zeolite pores, have been proposed as active sites in NO_x and N_2O decomposition processes. Rutkowska et al. proposed that monomeric Fe^{3+} cations and small Fe_xO_y oligomers in Fe-beta zeolites are more active than larger iron oxide aggregates in the catalytic decomposition and reduction of N_2O [16]. Kim et al. pointed to the activity of small Fe^{2+}-enriched iron oxide clusters in zeolite pores, maximized by reduction with hydrogen, in the same reaction [17]. Similar oligonuclear Fe_xO_y clusters, coupled to isolated Fe^{3+} ions, were reported to be active in Fe-containing AlPO-5, where N_2O SCR was carried out with CH_4 in the presence of water vapor and oxygen [18]. The authors also pointed out that larger Fe_2O_3 particles are able to activate both SCR and methane combustion only at temperatures higher than 500 °C. Fe-exchanged MOR, FER, BETA, and ZSM-5 prepared with iron (III) acetylacetonate precursor were shown to present a mixture of Fe^{3+} and Fe^{2+}, with catalytic activity in the NO reduction with C_3H_6 dependent on the Fe^{2+}/Fe^{3+} ratio, especially at low temperatures [19]. Fe/MOR showed the highest NO_x adsorption capacity and catalytic activity in the 200 to 400 °C temperature range. These few (not exhaustive) examples give a taste about the complexity of iron species that can be formed within zeolite pores. Zeolite's inner (and external) surface is indeed intrinsically rich in sites (negatively charged framework oxygen atoms, defective silanols, and silanol nests) able to stabilize Fe sites with a variety of coordination, nucleation, and oxidation states, strongly affecting their catalytic activity and selectivity [20].

Our team has earlier reported the characterization of a purified natural mordenite from Palmarito de Cauto deposit (Cuba), ion-exchanged by hydrothermal treatments with Fe^{2+} and Fe^{3+} salts. The three samples (parent ZP, Fe^{2+}ZP, and Fe^{3+}ZP, see the experimental section for details) were characterized by diffuse reflectance UV-Vis and Mössbauer spectroscopies [21]. These techniques showed the presence of a large variety of iron species already in the not-exchanged parent material: Highly dispersed Fe ions in extra framework positions, octahedral Fe ions in oligomeric clusters of the Fe_xO_y type inside the channels, and larger iron oxide/hydroxide clusters and Fe_2O_3 particles located on the external surface of the zeolite crystals. After the hydrothermal exchange, the presence of Fe^{2+} and Fe^{3+} in the Fe^{2+}ZP and Fe^{3+}ZP samples, respectively, was confirmed by Mössbauer and UV-Vis. Moreover, iron oxide agglomerates are favored in Fe^{3+}ZP, and dispersed cationic species in Fe^{2+}ZP. Together with a plethora of oxy-hydroxides clusters/aggregates, Fe^{2+} were proposed to be present in extra-framework positions as a charge compensating cations. The formation of Brønsted sites, likely formed by insertion of Fe^{3+} ions in framework positions during hydrothermal treatment, was moreover assessed by infrared spectroscopy in Fe^{3+}ZP.

The use of natural zeolites as catalyst has an intrinsic limitation related to phase purity, which cannot be controlled as carefully as for synthetic ones [22–25]. This also implies that the interpretation of the results can be very complex. On the other hand, these are relatively low-cost catalyst materials and importantly, no templates need to be used for their preparation, thus lowering the environmental impact of the process. With this respect, we here report about a detailed physico-chemical characterization of the set of samples ZP, Fe^{2+}ZP, and Fe^{3+}ZP, trying to correlate the nature and distribution of present Fe species to the NO storage and conversion in the CO/C_3H_6 SCR of NO, which is used as a test reaction.

2. Results

2.1. Elemental Analysis

The iron content of the starting material ZP, and its exchanged forms Fe^{2+}ZP and Fe^{3+}ZP, are 1.55%, 3.06%, and 2.68%, respectively. As discussed with more detail in our earlier paper [21], iron is already present in the natural zeolite sample. However, after the hydrothermal exchange treatments, the amount of iron increases, as a result of ion exchange of Fe^{2+} and Fe^{3+} with cations present in the natural mordenite, mainly Ca^{2+} and Na^{+}.

2.2. Catalytic Activity in HC-SCR

The catalytic activity of the set of samples in the NO reduction with CO/C_3H_6 in the presence of oxygen was measured as a function of temperature, as summarized in Figures 1 and 2. Figure 1 shows the change in the amount of NO and CO_2 in the gas output, monitored from 25 up to 500 °C. Figure 2 reports the NO conversion, which is negligible below 170 °C for all samples. Before carrying out the catalytic test, a steady state concentration of the gas in the flow was reached, around 700 ppm for NO at the reactor outlet. Once this condition was reached, the increase of temperature caused the desorption of the NO absorbed on the catalyst's surface. This can be seen in Figure 1a as a positive peak between RT and 200 to 250 °C. In this interval, no NO conversion is thus observed in Figure 2. Above this temperature, the negative peaks in Figure 1a correspond to NO conversion, as described below. It is well known that adsorption is the primary step in a catalytic reaction. Accordingly, studies reported on the catalytic test with Fe-MOR [19] and Cu-MFI [26] showed that these catalysts both have great capacities for NO adsorption and high catalytic activity for the selective reduction of NO. In agreement with these reports, the amount of NO absorbed/desorbed during the first step of NO interaction (from RT to 200 °C, Figure 1a) is related to the adsorption capacity of the materials. Thus, Fe^{2+}ZP shows a higher NO adsorption capacity with respect to Fe^{3+}ZP and, even more, with the ZP sample (full, dashed and dotted lines, respectively).

Coming to the different catalytic performances of the samples, ZP shows the lowest activity, with a low conversion (8%) starting from 170 °C, as shown in Figure 2. On the contrary, the highest activity is observed for Fe^{2+}ZP sample (34%), followed by Fe^{3+}ZP (19%), indicating the importance of the ion exchange hydrothermal process in order to increase the population of active sites. The conversion of Fe^{2+}ZP is comparable to literature data obtained in similar conditions [19]. Moreover, Fe^{2+}ZP shows a reaction onset below 190 °C while Fe^{3+}ZP only starts to be significantly active in NO reduction at around 240 °C.

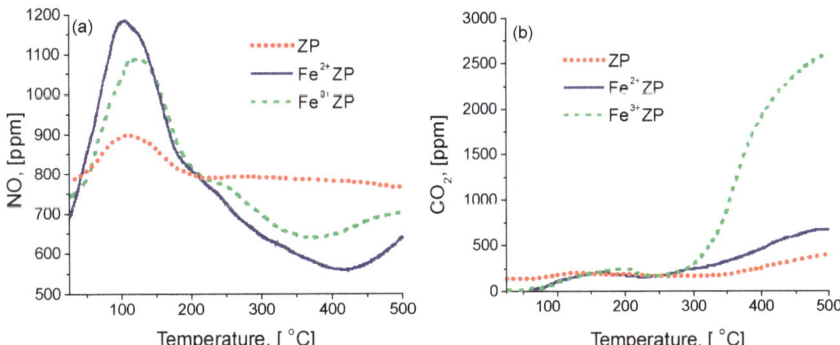

Figure 1. Behavior of the catalytic desorption–reduction of NO with CO/C_3H_6 and O_2 (**a**) and corresponding CO_2 formation (**b**) over the parent zeolite (ZP) and exchanged samples as function of temperature.

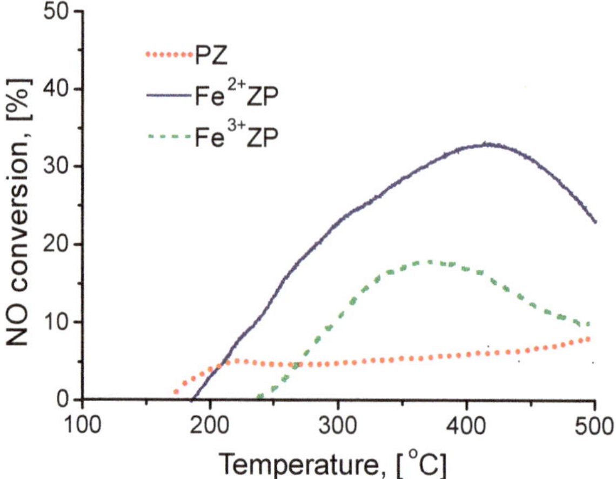

Figure 2. Dependence of NO conversion (%) with CO/C_3H_6 on the parent zeolite (ZP) and the exchanged samples as function of reaction temperature.

The formation of CO_2 during NO reduction with CO/C_3H_6 was also followed, and is presented in Figure 1b. The highest CO_2 production is observed on sample Fe^{3+}ZP, while Fe^{2+}ZP and ZP have similar CO_2 productivity. This trend is not correlated to the zeolite's activity in NO reduction, and can be explained by a direct oxidation of CO and C_3H_6 by O_2. This suggests that sample Fe^{3+}ZP, only mildly active in NO reduction, is instead a good oxidation catalyst. This reactivity could be related to the presence of small Fe_xO_y aggregates, providing oxygen atoms and reversible electron transfer involving Fe^{3+}/Fe^{2+} ions. The formation of these species after Fe^{3+} insertion in parent ZP was observed by UV-Vis spectroscopy in our previous report [21]. Thus, CO and/or C_3H_6 oxidation may be one of the reasons for the lower activity in NO reduction of Fe^{3+}ZP with respect to Fe^{2+}ZP.

Although the NO conversion levels achieved by using natural zeolite-based catalysts (included those studied in this work) are not elevated, the related studies provide important knowledge on the iron speciation in the samples and on their reactivity, as discussed in the following where a detailed physico-chemical characterization of the materials is reported.

2.3. XRD Analysis

The XRD patterns of the samples are shown in Figure 3. In parent ZP, the diffraction peaks of the mordenite structure are present, as can be seen in the figure where they are indexed to the (hkl) corresponding crystallographic planes. The diffraction pattern also shows peak related to impurity phases (quartz and clinoptinolite), as labeled in Figure 3. A small change in the intensity of the peaks is observed on Fe^{2+}ZP and Fe^{3+}ZP, as is often observed after ion exchange as a consequence of changes in the location and concentration of extra-framework cations in the zeolite channels [27,28]. This suggests that the crystalline structure of mordenite was not altered by the treatment with acid solutions, and that the zeolite is stable under applied conditions.

Besides the changes in intensity mentioned above, a small negative shift (about 0.14 degrees in average) with respect to parent ZP is observed for the position of most of the peaks of sample Fe^{3+}ZP, which is instead negligible for Fe^{2+}ZP (Table 1). This corresponds to an increase of the distances between crystallographic planes, and can be associated to an expansion of the zeolite matrix due to the incorporation of iron into the mordenite framework [29,30]. Indeed, this observation is in very good agreement with infrared evidences obtained for the formation of Brønsted Fe(OH)Si sites after hydrothermal treatment with Fe^{3+} [21].

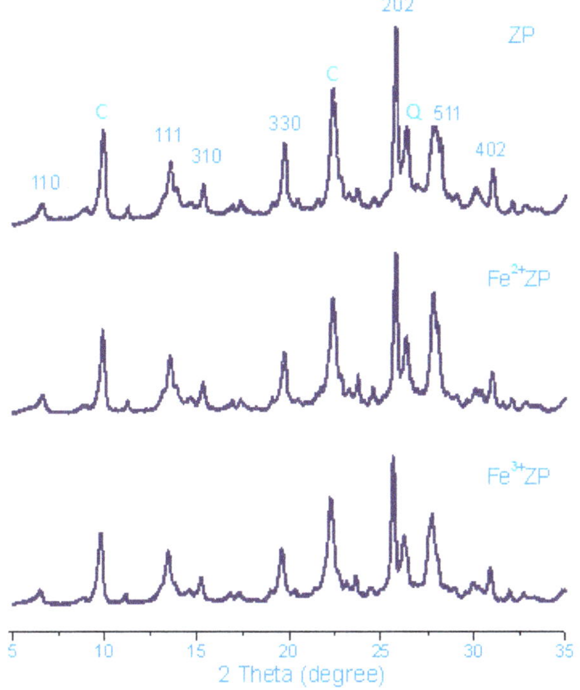

Figure 3. XRD patterns of the samples, with hkl labels of the mordenite structure. Q: quartz, C: clinoptilolite.

Table 1. Peaks position analysis of sample ZP and its exchange form, with the measured negative shift in brackets.

Plane		Peak Position (Shift)	
	ZP	Fe^{2+}ZP	Fe^{3+}ZP
111	13.57	13.55 (0.02)	13.44 (0.11)
310	15.41	15.40 (0.01)	15.27 (0.13)
330	19.77	19.74 (0.03)	19.59 (0.15)
202	25.76	25.76	25.63 (0.13)
402	31.06	31.04 (0.02)	30.89 (0.15)

2.4. Textural Properties

The textural properties of the three samples, determined by low-temperature nitrogen sorption measurements, are summarized in Table 2. A small increase in the specific surface area is observed passing from parent ZP to its exchanged forms. This trend can be related to the used exchange process, likely removing impurities or extra-phases (e.g., amorphous volcanic glass) from the zeolite particles' external surface. Moreover, the small increase in total surface area is accompanied by a mild decrease in the micropores' surface area and volume (see last columns of Table 2), in agreement with the formation of low solubility iron oxy-hydroxides phases inside the mordenite channels, as testified by our previous work [21]. This could be explained by a reaction of exchanged iron cations with atmospheric oxygen and OH- groups retained (adsorbed or occluded) in the zeolite pores, which are liberated during the ion exchange. Numerous studies are available in the literature on the occurrence of these processes, mainly during ion exchange [31–35]. Also, Fe^{3+} hydrolysis is expected to occur to a greater extent than that of Fe^{2+} [36,37]. However, we acknowledge that the observed changes are very small, being very close to the experimental error of 5%, so that formation of occluded particles should be limited to a restricted amount of small clusters/aggregates, in agreement with the absence of extra peaks in the XRD patterns of exchanged samples. No evidence for the formation of mesoporosity (often observed after acid treatment of zeolites) was observed, in agreement with the high stability of mordenite in these conditions.

Table 2. Textural properties of the studied samples.

Sample	SSA_{BET} (m^2/g)	SSA_{Lang} (m^2/g)	SSA_{ext} (m^2/g)	V_{mic} (cm^3/g)	A_{mic} (m^2/g)
ZP	244	312	3	0.109	241
Fe^{2+}ZP	259	331	26	0.105	233
Fe^{3+}ZP	256	328	26	0.104	230

SSA_{BET} = Specific surface area calculated using the standard Brunauer–Emmet–Teller (BET) method. SSA_{Lang} = Specific surface area, calculated from Langmuir method. SSA_{ext} = External surface area, calculated using the t-plot method. V_{mic} = Micropore volume, calculated using the t-plot method. A_{mic} = Micropore surface area, calculated using the t-plot method.

2.5. HRTEM Analysis

High-resolution TEM was used to investigate the effect of hydrothermal ion exchange on the morphology of the parent natural purified mordenite ZP. Measurements were carried out at 100 kV in order to avoid beam-induced damaging of the zeolite particles [38]. Figure 4 shows the TEM images of ZP sample, which is composed of particles of irregular shape, with sizes in the range of 200 to 400 nm, formed by the agglomeration of smaller particles. The left-hand panel of Figure 4 shows a magnification, allowing us to appreciate the crystalline nature of the zeolite. The measured lattice distance is 0.38 nm, in agreement with the (241) plane of mordenite [39].

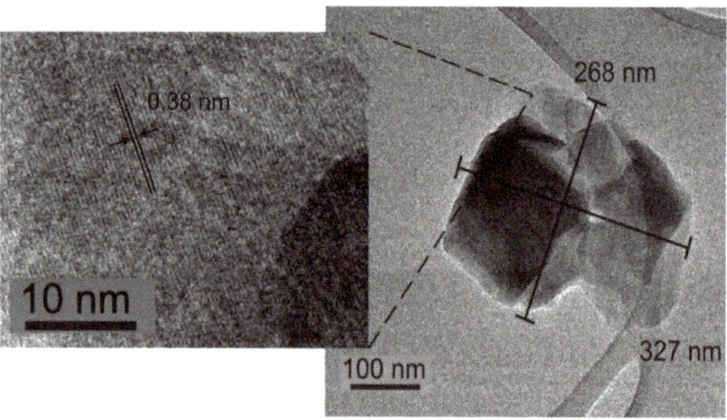

Figure 4. HRTEM images of ZP, with magnification of a particle border allowing appreciation of the mordenite lattice planes.

Representative TEM images of Fe^{2+}ZP and Fe^{3+}ZP samples are shown in Figures 5 and 6, respectively. In both cases, the morphology of the main particles is similar to that of ZP sample, with relatively large agglomerates of regular crystalline particles. Smaller particles (1 to 5 nm) can be appreciated in the figure magnifications, which could be related to the clinoptilolite extra phase on the basis of the measured crystallographic fringes.

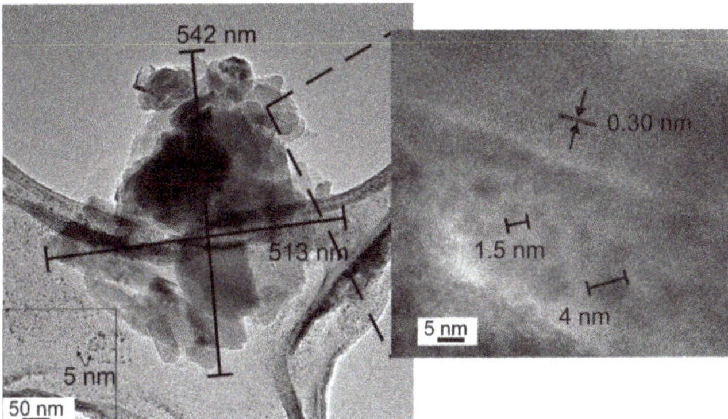

Figure 5. HRTEM images of Fe^{2+}ZP sample. Arrows in the bottom part of left-hand panel indicate the presence of small particles of extra phases. Some of these can be also seen in the right-hand magnification, where particles' size and crystalline fringes' distances were measured.

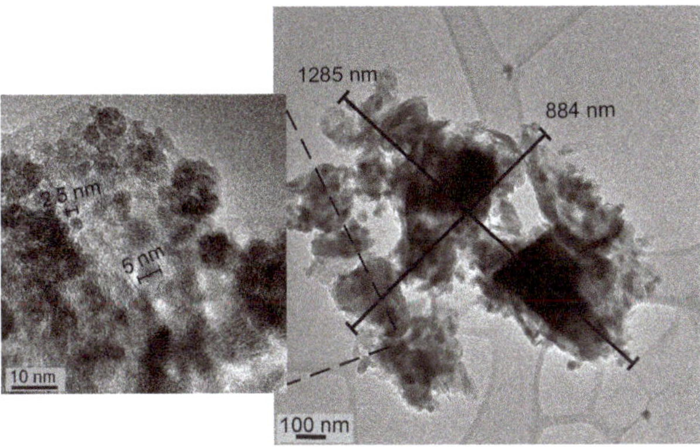

Figure 6. HRTEM images of Fe^{3+}ZP sample. The left-hand panel reports magnifications of the central image, showing the presence of nanometric-sized particles, likely related to the clinoptilolite extra phase.

EDS microanalysis carried out at different magnifications indicates a large heterogeneity in the iron distribution in all samples. Moreover, this analysis (spectra not reported) shows a difference in the cation exchange capability: While ZP and Fe^{3+}ZP samples show the presence of Mg, Na, K, Ca, and Fe, the Fe^{2+}ZP sample only presents Ca and Fe, in agreement with previous reports [21].

2.6. NO Adsorption Studies Followed by FTIR

The adsorption of NO or NO_2 on acid- and metal-exchanged zeolites followed by infrared spectroscopy is a well-known method for the characterization of exposed ions, through the formation of adsorbed complexes with characteristic vibrational fingerprints. Moreover, indirect indications about the Brønsted or metal ions' reactivity can be inferred by observing the formation of adsorbed NO/NO_2 reaction products, such as nitrosonium ions NO^+, nitrates NO_3^-, nitrites NO_2^-, and/or water [40–44].

NO adsorption experiments were carried out at room temperature (RT) on ZP, Fe^{2+}ZP, and Fe^{3+}ZP, after oxidation and reduction treatments. Modifications in the spectra upon dosing NO were observed in the OH stretching (υOH, from 3800 to 3000 cm^{-1}), in the nitrosonium and nitrosyl (from 2300 to 1700 cm^{-1}), and in the water bending $\delta(H_2O)$ and nitrate (from 1750 to 1300 cm^{-1}) regions. Spectra discussion will be mainly focused on the nitrosonium/nitrosyl region, showing the most intense and informative bands.

2.6.1. ZP Sample

The results obtained on the oxidized and reduced ZP sample are shown in Figure 7A and B, respectively. On oxidized ZP, three groups of bands are present: One centered at 2265 cm^{-1}, a second between 2200 and 1920 cm^{-1}, and a third one between 1920 and 1700 cm^{-1}. Some of these are also present in the reduced sample (bottom), though with lower intensity. The first group is formed by an intense band at 2265 cm^{-1}, with shoulders at 2282 and 2251 cm^{-1}, which can be assigned as N_2O formed upon reaction with NO on different Fe^{x+} ($x = 2, 3$) surface sites on small iron oxide particles [43,45–47].

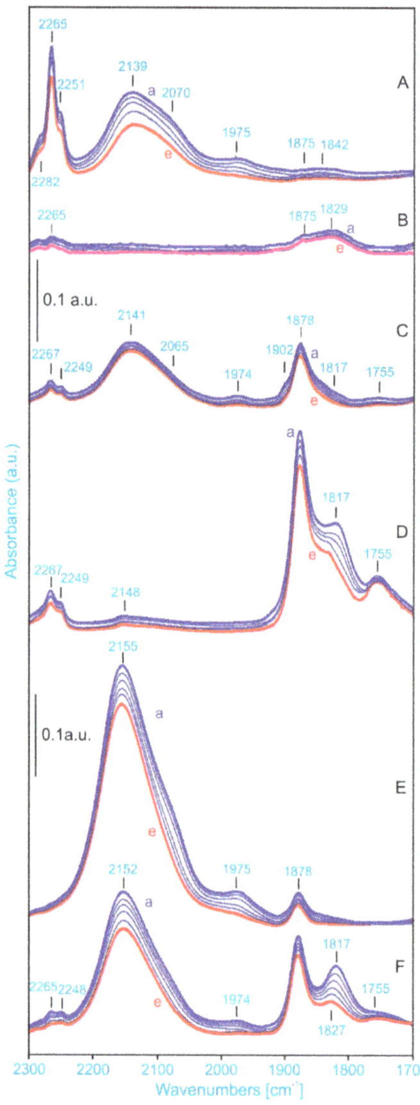

Figure 7. Infrared spectra in the nitrosonium/ nytrosyl region of NO adsorbed at RT on oxidized (**A**) and reduced (**B**) ZP sample, oxidized (**C**) and reduced (**D**) Fe^{2+}ZP sample, and oxidized (**E**) and reduced (**F**) Fe^{3+}ZP sample. Equilibrium P_{NO} = 0.4 mBar (a) and subsequent stepwise evacuation up to dynamic vacuum (e).

Coming to the second group of bands, it is composed of a broad absorption, with a maximum at 2139 cm^{-1} and a shoulder at 2070 cm^{-1}, with a very broad band centered at 1975 cm^{-1}, which are not present on the reduced sample. Similar bands have been assigned by different authors to NO^+ [48–50], formed by the reaction of NO with adsorbed oxygen atoms, and with the formation of NO_2, which then reacts with protons to give NO^+ and water [51]. Similarly, Rivallan et al. proposed that NO_2 is transformed into physisorbed HNO_3, which mediates the formation of NO^+ ions that exchange zeolite protons [44]. This interpretation is in agreement with the absence of these bands in the reduced sample (B). On the whole, these spectra indicate the activity of oxidized ZP sample in oxidizing NO, thanks to the presence of oxygen atoms, which are likely adsorbed on the surface of small iron oxides.

Finally, the third group of bands in the 1950 to 1700 cm^{-1} region is readily assigned to nitrosylic adducts on Fe^{2+}/Fe^{3+} extra-framework sites [44–46]. These bands are very weak on both oxidized and reduced ZP samples and will be described in more detail in the following. Since they are typical of highly dispersed extra-framework iron sites, their low intensity indicates that most iron in this sample is present as large iron oxide particles.

2.6.2. Fe^{2+}ZP and Fe^{3+}ZP Samples

The infrared spectra measured upon NO adsorption on Fe^{2+}ZP and Fe^{3+}ZP are reported in Figure 7 (C,D and E,F, respectively). In both samples, the set of bands assigned to N_2O adsorbed on Fe^{x+} (x = 2, 3) sites is weaker with respect to that observed on the oxidized ZP sample, in agreement with their assignment to surface sites present on small iron oxide clusters. As for the bands assigned to NO^+ ions, they have different position and intensity depending on the sample and activation treatment (oxidized or reduced, C,E or D,F, respectively). In detail, the oxidized Fe^{2+}ZP sample shows a main component centered at 2141 cm^{-1} (shoulder at 2065 cm^{-1}), which is similar to that observed on ZP). As observed on the ZP sample, this band almost disappears after the reduction treatment. On the contrary, a very intense band is observed on oxidized Fe^{3+}ZP at 2155 cm^{-1}, which decreases but still shows considerable intensity after reduction (spectra E and F). This suggests that the reactivity of NO with adsorbed oxygen strongly depends on the nature of the inserted iron species. Namely, this reaction takes place mostly in Fe^{3+}ZP, while oxidized ZP and Fe^{2+}ZP have a similar lower activity. The different positions of the NO^+ band suggest different local environments, which could be due to vicinal iron and Brønsted sites. This is in agreement with our previous results, showing the presence of Fe(OH)Si Brønsted sites in Fe^{3+}ZP [21].

Coming to the nitrosyl region, the oxidized Fe^{2+}ZP sample shows a relatively intense peak at 1878 cm^{-1}, with shoulders at 1902 cm^{-1} and 1817 cm^{-1} and a very weak band at 1755 cm^{-1} (curves C). A much higher intensity is observed after reduction (curves D). A similar trend with a lower intensity is observed on sample Fe^{3+}ZP (curves E). Bands in this spectral region have often been observed on synthetic Fe-zeolites, and are mainly ascribed to highly dispersed Fe^{2+} and Fe^{3+} sites [19,44,45,52,53].

3. Discussion

Several types of iron active sites, such as isolated iron cations, binuclear oxygen-bridged Fe species, iron oxide clusters, hydrocyanic acid, isocyanate species, etc., have been reported for the selective catalytic reduction of NO_x [23,54–56], but their precise assignment remains a controversial issue. We consider that such iron active species are heavily dependent on the applied method to incorporate iron, the used iron source, and the homogeneity of the resulting sample. In this report, the most active sample is the Fe^{2+}ZP sample, which also shows the highest dispersity of inserted Fe ions, as probed by NO (intensity of the nitrosyl bands in the 1950 to 1700 cm^{-1}).

The mechanism and the nature of intermediate species in producing N_2 in the selective catalytic reduction of NO_x is also a topic of controversy. Some intermediate species, such as NO_2, nitrosyl, nitro, nitrate, NO-NO triplet species, and dissociatively chemisorbed NO, have been proposed. Several studies have outlined that NO_2 in particular plays an important role in the reduction of NO_x to N_2. It has been reported that NO is oxidized to NO_2 with oxygen and then the resulting NO_2 reacts with hydrocarbons to form an active intermediate [53,57–59]. In our case, NO adsorption shows nitrosyl complexes as primary compounds formed in the most active Fe^{2+}ZP, NO^+ in Fe^{3+}ZP, and N_2O in the least active ZP. We can thus tentatively draw a correlation between the abundance of dispersed sites able to add NO as ligand(s) without its disproportionation at RT (bands in the 1950–1700 cm^{-1} region), and the catalytic activity in NO conversion. On the contrary, neither NO^+ nor N_2O formation at RT can be directly related to the samples' catalytic activity. In both cases, these features can be associated with the presence of agglomerated particles, in line with early studies by UV-Vis diffuse reflectance and Mössbauer spectroscopies [21].

On the other hand, it is observed that the maximum reduction of NO on Fe^{2+}ZP takes place at 410 °C. After this temperature, NO reduction starts to decrease. According to [53], such a decrement can be associated to a shift in the NO adsorption equilibrium (adsorption is usually an exothermic process) and to the properties of iron species active in NO reduction at higher temperatures.

Finally, the reported results also allow for some semi-quantitative considerations about the extinction coefficients of the bands related to different adsorbed species formed upon NO adsorption. As mentioned above, the sample with higher activity (Fe^{2+}ZP) is not only characterized by the highest nitrosyl band intensity, but also the highest NO sorption capacity. This should be, at first approximation, related to the overall integrated area of the adsorbed specie. However, by comparing the infrared results of Figure 7, the highest integrated area among the three samples is observed in Fe^{3+}ZP, particularly in the nitrosonium region. Since this does not correspond in a higher NO sorption capacity, we can conclude that NO^+ species are characterized by a higher extinction coefficient with respect to nitrosyl bands.

4. Materials and Methods

A purified zeolite material was obtained from the zeolitic rock of Palmarito de Cauto deposit, via magnetic and gravimetric purifications [60]. This zeolite is a mixture of ~70% of mordenite with other phases (clinoptilolite-heulandite, montmorillonite, quartz, feldspar, and iron oxides). The chemical composition of the sample in oxide form, with the balance as water, is 61.93% SiO_2; 13.42% Al_2O_3; 4.95% CaO; 2.53% Fe_2O_3; 1.14% MgO; 1.52% K_2O; 2.33% Na_2O; and 1.22% FeO. This purified zeolite is referred to as the natural mordenite (ZP).

In order to obtain natural mordenite forms modified with iron, ZP samples with a particle size class of +38–74 μm, were exchanged at pH = 2 with $FeSO_4$ and $Fe(NO_3)_3$ 0.05 N solutions. The exchanges were carried out with a solid/liquid ratio of 1 g/20 mL at 80 °C while stirring for 24 h and replacing exchange solutions after 12 h. The resulting samples were washed (firstly with acidified distilled water and then with distilled water) until nitrate and sulfate anions were totally removed and pH was neutral. They were then oven dried at 100 °C and stored in a desiccator. The ZP forms modified with Fe^{2+} and Fe^{3+} were designated as Fe^{2+}ZP and Fe^{3+}ZP, respectively.

Elemental analysis of the samples was obtained by flame photometry (Na and K) and atomic emission spectrometry with inductively coupled plasma (Si, Al, Fe, Ca, and Mg) using a Corning 400 photometer and an EPECTROFLAME Modula F spectrometer, respectively. To this aim, the samples were firstly dissolved using a mixture of fluoridric and perchloric acids. Powder X-ray diffraction (XRD) patterns of ZP, Fe^{2+}ZP, and Fe^{3+}ZP samples were measured with a D8 Advance Bruker X-ray diffractometer (Bruker AXS GmbH, Berlin, Germany) with Cu-Kα monochromatic radiation (1.5406 Å) with 40 kV and a current of 20 mA. The XRD patterns were registered in the 2θ range from 10 to 80° at room temperature (RT). N_2 gas adsorption/desorption isotherms were measured by means of a Micromeritics TriStar II device (Norcross, GA, USA) at liquid nitrogen temperature (−196 °C, LNT). Before the sorption measurements, the samples were pretreated in vacuum for 8 h at 350 °C. High resolution transition electron microscope (HRTEM) images of the samples were registered with a JEOL 3010-UHR microscope (Tokyo, Japan) with an acceleration potential of 100 kV, by dispersing the powdered samples on a copper grid covered with a lacey carbon film.

Infrared spectra were recorded on a BRUKER FTIR-66 spectrometer with a resolution of 2 cm^{-1}, using an MCT detector. Measurements were carried out using a home-made cell, allowing in situ thermal treatment, gas dosage, and measurement at RT. Thin self-supporting pellets for transmission measurements (around 10 mg/cm^2) were prepared with a hydraulic press. Oxidation treatments were carried out by dosing 60 Torr of O_2 at 550 °C for 1 h after a heating ramp in vacuum. Similarly, reduction activation was carried out at 400 °C for 1 h with 60 Torr of H_2. In both cases, the samples were evacuated before cooling down to RT. NO, carefully distilled, was dosed on the samples at RT after measuring the zeolite-activated reference spectra. Spectra were measured following a stepwise NO pressure (P_{NO}) reduction.

Catalytic activity tests of the samples in the selective catalytic reduction of NO with CO/C_3H_6 were done in a quartz flow reactor, using 100 mg of catalysts, within a 25 to 500 °C temperature interval with a ramp rate of 5 °C/min. The NO (0.09%), C_3H_6 (0.22%), O_2 (0.46%), and CO (1.179%) reaction mixture was prepared by mixing individual flows with mass-flow controllers. A total flow of 55 mL/min was obtained using N_2 as a diluent. Effluent gases were analyzed with a CAI ZRE gas analyzer. Before the catalytic test, the samples were pretreated in oxygen flow (0.5% in N_2) up to 350 °C with ramp rate of 5 °C/min. The temperature was then decreased to 25 °C and the flow was switched to the reaction mixture.

5. Conclusions

Iron-mordenite forms (Fe^{2+}ZP and Fe^{3+}ZP) were obtained from natural purified mordenite from Palmarito de Cauto (ZP) deposit, Cuba, by hydrothermal ion-exchange processes in acid medium with Fe^{2+} and Fe^{3+} cations. Only small variations were observed in the samples' crystallinity and textural properties after ion exchange, the latter ascribed to the dissolution in the acidic environment of large iron oxide particles from the zeolite external surface with parallel formation of small Fe_xO_y clusters within the zeolite pores. In Fe^{3+}ZP, a shift in the diffraction peaks' angular position was observed, which corresponds to an increment in the inter-planar distance and expansion of the zeolite matrix due to the incorporation of Fe^{3+} into the mordenite framework.

The activity in HC-SCR (used as a reaction test) was found to be dependent upon the ion exchange process, with Fe^{2+}ZP showing the highest catalytic activity, followed by Fe^{3+}ZP and ZP samples. The same order Fe^{2+}ZP >> Fe^{3+}ZP >> ZP was also observed in the NO sorption capacity of the samples at low temperatures (<170–200 °C).

In agreement with these results, FTIR-NO studies showed a higher intensity of the nitrosyl bands formed upon NO adsorption on the Fe^{2+}ZP sample. These bands (present on both oxidized and reduced samples) are related to isolated/highly dispersed Fe^{2+}/Fe^{3+} ions in extra-framework positions, which are more abundant in Fe^{2+}ZP and are responsible for the high sorption capacity (and likely activity) of this sample. On the contrary, the intensity of nitrosonium ions NO^+, formed upon reaction of NO with oxygen and likely protons, and more abundant on Fe^{3+}ZP sample, could not be related to neither NO sorption capacity or activity.

Author Contributions: Conceptualization, F.C.R. and G.B.; investigation F.C.R., D.T.F. and B.C.-R.; resources V.P.; data curation, F.C.R., I.R.-I. and G.B.; writing—original draft preparation, F.C.R., I.R.-I. and G.B.; writing—review and editing, F.C.R., I.R.-I. and G.B.

Funding: This research was partially supported by UNAM-PAPIIT through grant IN107817 and grant Russia-Cuba 18-53-34004.

Acknowledgments: F.C.R. acknowledges support from COFAA-IPN-México as well as technical support from Marco Fabbiani of the Department of Chemistry in Torino, Italy. Thanks are given to E. Aparicio, I. Gradilla, J. Peralta, J. Gonzalez and E. Flores for technical assistance.

Conflicts of Interest: The authors declare no conflict of interest.

References

1. Heinrich, F.; Schmidt, C.; Löffler, E.; Grünert, W. A highly active intra-zeolite iron site for the selective catalytic reduction of NO by isobutene. *Catal. Commun.* **2001**, *2*, 317–321. [CrossRef]
2. Frey, A.M.; Mert, S.; Due-Hansen, J.; Fehrmann, R.; Christensen, C.H. Fe-BEA Zeolite Catalysts for NH_3-SCR of NO_x. *Catal. Lett.* **2009**, *130*, 1–8. [CrossRef]
3. Krishna, K.; Makkee, M. Preparation of Fe-ZSM-5 with enhanced activity and stability for SCR of NO_x. *Catal. Today* **2006**, *114*, 23–30. [CrossRef]
4. Brandenberger, S.; Kröcher, O.; Tissler, A.; Althoff, R. The determination of the activities of different iron species in Fe-ZSM-5 for SCRof NO by NH_3. *Appl. Catal. B Environ.* **2010**, *95*, 348–357. [CrossRef]
5. Colombo, M.; Nova, I.; Tronconi, E. A comparative study of the NH3-SCR reactions over a Cu-zeoliteand a Fe-zeolite catalys. *Catal. Today* **2010**, *151*, 223–230. [CrossRef]
6. Liu, Z.; Millington, P.J.; Bailie, J.E.; Rajaram, R.R.; Anderson, J.A. A comparative study of the role of the support on the behaviourof iron based ammonia SCR catalysts. *Microporous Mesoporous Mater.* **2007**, *104*, 159–170. [CrossRef]
7. Høj, M.; Beier, M.J.; Grunwaldt, J.D.; Dahl, S. The role of monomeric iron during the selective catalytic reduction of NO_x by NH_3 over Fe-BEA zeolite catalysts. *Appl. Catal. B Environ.* **2009**, *93*, 166–176. [CrossRef]
8. Li, J.; Zhu, R.H.; Cheng, Y.S.; Lambert, C.K.; Yang, R.T. Mechanism of Propene Poisoning on Fe-ZSM-5 for Selective Catalytic Reduction of NO_x with Ammonia. *Environ. Sci. Technol.* **2010**, *44*, 1799–1805. [CrossRef]
9. Kucherov, A.; Montreuil, C.; Kucherova, T.; Shelef, M. In situ high-temperature ESR characterization of FeZSM-5 and FeSAPO-34 catalysts in flowing mixtures of NO, C_3H_6, aand O_2. *Catal. Lett.* **1998**, *56*, 173–181. [CrossRef]
10. Brosius, R.; Martens, J.A. Reaction Mechanisms of Lean-Burn Hydrocarbon SCR over Zeolite Catalysts. *Top. Catal.* **2004**, *28*, 119–130. [CrossRef]
11. Brosius, R.; Martens, J.A. Analysis of the structural parameters controlling the temperature window of the process of SCR-NO_x by low paraffins over metal-exchanged zeolites. *Catal. Today* **2002**, *75*, 347–351.
12. Chen, H.Y.; Wang, X.; Sachtler, W.M.H. Reduction of NOx over zeolite MFI supported iron catalysts: Nature of active sites. *Phys. Chem. Chem. Phys.* **2000**, *2*, 3083–3090. [CrossRef]
13. Janas, J.; Gurgul, J.; Socha, R.P.; Shishido, T.; Che, M.; Dzwigaj, S. Selective catalytic reduction of NO by ethanol: Speciation of iron and "structure–properties" relationship in FeSiBEA zeolite. *Appl. Catal. B Environ.* **2009**, *91*, 113–122. [CrossRef]
14. Mrad, R.; Aissat, A.; Cousin, R.; Courcot, D.; Siffert, S. Catalysts for NO_x selective catalytic reduction by hydrocarbons (HC-SCR). *Appl. Catal. A Gen.* **2015**, *504*, 542–548. [CrossRef]
15. Gómez-García, M.A.; Pitchon, V.; Kiennemann, A. Pollution by nitrogen oxides: An approach to NO(x) abatement by using sorbing catalytic materials. *Environ. Int.* **2005**, *31*, 445–467. [CrossRef]
16. Rutkowskaa, M.; Chmielarz, L.; Macina, D.; Piwowarska, Z.; Dudek, B.; Adamski, A.; Witkowski, S.; Sojka, Z.; Obalová, L.; van Oers, C.J.; et al. Catalytic decomposition and reduction of N_2O over micro-mesoporous materials containing Beta zeolite nanoparticles. *Appl. Catal. B Environ.* **2014**, *146*, 112–122. [CrossRef]
17. Kim, M.Y.; Lee, K.W.; Park, J.H.; Shin, C.H.; Lee, J.; Seo, G. Catalytic decomposition of nitrous oxide over Fe-BEA zeolites:Essential components of iron active sites. *Korean J. Chem. Eng.* **2010**, *27*, 76–82. [CrossRef]
18. Cheng, D.G.; Chen, F.; Zhan, X. Characterization of iron-containing AlPO-5 as a stable catalyst for selective catalytic reduction of N_2O with CH_4 in the presence of steam. *Appl. Catal. A Gen.* **2012**, *435*, 27–31. [CrossRef]
19. Pan, H.; Guo, Y.; Bi, H.T. NO_x adsorption and reduction with C_3H_6 over Fe/zeolite catalysts: Effect of catalyst support. *Chem. Eng. J.* **2015**, *280*, 66–73. [CrossRef]
20. Zecchina, A.; Rivallan, M.; Berlier, G.; Lamberti, C.; Ricchiardi, G. Structure and nuclearity of active sites in Fe-zeolites: Comparison with iron sites in enzymes and homogeneous catalysts. *Phys. Chem. Chem. Phys.* **2007**, *9*, 3483–3499. [CrossRef]
21. Tito-Ferro, D.; Rodríguez-Iznaga, I.; Concepción-Rosabal, B.; Berlier, G.; Chávez-Rivas, F.; Penton-Madrigal, A.; Castillón-Barraza, F.; Petranovskii, V. Iron exchanged natural mordenite: UV–Vis diffuse reflectance and Mössbauer spectroscopy characterisation. *Int. J. Nanotechnol.* **2016**, *3*, 1–3.
22. Gelves, J.F.; Dorkis, L.; Márquez, M.A.; Álvarez, A.C.; González, L.M.; Villa, A.L. Activity of an iron Colombian natural zeolite as potential geo-catalyst for NH_3-SCR of NO_x. *Catal. Today* **2019**, *320*, 112–122. [CrossRef]

23. Ates, A.; Reitzmann, A.; Hardacre, C.; Yalcin, H. Abatement of nitrous oxide over natural and iron modified natural zeolites. *Appl. Catal. A Gen.* **2011**, *407*, 67–75. [CrossRef]
24. Maisuradze, G.; Sidamonidze, S.; Akhalbedashvili, L.; Kvatashidze, R. Modified Natural Zeolites as Catalysts for Catalytic Reduction of NO with CO—Main Components of Exhaust Gases. *J. Environ. Sci. Eng. B* **2015**, *4*, 574–582. [CrossRef]
25. Moreno-Tost, R.; Santamaría-González, J.; Rodríguez-Castellón, E.; Jiménez-López, A.; Autié, M.A.; Carreras-Glacial, M.; Autié-Castro, G.; Guerra, M. Selective Catalytic Reduction of Nitric Oxide by Ammonia over Ag and Zn-Exchanged Cuban Natural Zeolites. *Z. Anorg. Allg. Chem.* **2005**, *631*, 2253–2257. [CrossRef]
26. Yahiro, H.; Iwamoto, M. Copper ion-exchanged zeolite catalysts in deNO$_x$ reaction. *Appl. Catal. A* **2001**, *222*, 163–181. [CrossRef]
27. Rodríguez-Iznaga, I.; Petranovskii, V.; Castillón-Barraza, F.; Concepción-Rosabal, B. Copper-Silver Bimetallic System on Natural Clinoptilolite: Thermal Reduction of Cu^{2+} and Ag^{+} Exchanged. *J. Nanosci. Nanotechnol.* **2011**, *11*, 1–7. [CrossRef]
28. Petrov, O.E. *Natural Zeolites 93: Occurrence, Properties, Use*; Ming, D.W., Mumpton, F.A., Eds.; ICNZ: Brockport, NY, USA, 1995.
29. Hadjiivanov, K.; Ivanova, E.; Kefirov, R.; Janas, J.; Plesniar, A.; Dzwigaj, S.; Che, M. Adsorption properties of Fe-containing dealuminated BEA zeolites as revealed by FTIR spectroscopy. *Microporous Mesoporous Mater.* **2010**, *131*, 1–12. [CrossRef]
30. Hajjar, R.; Millot, Y.; Man, P.P.; Che, M.; Dzwigaj, S. Two Kinds of Framework Al Sites Studied in BEA Zeolite by X-ray Diffraction, Fourier Transform Infrared Spectroscopy, NMR Techniques, and V Probe. *J. Phys. Chem. C* **2008**, *112*, 20167–20175. [CrossRef]
31. Rodríguez-Iznaga, I.; Fuentes, G.R.; Aguilar, A.B. The role of carbonate ions in the ion-exchange Ni^{2+}=2NH_4^{+} in natural clinoptilolite. *Microporous Mesoporous Mater.* **2000**, *41*, 129–136. [CrossRef]
32. Rodríguez-Iznaga, I.; Gómez, A.; Rodríguez-Fuentes, G.; Aguilar, A.B.; Ballan, J.S. Natural clinoptilolite as an exchanger of Ni^{2+} and NH_4^{+} ions under hydrothermal conditions and high ammonia concentration. *Microporous Mesoporous Mater.* **2002**, *53*, 71–80. [CrossRef]
33. Breck, D.W. *Zeolite Molecular Sieves: Structure, Chemistry and Use*; John Wiley & Sons: New York, NY, USA, 1974.
34. Inglezakis, V.J.; Zorpas, A.A.; Loizidou, M.D.; Grigoropoulou, H.P. Simultaneous removal of metals Cu^{2+}, Fe^{3+} and Cr^{3+} with anions SO_4^{2-} and HPO_4^{2-} using clinoptilolite. *Microporous Mesoporous Mater.* **2003**, *61*, 167–171. [CrossRef]
35. Yamamoto, S.; Sugiyama, S.; Matsuoka, O.; Honda, T.; Banno, Y.; Nosoye, H. AFM imaging of the surface of natural heulandite. *Microporous Mesoporous Mater.* **1998**, *21*, 1–6. [CrossRef]
36. Barrett, J. *Inorganic Chemistry in Aqueous Solutions*; The Royal Society of Chemistry: Cambridge, UK, 2003.
37. Cotton, F.A.; Wilkinson, G.; Murrillo, C.A.; Bochmann, M. *Advanced Inorganic Chemistry*, 6th ed.; John Wiley & Sons: New York, NY, USA, 1999.
38. Ugurlu, O.; Haus, J.; Gunawan, A.A.; Thomas, M.G.; Maheshwari, S.; Tsapatsis, M.; Mkhoyan, K.A. Radiolysis to knock-on damage transition in zeolites under electron beam irradiation. *Phys. Rev. B* **2011**, *83*, 113408. [CrossRef]
39. Baerlocher, C.; Meier, W.M.; Olson, D.H. Framework Type MOR (Material: Mordenite), Database of Zeolite Structures. 2013. Available online: http://www.iza-structure.org/databases/ (accessed on 3 April 2019).
40. Skarlis, S.A.; Berthout, D.; Nicolle, A.; Dujardin, C.; Granger, P. Combined IR spectroscopy and kinetic modeling of NOx storage and NO oxidation on Fe-BEA SCR catalysts. *Appl. Catal. B Environ.* **2014**, *148*, 446–465. [CrossRef]
41. Ahrens, M.; Marie, O.; Bazin, P.; Daturi, M. Fe-H-BEA and Fe-H-ZSM-5 for NO_2 removal from ambient air —A detailed in situ and operando FTIR study revealing an unexpected positive water-effect. *J. Catal.* **2010**, *271*, 1–11. [CrossRef]
42. Rivallan, M.; Ricchiardi, G.; Bordiga, S.; Zecchina, A. Adsorption and reactivity of nitrogen oxides (NO_2, NO, N_2O) on Fe–zeolites. *J. Catal.* **2009**, *264*, 104–116. [CrossRef]
43. Berlier, G.; Lamberti, C.; Rivallan, M.; Mul, G. Characterization of Fe sites in Fe-zeolites by FTIR spectroscopy of adsorbed NO: Are the spectra obtained in static vacuum and dynamic flow set-ups comparable? *Phys. Chem. Chem. Phys.* **2010**, *12*, 358–364. [CrossRef]

44. Berlier, G.; Pourny, M.; Bordiga, S.; Spoto, G.; Zecchina, A.; Lamberti, C. Coordination and oxidation changes undergone by iron species in Fe-MCM-22 upon template removal, activation and red–ox treatments: An in situ IR, EXAFS and XANES study. *J. Catal.* **2005**, *229*, 45–54. [CrossRef]
45. Hadjiivanov, K.; Tsyntsarski, B.; Nikolova, T. Stability and reactivity of the nitrogen-oxo species formed after NO adsorption and NO+O_2 coadsorption on Co-ZSM-5: An FTIR spectroscopic study. *Phys. Chem. Chem. Phys.* **1999**, *1*, 4521–4528. [CrossRef]
46. Gankanda, A.; Grassian, V.H. Nitrate Photochemistry in NaY Zeolite: Product Formation and Product Stability under Different Environmental Conditions. *J. Phys. Chem. A* **2013**, *117*, 2205–2212. [CrossRef] [PubMed]
47. Wacław, A.; Nowinska, K.; Schwieger, W.; Zielinska, A. N_2O decomposition over iron modified zeolites ZSM-5. *Catal. Today* **2004**, *90*, 21–25. [CrossRef]
48. Hadjiivanov, K.; Penkova, A.; Daturi, M.; Saussey, J.; Lavalley, J.C. FTIR spectroscopic study of low-temperature co-adsorption of NO and O_2 on H-ZSM-5: Evidence of formation of $[ONNO]^+$ species. *Chem. Phys. Lett.* **2003**, *377*, 642–646. [CrossRef]
49. Hadjiivanov, K.; Saussey, J.; Freysz, J.L.; Lavalley, J.C. FT-IR study of NO + O_2 co-adsorption on H-ZSM-5: Re-assignment of the 2133 cm^{-1} band to NO^+ species. *Catal. Lett.* **1998**, *52*, 103–108. [CrossRef]
50. Ma, X.; Wang, X.; Bi, R.; Zhao, Z.; He, H. Defect of HY as catalyst for selective catalytic reduction of NO in comparison with the pentasil zeolites. *J. Mol. Catal. A* **2009**, *303*, 90–95. [CrossRef]
51. Henriques, C.; Marie, O.; Thibault-Starzyk, F.; Lavalley, J.C. NO^+ ions as IR probes for the location of OH groups and Na^+ ions in main channels and side pockets of mordenite. *Microporous Mesoporous Mater.* **2001**, *50*, 167–171. [CrossRef]
52. Spoto, G.B.G.; Bordiga, S.; Richiardi, G.; Fisicaro, P.; Zecchina, A.; Rossetti, I.; Selli, E.; Forni, L.; Giamello, E.; Lamberti, C. Evolution of Extraframework Iron Species in Fe Silicalite; 1. Effect of Fe Content, Activation Temperature, and Interaction with Redox Agents. *J. Catal.* **2002**, *208*, 64–82.
53. Lezcano, M.; Kovalchuk, V.I.; d'Itri, J.L. FTIR Study of the Interaction of Nitric Oxide with Fe-ZSM-51. *Kinet. Catal.* **2001**, *42*, 104–111. [CrossRef]
54. Traa, Y.; Burger, B.; Weitkamp, J. Zeolite-based materials for the selective catalytic reduction of NO_x with hydrocarbons. *Microporous Mesoporous Mater.* **1999**, *30*, 3–41. [CrossRef]
55. Long, J.; Zhang, Z.; Ding, Z.; Ruan, R.; Li, Z.; Wang, X. Infrared Study of the NO Reduction by Hydrocarbons over Iron Sites with Low Nuclearity: Some New Insight into the Reaction Pathway. *J. Phys. Chem. C* **2010**, *114*, 15713–15727. [CrossRef]
56. Battiston, A.A.; Bitter, J.H.; Koningsberger, D.C. Reactivity of binuclear Fe complexes in over-exchanged Fe/ZSM5, studied by in situ XAFS spectroscopy 2. Selective catalytic reduction of NO with isobutene. *J. Catal.* **2003**, *218*, 163–177. [CrossRef]
57. Ren, L.; Zhang, T.; Liang, D.; Xu, C.; Tang, J.; Lin, L. Effect of addition of Zn on the catalytic activity of a Co/HZSM-5 catalyst for the SCR of NO_x with CH_4. *Appl. Catal. B Environ.* **2002**, *35*, 317–321. [CrossRef]
58. de Oliveira, M.L.M.; Silva, C.M.; Moreno-Tost, R.; Lopes-Farias, T.; Jiménez-López, A.; Rodríguez-Castellón, E. A study of copper-exchanged mordenite natural and ZSM-5 zeolites as SCR–NOx catalysts for diesel road vehicles: Simulation by neural networks approach. *Appl. Catal. B Environ.* **2009**, *88*, 420–429. [CrossRef]
59. Biglino, D.; Li, H.; Erickson, R.; Lund, A.; Yahiro, H.; Shiotani, M. EPR and ENDOR studies of NO_x and Cu^{2+} in zeolites: Bonding and diffusion. *Phys. Chem. Chem. Phys.* **1999**, *1*, 2887–2896. [CrossRef]
60. Tito-Ferro, D.; Rodríguez-Iznaga, I.; Concepción-Rosabal, B.; Chávez-Rivas, F. El hierro en la roca zeolitizada del yacimiento de Palmarito de Cauto: Separación y caracterización de fases magnéticas, V. Córdova-Rodríguez, R. Rizo-Beyra. *Rev. Min. Y Geol.* **2011**, *27*, 22–37.

Article

Oxidative Degradation of Trichloroethylene over Fe_2O_3-Doped Mayenite: Chlorine Poisoning Mitigation and Improved Catalytic Performance

Raffaele Cucciniello [1,†], Adriano Intiso [1,†], Tiziana Siciliano [2], Antonio Eduardo Palomares [3,*], Joaquín Martínez-Triguero [3], Jose Luis Cerrillo [3], Antonio Proto [1] and Federico Rossi [4,*]

1 Department of Chemistry and Biology, University of Salerno, via Giovanni Paolo II, 132, 84084 Fisciano (SA), Italy
2 DiSTeBa Department, University of Salento, 73100 Lecce (LE), Italy
3 Instituto de Tecnología Química, Universitat Politècnica de València-CSIC, Valencia 46022, Spain
4 Department of Earth, Environmental and Physical Sciences, DEEP Sciences, Pian dei Mantellini 44, 53100 Siena, Italy
* Correspondence: apalomar@iqn.upv.es (A.E.P.); federico.rossi2@unisi.it (F.R.)
† These authors contributed equally to this work.

Received: 14 August 2019; Accepted: 31 August 2019; Published: 5 September 2019

Abstract: Mayenite was recently successfully employed as an active catalyst for trichloroethylene (TCE) oxidation. It was effective in promoting the conversion of TCE in less harmful products (CO_2 and HCl) with high activity and selectivity. However, there is a potential limitation to the use of mayenite in the industrial degradation of chlorinated compounds—its limited operating lifespan owing to chlorine poisoning of the catalyst. To overcome this problem, in this work, mayenite-based catalysts loaded with iron (Fe/mayenite) were prepared and tested for TCE oxidation in a gaseous phase. The catalysts were characterized using different physico-chemical techniques, including XRD, ICP, N_2-sorption (BET), H_2-TPR analysis, SEM-EDX, XPS FESEM-EDS, and Raman. Fe/mayenite was found to be more active and stable than the pure material for TCE oxidation, maintaining the same selectivity. This result was interpreted as the synergistic effect of the metal and the oxo-anionic species present in the mayenite framework, thus promoting TCE oxidation, while avoiding catalyst deactivation.

Keywords: trichloroethylene; mayenite; catalytic oxidation; iron; chlorine poisoning

---

1. Introduction

Trichloroethylene (TCE) is a chlorinated volatile organic solvent belonging to the class of dense non aqueous phase liquids (DNAPL) pollutants [1–3]. Several strategies have been considered for TCE remediation, including the use of CaO [4], bioremediation [5,6] and adsorption processes with activated charcoal or zeolites [7,8]. In addition, as TCE is highly volatile, it can be easily stripped from the remediation media (water, surfactant solutions, removed soils, etc.) with air flux and directed to further treatments in gas phase [9,10]. In this respect, catalytic heterogeneous oxidation is becoming a popular alternative to thermal incineration for treating exhausted gases rich in TCE, as catalysts lower operative temperatures and improve selectivity of the reaction towards less harmful products, with high benefits in terms of energy consumption and environmental impact.

Several heterogeneous catalysts have been developed and tested for gaseous TCE oxidation. Catalytic systems based on noble metals, particularly Pt and Pd, have been extensively employed, showing good results in terms of activity and selectivity, as reported by Gonzalez-Velasco and co-workers in a recently published review [11]. Less-expensive catalysts, based on metallic oxides, have also been prepared as uniform catalyst or supported on high surface materials (e.g., γ-Al_2O_3) [12,13]. Blanch-Raga et al. reported the oxidation of TCE over different mixed oxides derived from hydrotalcites [14], with the Co(Fe/Al) catalyst being the most active ($T_{50\%}$ = 280 °C and $T_{90\%}$ = 340 °C at Gas Hourly Space Velocity, GHSV = 15,000 h^{-1} and [TCE] = 1000 ppm) due to its acidic and oxidative properties. Zeolites also represent an important type of active catalysts for the oxidation of TCE and many papers have reported on the synergic effect of acidic sites in zeolites [15] with metal catalysts in order to improve the performance of the whole catalytic system. Romero-Saez et al. [16], studied the performance of iron-doped ZSM-5 zeolite for TCE oxidation, finding that a ZSM-5 containing 2 wt% of Fe quantitatively oxidizes 1000 ppm of TCE at 500 °C and GHSV = 13,500 h^{-1}. This paper shows that the formation of active iron (III) species, as Fe_2O_3 nanoparticles, was most likely responsible for the enhanced catalytic performance of the zeolite. Nevertheless the catalyst suffers some deactivation after 16 h of reaction due to the formation of $FeCl_3$ [16]. Recently, Palomares et al. reported a remarkably high selectivity towards CO_2 during TCE oxidation with Cu and Co-doped beta zeolites. The best results ($T_{50\%}$ = 310 °C and $T_{90\%}$ = 360 °C at GHSV = 15,000 h^{-1} and [TCE] = 1000 ppm) were obtained by using Cu-doped zeolite, which combined the acid sites of the zeolite with the redox properties of the copper ions [17]. Notwithstanding, zeolites-based catalysts suffer some drawbacks, which include coke formation, deactivation, and formation of chlorinated by-products [14].

In previous works, we reported about the oxidation of TCE by using the mesoporous calcium aluminate mayenite ($Ca_{12}Al_{14}O_{33}$) as a catalyst [18–22]; mayenite had a good overall performance, showing high activity and selectivity towards nontoxic compounds, and fair thermal stability and recyclability. As a main drawback, the material shows a certain tendency towards chlorine poisoning, leading to slow deactivation of the catalyst. Mayenite has a zeolite-type structure with interconnected cages and a positive electric charge per unit cell that is balanced by O^{2-} ions (free oxygen ions) [23]. The free oxygen ions can be substituted by other species (Cl^-, H^-, NH_2^-, etc.) [24,25] and can migrate from the bulk to the surface at temperatures higher than 400 °C [26], thus conferring to the mayenite oxidative properties exploited for many applications [27–29]; for instance, as Ni support for the catalytic reforming of tar [30–32].

With the aim of further improving mayenite activity for TCE oxidation and mitigating deactivation of the material, in this work, we use a mayenite containing iron that is employed for the catalytic oxidation of TCE. The performance of the system has been evaluated by means of the light-off curve and the structural properties of the material have been characterized, before and after the reaction, by means of different physico-chemical techniques, including XRD, ICP analysis, N_2-sorption (BET), H_2-TPR analysis, SEM-EDX, FESEM-EDS, XPS, and Raman spectroscopy. We have prepared catalysts with different iron content and compared the activity and stability of this material with those of pure mayenite.

2. Results and Discussion

Figure 1 shows the XRD patterns of mayenite and Fe/Mayenite loaded with 2% of Fe_2O_3 (1.5% Fe/mayenite has an identical spectrum). The mayenite XRD pattern presented typical peaks of mayenite around 2θ = 18.1°, 30°, 33.4°, 36.7°, 41.2°, 46.7°, 55.2° and 57.4°. $Ca_3Al_2O_6$ (●) and $CaAl_2O_4$ (Δ) were also found as impurities formed during the mayenite preparation processes [33].

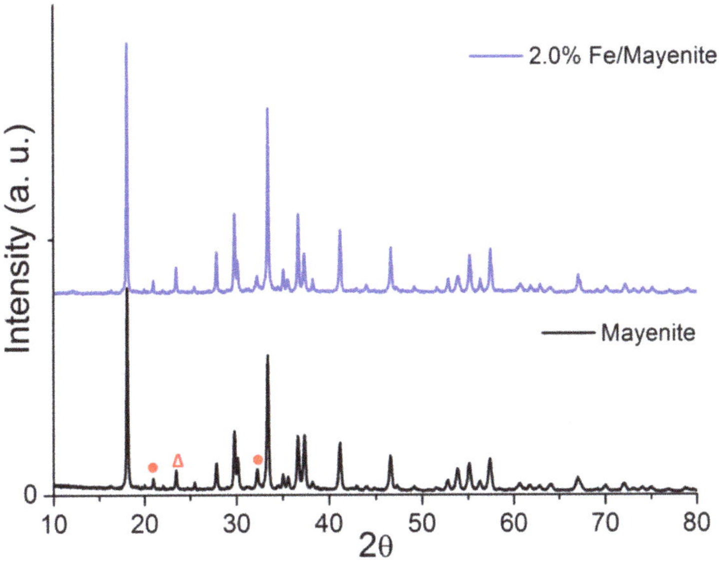

Figure 1. XRD patterns of mayenite (black) and 2.0% Fe/mayenite (blue).

The Fe/mayenite catalyst clearly maintains the crystalline structure of mayenite. Furthermore, no peaks associated with iron oxides were observed in the Fe/mayenite samples; this is due to the low metal loading in the mayenite and their good dispersion on the mayenite support [34].

Table 1 shows the metal loading and specific surface area of the catalysts. The BET surface area of mayenite was 11.7 m^2/g, in line with data reported for this type of material [27]. The Fe/mayenite samples had BET surface area values similar to that of mayenite, showing that the incorporation of iron does not modify its textural properties. ICP analysis confirmed that the iron content was close to the nominal value.

Table 1. Iron content and BET surface area of the catalysts.

Catalyst	Fe Content (wt.%)	BET Surface Area (m^2/g)
Mayenite	-	11.7
1.5% Fe/mayenite	1.72	11.5
2.0% Fe/mayenite	2.30	11.2

Fe/mayenite is a porous material (view SI, Figure S1), characterized by large pores with dimensions of μm (macropores) and nm (mesopores) composed of calcium, aluminium, oxygen, and iron with an approximate content of 34%, 40%, 26%, and 2 wt.%, respectively. The structure and composition of the synthetized materials were also characterised by FESEM-EDS analysis, which yielded similar results, but allowed a detailed distribution of the atomic content. Figure 2 shows the results obtained, observing the most abundant elements in mayenite, i.e., aluminium, calcium, and oxygen. Iron atoms appear in low quantity and with a homogenous distribution in the mayenite, demonstrating good iron dispersion on the mayenite support. Moreover, the FESEM images of Fe/mayenite and pure mayenite (not shown) showed no significant differences in terms of morphology, maintaining the typical morphology of mayenite in both cases.

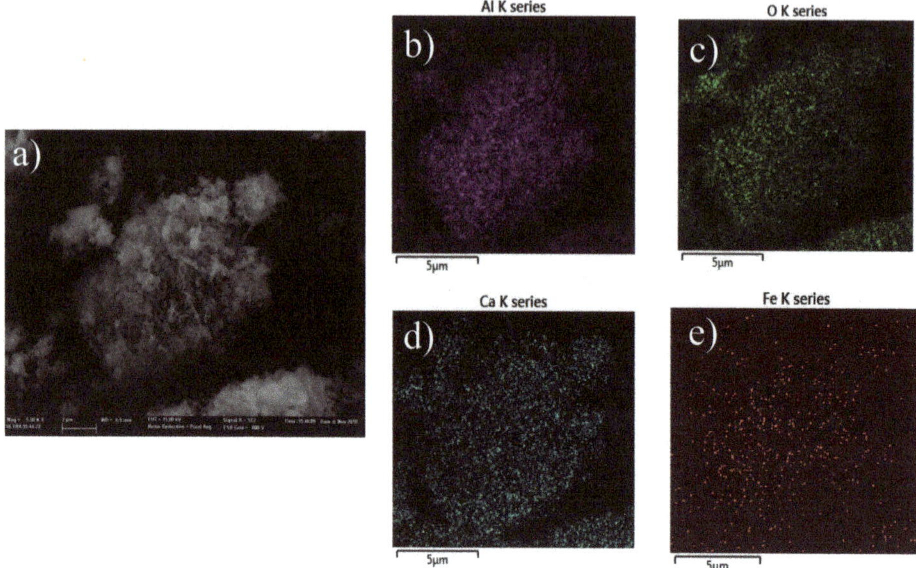

Figure 2. (**a**) FESEM micrograph (magnification of × 5500) of Fe/mayenite; atomic mapping of Fe/mayenite: (**b**) Al (purple); (**c**) O (green); (**d**) Ca (blue); (**e**) Fe (red).

TPR study of the catalysts is reported in Figure 3. All samples have similar hydrogen consumption, but present different TPR profiles. As can be seen, two peaks are observed for mayenite: the first one is a smaller band that appears around 550 °C, while the second is more intense and has the maximum at 620 °C. The first corresponds to the dissociative adsorption of H_2 in a heterolitic fashion [35] and the second to the reaction of these species with extra framework O_x^- and O_2^{2-} anions [36]. Iron-containing mayenites show a different profile with a unique band centred at 550 °C for the sample with 1.5% Fe or at 530 °C for the sample containing 2% Fe. These bands are assigned to the reduction of extra framework O_x^- and O_2^{2-} anions that in these catalysts are coincident with the band assigned to the dissociative adsorption of H_2 with consequential water formation, as previously reported for iron oxide-based catalysts [37]. Also, a small shoulder at 400–450 °C can be observed in the sample with higher iron content. This shoulder is assigned to the reduction of Fe^{3+} to Fe^{2+}, as reported by Romero-Saez et al. for Fe/zeolite samples [16]. Quantification of the hydrogen consumption shows similar results for the different samples, as the main species reduced in all the catalysts are the anionic oxygens present in the mayenite. The low iron content of the Fe/mayenite and the only partial reaction of the Fe^{3+} species results in a negligible consumption of hydrogen compared to that necessary for the oxygen species reduction. The results also show that there is a relationship between the content of iron and the shift towards lower temperatures of the extra framework O_x^- and O_2^{2-} reduction peak. This indicates that there is an interaction between iron species and anionic oxygen, which improves the redox properties of the catalysts containing iron.

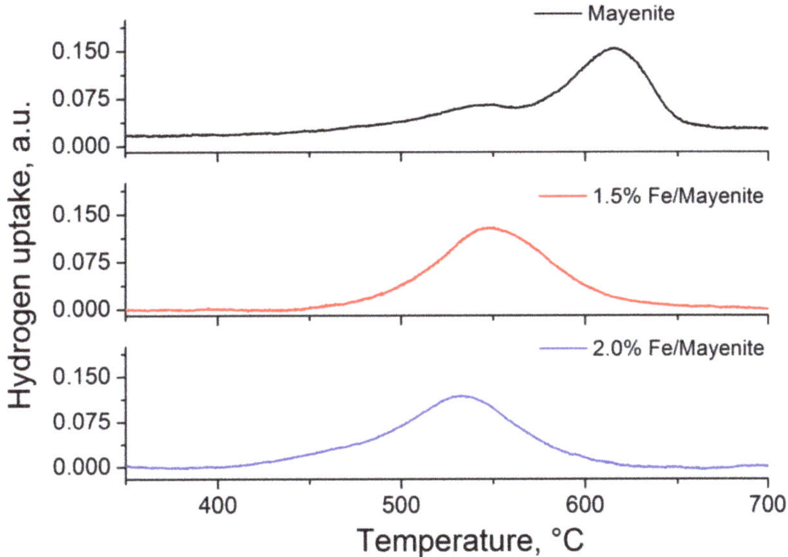

Figure 3. H_2-TPR profiles of mayenite (black), 1.5% Fe/mayenite (red), and 2.0% Fe/mayenite (blue).

The synthetized catalysts have been evaluated for oxidation of trichloroethylene by monitoring the conversion percentage calculated by means of Equation (1) as a function of the temperature (light-off curve). As shown in Figure 4, 2.0% Fe/mayenite has the best conversion rate among the different catalysts tested.

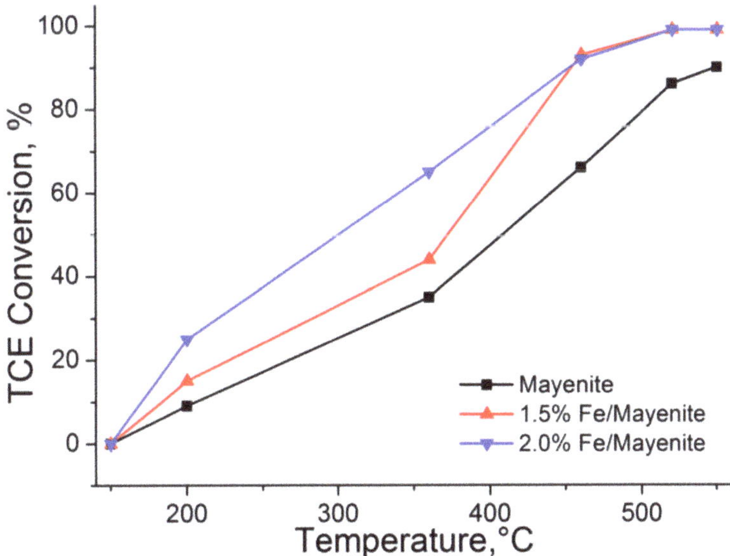

Figure 4. TCE conversion in wet conditions for mayenite (black), 1.5% (red), and 2.0% (blue) Fe/mayenite.

Blank experiments performed without a catalyst showed no significant TCE conversion below 550 °C. T_{50} and T_{90} (temperature at which 50% and 90% of the TCE was depleted, respectively) were found significantly lower for Fe/mayenite, with respect to pure mayenite (Table 2). In particular, pure mayenite, used as a reference, showed T_{50} = 410 °C and T_{90} = 550 °C; 2.0% Fe/mayenite has a T_{50} = 300 °C and T_{90} = 460 °C and for 1.5% Fe/mayenite T_{50} was found slightly higher (375 °C), whilst T_{90} was the same as 2.0% iron-loaded mayenite. Conversion rates at 365 °C were also calculated according to Equation (2) for the three catalysts and reported in Table 2. Preliminary experiments revealed that an iron loading higher than ~2% in the mayenite did not significantly improve the conversion performance.

Table 2. Catalysts activity in terms of T_{50} and T_{90} and conversion rate at 365 °C.

Catalyst	T_{50} (°C)	T_{90} (°C)	Conversion Rate (mol g^{-1} s^{-1})
Mayenite	410	550	5.58×10^{-6}
1.5% Fe/mayenite	375	460	7.14×10^{-6}
2.0% Fe/mayenite	300	460	1.05×10^{-5}

The results show an increased catalytic activity of mayenite in the presence of iron. This can be explained by the synergistic effect of dispersed iron with the ionic oxygen species (O^{2-} and $O_2{}^{2-}$) present in the mayenite framework. In fact, iron species have a twofold role: *i*) they are active oxidants [11,16] and, *ii*) they make easier the interaction of the atmospheric O_2 to form O^{2-} and $O_2{}^{2-}$, responsible for the TCE oxidation [26]. Both roles result in improved redox and catalytic properties for Fe/mayenite, when compared with pure mayenite, as the TPR and catalytic results have shown.

The results of selectivity tests showed that, for all the catalysts, CO_2 and CO were the main oxidation reaction products (CO/CO_2 ratio was 50:50 for all the investigated temperatures and catalysts), while only HCl was detected as a chlorinated product.

2% Fe/mayenite stability was evaluated for several hours (12 h) at stationary conditions (T = 460 °C, 1700 ppm TCE, GHSV = 6000 h^{-1}, 0.8 g of catalyst). As highlighted in Figure 5, the presence of iron on the mayenite surface drastically improved the stability of the material. Indeed, after 2 h of reaction, pure mayenite showed a significant loss in activity, whilst 2.0% Fe/mayenite retained catalytic activity. Only a partial deactivation of the 2% Fe/mayenite catalyst was observed after 12 h of reaction, decreasing the conversion from 95% to 80%.

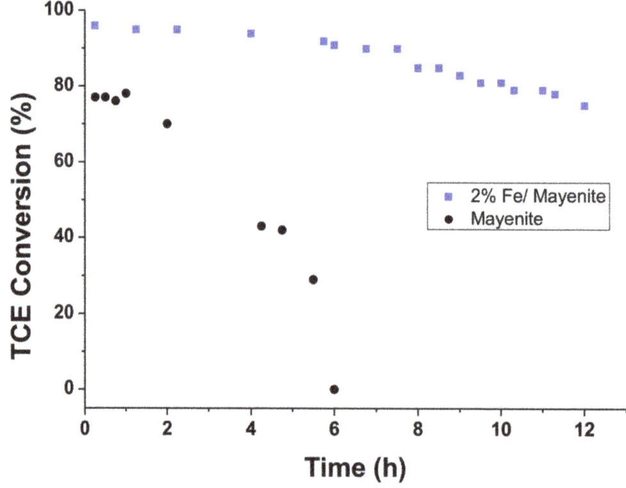

Figure 5. Comparison of catalyst stability (T = 460 °C, 1700 ppm TCE, GHSV = 6000 h^{-1}, 0.8 g of catalyst).

These results are interesting because stability of the catalyst is essential for possible commercial use of the material. As reported in previous works, the deactivation mechanism of mayenite catalyst is related with the displacement of active oxygen species by chloride ions [19,20,24,25]. In particular, the formation of HCl during the reaction caused a partial irreversible substitution of the anionic oxygen species (O_x^{x-}) by chloride ions forming chloromayenite [19,22], with consequent deactivation of the catalyst. It is observed that the presence of iron in mayenite modifies this mechanism, probably because the Fe species catalyze the interaction of the atmospheric O_2 to form O^{2-} and O_2^{2-}, making reversible the substitution of chloride ions by oxygen and avoiding or diminishing the formation of chloromayenite. In order to understand the role of iron in improving the stability of the material, XRD, SEM-EDX, FESEM-mapping, XPS, and Raman analyses of fresh and used catalysts were performed. XRD patterns and SEM images did not show important differences before and after the reaction, indicating a high structural stability of the material both in the absence and in the presence of iron (see Supplementary Materials for details, Figures S2 and S3a). Nevertheless, the EDX spectrum (Figure S3b) of the 2% Fe/mayenite sample after the stability test shows the presence of chloride. This is clearly observed in Figure 6 with the atomic mapping of Fe/mayenite after reaction. Comparing these images with those of the catalyst before reaction (Figure 2), it is observed that after the 12 h reaction, mayenite has the same morphology and distribution as Ca, Al, O and Fe atoms, revealing high stability of Fe/mayenite. It is observed that iron atoms do not agglomerate after reaction, indicating high stability of the metal supported on the mayenite. The main difference with the material before reaction is the presence of Cl atoms (orange points). The Cl mapping shows that this element has the same distribution as Ca atoms, suggesting that Cl is mainly interacting with Ca, because the basic properties of this element favour interaction with an acidic molecule as HCl. Different distribution of Cl and Fe atoms suggests that $FeCl_3$ was not formed, as it occurs in zeolites [16], indicating a high stability of the Fe species not poisoned by the chloride present on the catalyst surface.

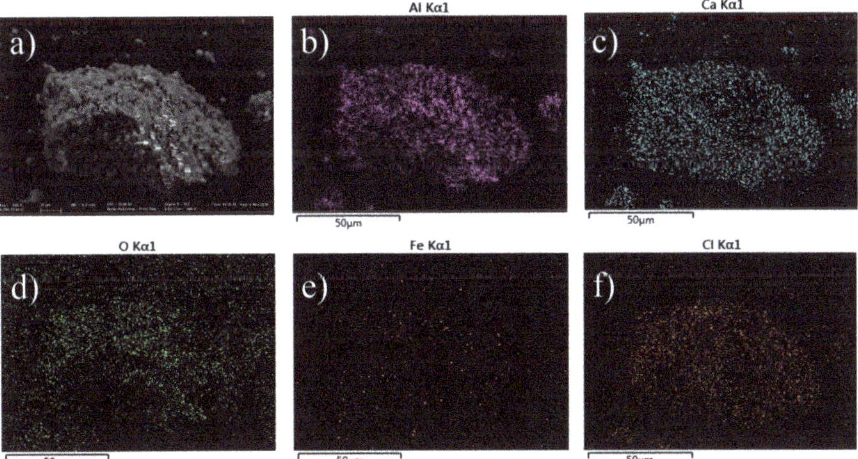

Figure 6. (**a**) FESEM micrograph (magnification of × 5500) of post-reaction Fe/mayenite; atomic mapping of post-reaction Fe/mayenite: (**b**) Al (purple); (**c**) O (green); (**d**) Ca (blue); (**e**) Fe (red); (**f**) Cl (orange).

The full XPS spectra of the mayenite containing iron before and after reaction are shown in Fig S4, where a peak at 200 eV is observed in the sample after reaction corresponding to the binding energy of Cl (2p) with two contributions at 200 and 189.7 eV, characteristic of Cl ($2p_{3/2}$) and Cl ($2p_{1/2}$ for ionic chlorine (Cl^-) [38]. The presence of this peak in the sample after reaction coincides with a decrease in the relative intensity of the O peak at 530.9 eV due to the oxygen surface species consumption during the reaction. On the other hand, Figure 7 shows a complex multiplet-split Fe 2p XPS spectra.

At least two contributions are observed: the first one is centred at 709.5 eV and the second, which is more intense, is centred at 711.8 eV. These peaks have been assigned to Fe^{2+} ($2p_{3/2}$) and Fe^{3+} ($2p_{3/2}$), respectively, indicating the formation of Fe_2O_3, probably together with Fe_3O_4 [38,39]. After reaction, an increase in the band centred at 711.8 eV, together with a small shift to the highest eV, is observed; it could be related with the disappearance of the Fe^{2+} contribution in the Fe_3O_4 phase or the formation of $FeCl_3$ [39]. These results reveal that after 12 h of reaction, part of the HCl formed in the reaction interacts with the mayenite and some chloride species are formed on the catalyst surface. Nevertheless, these species are not irreversibly adsorbed on the ionic vacancies (as in the case with pure mayenite [19,22]) and oxidant centres (anionic oxygen) are present after reaction. This is due to a synergic action of the iron present in the mayenite surface, which catalyses O_x^{x-} regeneration with the gas-phase oxygen present in the reaction media.

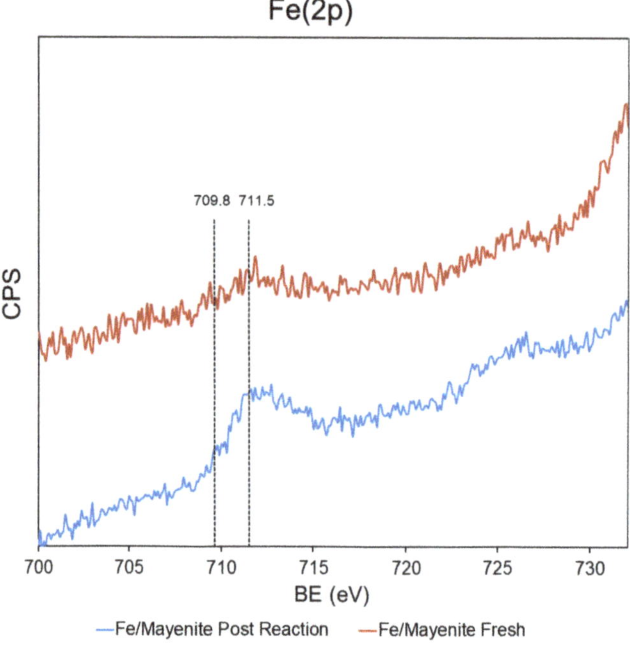

Figure 7. XPS spectra for the Fe (2p) of the Fe/mayenite (2% Fe) before and after reaction.

To understand this process better, Raman spectroscopy studies were conducted; the results are given in Figure 8. It is shown that in Fe/mayenite, the signal for oxygen O_2^- at 1075 cm^{-1} is preserved after the catalytic test. Similar results were obtained in the XPS analysis of the samples after reaction (see Supplementary Materials). This does not occur with pure mayenite [19,22], demonstrating that the redox properties of iron catalyze the regeneration of anionic oxygen by the O_2 present in the gas feed and avoid the irreversible chlorine poisoning of mayenite.

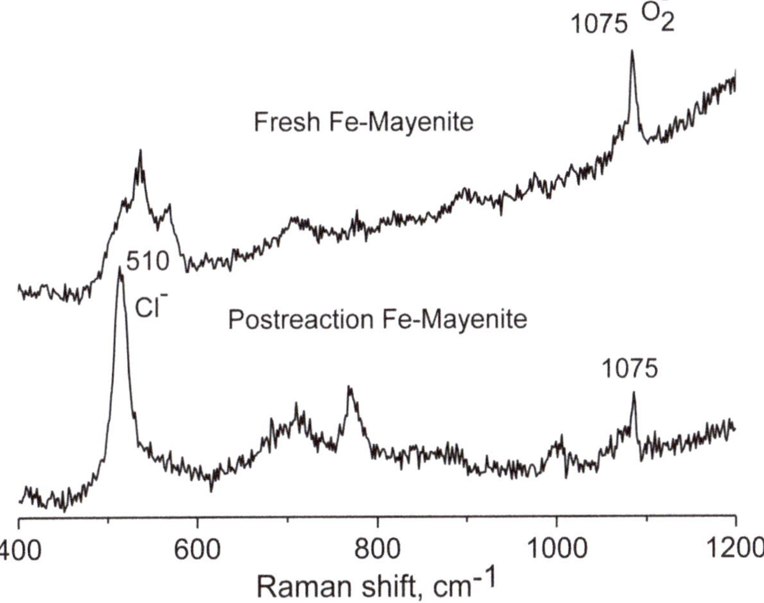

Figure 8. Raman spectra of Fe/mayenite (2.0%), fresh (top) and after reaction (bottom).

These results clearly show that the presence of well-dispersed iron on the mayenite surface improves the redox properties of the mayenite together with its stability, due to the ability of iron to catalyse the regeneration of the anionic oxygen responsible for TCE oxidation.

3. Materials and Methods

3.1. Materials

Trichloroethylene, calcium hydroxide, aluminium hydroxide, iron (III) acetylacetonate ($Fe(acac)_3$), toluene, and nitric acid were purchased from Sigma Aldrich and used without further purification.

3.2. Catalyst Preparation and Characterization

Details on the experimental methods are reported in the Supplementary Materials; here, we briefly sketch the basic procedures employed in this work. Mayenite was prepared by following the ceramic method described by Li et al. [40], starting with a mixture of calcium and aluminium hydroxide. Successively, mayenite was loaded with two different amounts of iron (1.5 and 2.0%) by using a solution of $Fe(acac)_3$ [41].

Diffraction patterns were recorded on a Bruker D8 Advance automatic diffractometer, operating with nickel-filtered CuKα radiation.

The iron content of Fe/mayenite catalyst was determined by ICP-OES analysis using a Perkin Elmer Optima 7000 DV after digestion of the sample in HNO_3. Three replicates for each sample were made.

The BET surface areas were determined with an 11-point BET analysis, after a degassing procedure in vacuum at 200 °C, by using a Nova Quantachrome 4200e instrument.

Temperature programmed reduction (TPR) of samples (10–20 mg) was measured with a TPD-TPR Autochem 2910 analyzer in the range of 25–800 °C.

The morphological and elemental analysis of the catalysts was performed by a scanning electron microscope (SEM, Tescan Vega LMU) equipped with a X-Ray energy dispersive microanalysis of elements (EDX, Bruker Quantax 800). Additionally, Fe/mayenite was characterized by field emission scanning electronic microscope (FESEM, ZEISS ULTRA 55).

X-ray photoelectron spectroscopy (XPS) was performed by using a SPECS spectrometer equipped with a Phoibos 150MCD-9 multichannel analyser. CasaXPS software was used for spectra treatment.

Raman spectra were recorded at RT with a 514 nm laser excitation on a Renishaw Raman Spectrometer (in via) equipped with a CCD detector. The laser power on the sample was 25 mW and a total of 20 acquisitions were taken for each spectra.

3.3. Experimental Setup and Catalytic Reactions

The experiments were performed in a stainless-steel fixed bed reactor, with the catalysts (0.8 g) placed between two quartz fiberglass stoppers. The cylindrical reactor (150 mm × 18 mm i.d., 22 mm o.d.) was placed in a furnace and the temperature was controlled via a K-thermocouple located inside the reactor.

Catalytic oxidations were monitored in the temperature range of 150–550 °C and the catalysts were kept at an operative temperature in air for 30 min before flow of TCE. The inlet gas was prepared by flowing a stream of air through pure TCE at room temperature; the final gas composition was [TCE] = 1700 ppm in wet air (RH = 60%). The Gas Hourly Space Velocity (GHSV) was set to 6000 h^{-1} by flowing the inlet gas at 110 mL/min. The residence time, based on the packing volume of the catalyst, was 0.87 s. Blank experiments were performed in the same conditions to evaluate the thermal oxidation.

3.4. Analytical Methods

Post-reaction gases were collected in a Tedlar sampling bag (SKC Inc., Eighty Four, PA, USA) for further analyses. To evaluate the conversion yield, the organic species were determined by means of GC-MS (Agilent 7890A) equipped with a DB 17-MS column (30 m × 0.25 mm, 0.25 μm). [CO] and [CO_2] were measured with an NDIRS on line system (Q-Track Plus IAQ Monitor, TSI), placed at the reactor outlet [23]. [Cl_2] and [HCl] were determined by tritation and ion chromatographic (IC) analyses of two aqueous solutions obtained by bubbling the reactor effluent gases into two water solutions (6×10^{-2} M solution of KI and 0.1 M of H_2SO_4 for Cl_2 and 2.6/0.76 mM solution of $NaHCO_3$/Na_2CO_3 for Cl^-). The production of Cl_2 was also assessed by SIM mode GC-MS analysis, using the same operative conditions described for the organic compounds [42]. The reproducibility of the results was checked by triplicate analyses (error < 5%). More details about products characterization are reported in our previous work [19].

The conversion yield was calculated as the ratio between the reacted TCE over the total TCE introduced into the reactor (1):

$$\text{TCE conversion } [\%] = \frac{m^i_{\text{TCE}} - m^o_{\text{TCE}}}{m^i_{\text{TCE}}} \times 100 \quad (1)$$

where m^i_{TCE} are the moles of TCE introduced into the reactor and m^o_{TCE} are the number of moles measured in the outlet gases; the reaction rate was calculated in terms of converted mass of TCE with respect to the catalyst mass and the residence time of the gas in the reactor (2):

$$\text{conversion rate} \left[\frac{mol}{g \times s}\right] = \frac{\text{moles of converted TCE}}{\text{catalyst mass } \times \text{ residence time}} \quad (2)$$

4. Conclusions

Iron-doped mayenite catalysts were prepared with different metal loading, obtaining active and selective catalysts able to quantitatively oxidize TCE in the gas phase. All the synthesized catalysts showed good performances for TCE oxidation, totally converted in CO_2, CO, and HCl. Mayenite loaded with 2% iron was found to be the best catalyst in terms of T_{50} (300 °C). This result was correlated with an optimum combination of the oxidative properties of the mayenite active support with the redox properties of iron, as TPR, XPS, and Raman results have shown.

2.0% Fe/mayenite showed good stability in terms of TCE conversion, respect to pure mayenite. The results only show some deactivation after 12 h of reaction, with a small decrease in the TCE conversion from 95% to 80%. In contrast, pure mayenite had significant deactivation (TCE conversion decrease from 80% to 40%) after 4 h of reaction. The atomic mapping of the samples shows that iron is well dispersed on the mayenite surface, minimizing catalyst deactivation and improving mayenite oxidant activity.

In conclusion, Fe/mayenite was found to be a promising catalyst due to several advantages such as: (i) total conversion of TCE in less-harmful products (CO_2, CO and HCl), (ii) absence of noble and heavy metals, thus reducing costs and the environmental impact, (iii) low cost of catalyst precursors and low deactivation.

Supplementary Materials: The following are available online at http://www.mdpi.com/2073-4344/9/9/747/s1, Figure S1 SEM and EDX analyses of 2.0% Fe/mayenite; Figure S2 XRD patterns of 2.0% Fe/mayenite; Figure S3 SEM and EDX spectrum of 2% Fe/mayenite after reaction; Figure S4 XPS survey spectra of the 2.0% Fe/mayenite.

Author Contributions: Conceptualization, R.C., A.I., A.P. and F.R.; Data curation, R.C., A.I. and J.M.-T.; Investigation, A.I., T.S., J.M.-T. and J.L.C.; Resources, A.E.P., A.P. and F.R.; Supervision, A.E.P., A.P. and F.R.; Writing—original draft, R.C. and A.I.; Writing—review & editing, A.E.P., J.M.-T. and F.R.

Funding: This work was supported by the grants ORSA167988 and ORSA174250 funded by the University of Salerno. AEP and JLC thank the Spanish Ministry of Economy and Competitiveness through RTI2018-101784-B-I00 and SEV-2016-0683 for the financial support. J.L. Cerrillo wishes to thank the Spanish Ministry of Economy and Competitiveness for the Severo Ochoa PhD fellowship (SVP-2014-068600).

Acknowledgments: The authors are thankful to Michele Napoli and Antonio Rea for technical assistance.

Conflicts of Interest: The authors declare no conflict of interest.

References

1. Schwille, F. *Dense Chlorinated Solvents in Porous and Fractured Media: Model Experiments*; Lewis Publishers: Boca Raton, FL, USA, 1988; ISBN 978-0-87371-121-0.
2. Russell, H.H.; Matthews, J.E.; Guy, W.S. TCE removal from contaminated soil and ground water. In *EPA Environmental Engineering Sourcebook*; Ann Arbor Press, Inc.: Ann Arbor, MI, USA, 1996.
3. Rossi, F.; Cucciniello, R.; Intiso, A.; Proto, A.; Motta, O.; Marchettini, N. Determination of the trichloroethylene diffusion coefficient in water. *AIChE J.* **2015**, *61*, 3511–3515. [CrossRef]
4. Ko, J.H.; Musson, S.; Townsend, T. Destruction of trichloroethylene during hydration of calcium oxide. *J. Hazard. Mater.* **2010**, *174*, 876–879. [CrossRef]
5. Ge, J.; Huang, S.; Han, I.; Jaffé, P.R. Degradation of tetra- and trichloroethylene under iron reducing conditions by Acidimicrobiaceae sp. A6. *Environ. Pollut.* **2019**, *247*, 248–255. [CrossRef]
6. Moccia, E.; Intiso, A.; Cicatelli, A.; Proto, A.; Guarino, F.; Iannece, P.; Castiglione, S.; Rossi, F. Use of Zea mays L. in phytoremediation of trichloroethylene. *Environ. Sci. Pollut. Res.* **2017**, *24*, 11053–11060. [CrossRef]
7. Meyer, C.; Borgna, A.; Monzon, A.; Garetto, T. Kinetic study of trichloroethylene combustion on exchanged zeolites catalysts. *J. Hazard. Mater.* **2011**, *190*, 903–908. [CrossRef]
8. Cucciniello, R.; Proto, A.; Rossi, F.; Marchettini, N.; Motta, O. An improved method for BTEX extraction from charcoal. *Anal. Methods* **2015**, *7*, 4811–4815. [CrossRef]
9. Intiso, A.; Miele, Y.; Marchettini, N.; Proto, A.; Sánchez-Domínguez, M.; Rossi, F. Enhanced solubility of trichloroethylene (TCE) by a poly-oxyethylene alcohol as green surfactant. *Environ. Technol. Innov.* **2018**, *12*, 72–79. [CrossRef]
10. Garza-Arévalo, J.I.; Intiso, A.; Proto, A.; Rossi, F.; Sanchez-Dominguez, M. Trichloroethylene solubilization using a series of commercial biodegradable ethoxylated fatty alcohol surfactants. *J. Chem. Technol. Biotechnol.* **2019**. [CrossRef]
11. Aranzabal, A.; Ayo, B.P.; González-Marcos, M.P.; González-Marcos, J.A.; López-Fonseca, R.; González-Velasco, J.R. State of the art in catalytic oxidation of chlorinated volatile organic compounds. *Chem. Pap.* **2014**, *68*, 1169–1186. [CrossRef]

12. Li, D.; Li, C.; Suzuki, K. Catalytic oxidation of VOCs over Al- and Fe-pillared montmorillonite. *Appl. Clay Sci.* **2013**, *77*, 56–60. [CrossRef]
13. Tian, W.; Fan, X.; Yang, H.; Zhang, X. Preparation of MnOx/TiO_2 composites and their properties for catalytic oxidation of chlorobenzene. *J. Hazard. Mater.* **2010**, *177*, 887–891. [CrossRef]
14. Blanch-Raga, N.; Palomares, A.E.; Martínez-Triguero, J.; Puche, M.; Fetter, G.; Bosch, P. The oxidation of trichloroethylene over different mixed oxides derived from hydrotalcites. *Appl. Catal. B Environ.* **2014**, *160*, 129–134. [CrossRef]
15. Taralunga, M.; Mijoin, J.; Magnoux, P. Catalytic destruction of 1,2-dichlorobenzene over zeolites. *Catal. Commun.* **2006**, *7*, 115–121. [CrossRef]
16. Romero-Sáez, M.; Divakar, D.; Aranzabal, A.; González-Velasco, J.; González-Marcos, J.; Marcos, J.A.G. Catalytic oxidation of trichloroethylene over Fe-ZSM-5: Influence of the preparation method on the iron species and the catalytic behavior. *Appl. Catal. B Environ.* **2016**, *180*, 210–218. [CrossRef]
17. Blanch-Raga, N.; Palomares, A.E.; Triguero, J.M.; Valencia, S. Cu and Co modified beta zeolite catalysts for the trichloroethylene oxidation. *Appl. Catal. B Environ.* **2016**, *187*, 90–97. [CrossRef]
18. Cucciniello, R.; Proto, A.; Rossi, F.; Motta, O. Mayenite based supports for atmospheric NOx sampling. *Atmos. Environ.* **2013**, *79*, 666–671. [CrossRef]
19. Cucciniello, R.; Intiso, A.; Castiglione, S.; Genga, A.; Proto, A.; Rossi, F. Total oxidation of trichloroethylene over mayenite ($Ca_{12}Al_{14}O_{33}$) catalyst. *Appl. Catal. B Environ.* **2017**, *204*, 167–172. [CrossRef]
20. Intiso, A.; Cucciniello, R.; Castiglione, S.; Proto, A.; Rossi, F. Environmental Application of Extra-Framework Oxygen Anions in the Nano-Cages of Mayenite. In *Advances in Bionanomaterials*; Lecture Notes in Bioengineering; Springer: Cham, Switzerland, 2018; pp. 131–139. ISBN 978-3-319-62026-8.
21. Intiso, A.; Martinez-Triguero, J.; Cucciniello, R.; Proto, A.; Palomares, A.E.; Rossi, F. A Novel Synthetic Route to Prepare High Surface Area Mayenite Catalyst for TCE Oxidation. *Catalysts* **2019**, *9*, 27. [CrossRef]
22. Intiso, A.; Martinez-Triguero, J.; Cucciniello, R.; Rossi, F.; Palomares, A.E. Influence of the synthesis method on the catalytic activity of mayenite for the oxidation of gas-phase trichloroethylene. *Sci. Rep.* **2019**, *9*, 425. [CrossRef]
23. Proto, A.; Cucciniello, R.; Rossi, F.; Motta, O. Stable carbon isotope ratio in atmospheric CO2 collected by new diffusive devices. *Environ. Sci. Pollut. Res.* **2013**, *21*, 3182–3186. [CrossRef]
24. Eufinger, J.-P.; Schmidt, A.; Lerch, M.; Janek, J. Novel anion conductors—Conductivity, thermodynamic stability and hydration of anion-substituted mayenite-type cage compounds $C_{12}A_7$:X (X = O, OH, Cl, F, CN, S, N). *Phys. Chem. Chem. Phys.* **2015**, *17*, 6844–6857. [CrossRef]
25. Schmidt, A.; Lerch, M.; Eufinger, J.-P.; Janek, J.; Tranca, I.; Islam, M.M.; Bredow, T.; Dolle, R.; Wiemhofer, H.-D.; Boysen, H.; et al. Chlorine ion mobility in Cl-mayenite ($Ca_{12}Al_{14}O_{32}C_{l2}$): An investigation combining high-temperature neutron powder diffraction, impedance spectroscopy and quantum-chemical calculations. *Solid State Ionics* **2014**, *254*, 48–58. [CrossRef]
26. Teusner, M.; De Souza, R.A.; Krause, H.; Ebbinghaus, S.G.; Belghoul, B.; Martin, M. Oxygen Diffusion in Mayenite. *J. Phys. Chem. C* **2015**, *119*, 9721–9727. [CrossRef]
27. Ruszak, M.; Inger, M.; Witkowski, S.; Wilk, M.; Kotarba, A.; Sojka, Z. Selective N_2O Removal from the Process Gas of Nitric Acid Plants over Ceramic $_{12}CaO{\cdot}_7Al_2O_3$ Catalyst. *Catal. Lett.* **2008**, *126*, 72–77. [CrossRef]
28. Proto, A.; Cucciniello, R.; Genga, A.; Capacchione, C. A study on the catalytic hydrogenation of aldehydes using mayenite as active support for palladium. *Catal. Commun.* **2015**, *68*, 41–45. [CrossRef]
29. Ye, T.-N.; Li, J.; Kitano, M.; Hosono, H. Unique nanocages of $_{12}CaO{\cdot}_7Al_2O_3$ boost heterolytic hydrogen activation and selective hydrogenation of heteroarenes over ruthenium catalyst. *Green Chem.* **2017**, *19*, 749–756. [CrossRef]
30. Li, C.; Hirabayashi, D.; Suzuki, K. A crucial role of O^{2-} and $O_2{}^{2-}$ on mayenite structure for biomass tar steam reforming over Ni/$Ca_{12}Al_{14}O_{33}$. *Appl. Catal. B Environ.* **2009**, *88*, 351–360. [CrossRef]
31. Di Carlo, A.; Borello, D.; Sisinni, M.; Savuto, E.; Venturini, P.; Bocci, E.; Kuramoto, K. Reforming of tar contained in a raw fuel gas from biomass gasification using nickel-mayenite catalyst. *Int. J. Hydrog. Energy* **2015**, *40*, 9088–9095. [CrossRef]
32. Mani, S.; Kastner, J.R.; Juneja, A. Catalytic decomposition of toluene using a biomass derived catalyst. *Fuel Process. Technol.* **2013**, *114*, 118–125. [CrossRef]
33. Lacerda, M.; Irvine, J.T.S.; Glasser, F.P.; West, A.R. High oxide ion conductivity in $Ca_{12}Al_{14}O_{33}$. *Nature* **1988**, *332*, 525–526. [CrossRef]

34. Li, J.; Kitano, M.; Ye, T.-N.; Sasase, M.; Yokoyama, T.; Hosono, H.; Ye, T. Chlorine-Tolerant Ruthenium Catalyst Derived Using the Unique Anion-Exchange Properties of $_{12}CaO{\cdot}_7Al_2O_3$ for Ammonia Synthesis. *ChemCatChem* **2017**, *9*, 3078–3083. [CrossRef]
35. Diwald, O.; Sterrer, M.; Knözinger, E. Site selective hydroxylation of the MgO surface. *Phys. Chem. Chem. Phys.* **2002**, *4*, 2811–2817. [CrossRef]
36. Ruszak, M.; Witkowski, S.; Sojka, Z. EPR and Raman investigations into anionic redox chemistry of nanoporous $_{12}CaO{\cdot}_7Al_2O_3$ interacting with O_2, H_2 and N_2O. *Res. Chem. Intermed.* **2007**, *33*, 689–703. [CrossRef]
37. Levasseur, B.; Kaliaguine, S. Effects of iron and cerium in La1−yCeyCo1−xFexO3 perovskites as catalysts for VOC oxidation. *Appl. Catal. B Environ.* **2009**, *88*, 305–314. [CrossRef]
38. Li, H.; Wang, S.; Tang, N.; Pan, S.; Hu, J. FeCl3-modified Co–Ce oxides catalysts for mercury removal from coal-fired flue gas. *Chem. Pap.* **2017**, *71*, 2545–2555. [CrossRef]
39. Grosvenor, A.P.; Kobe, B.A.; Biesinger, M.C.; McIntyre, N.S.; Grosvenor, A. Investigation of multiplet splitting of Fe 2p XPS spectra and bonding in iron compounds. *Surf. Interface Anal.* **2004**, *36*, 1564–1574. [CrossRef]
40. Li, C.; Hirabayashi, D.; Suzuki, K. Synthesis of higher surface area mayenite by hydrothermal method. *Mater. Res. Bull.* **2011**, *46*, 1307–1310. [CrossRef]
41. Molina, R.; Poncelet, G. α-Alumina-Supported Nickel Catalysts Prepared from Nickel Acetylacetonate: A TPR Study. *J. Catal.* **1998**, *173*, 257–267. [CrossRef]
42. Méndez, M.; Ciuraru, R.; Gosselin, S.; Batut, S.; Visez, N.; Petitprez, D. Reactivity of chlorine radical with submicron palmitic acid particles: Kinetic measurements and products identification. *Atmospheric Chem. Phys. Discuss.* **2013**, *13*, 16925–16960. [CrossRef]

MDPI
St. Alban-Anlage 66
4052 Basel
Switzerland
Tel. +41 61 683 77 34
Fax +41 61 302 89 18
www.mdpi.com

Catalysts Editorial Office
E-mail: catalysts@mdpi.com
www.mdpi.com/journal/catalysts

www.ingramcontent.com/pod-product-compliance
Lightning Source LLC
LaVergne TN
LVHW070101170726
843515LV00007B/1588

* 9 7 8 3 0 3 6 5 1 9 8 0 7 *